Interdisciplinary Approaches to Transportation and Urban Planning

Görkem Gülhan
Pamukkale University, Turkey

Olcay Polat
Pamukkale University, Turkey

IGI Global
Publishing Tomorrow's Research Today

Published in the United States of America by
IGI Global Scientific Publishing
701 East Chocolate Avenue
Hershey, PA, 17033, USA
Tel: 717-533-8845
Fax: 717-533-8661
E-mail: cust@igi-global.com
Website: https://www.igi-global.com

Library of Congress Cataloging-in-Publication Data

Names: Gülhan, Görkem, editor. | Polat, Olcay, 1983- editor.
Title: Interdisciplinary approaches to transportation and urban planning / edited by, Görkem Gülhan, Olcay Polat.
Description: Hershey, PA : IGI Global Scientific Publishing, [2025] | Includes bibliographical references and index. | Summary: "This book delves into the intricate relationship between transportation systems and urban planning strategies, emphasizing the need for interdisciplinary collaboration to address the multifaceted challenges of contemporary urban environments"-- Provided by publisher.
Identifiers: LCCN 2024052887 (print) | LCCN 2024052888 (ebook) | ISBN 9798369366950 (hardcover) | ISBN 9798369366967 (paperback) | ISBN 9798369366974 (ebook)
Subjects: LCSH: Urban transportation. | City planning--Environmental aspects. | Intelligent transportation systems.
Classification: LCC TA1205 .I56 2025 (print) | LCC TA1205 (ebook) | DDC 307.1/216--dc23/eng/20241216
LC record available at https://lccn.loc.gov/2024052887
LC ebook record available at https://lccn.loc.gov/2024052888

British Cataloguing in Publication Data
A Cataloguing in Publication record for this book is available from the British Library.

Table of Contents

Preface xiv

Chapter 1
The Role of Transportation in Environmental Sustainability and Climate Change Mitigation 1
Aylin Tepecik, Erciyes University, Turkey

Chapter 2
The Role of Smart Cities in Creating Sustainable Urban Environments Through Strategies and Innovative Solutions 23
Jorge A. Ruiz-Vanoye, Universidad Politécnica de Pachuca, Mexico
Ocotlán Díaz-Parra, Universidad Politécnica de Pachuca, Mexico
Miguel A. Ruiz-Jaimes, Universidad Politécnica del Estado de Morelos, Mexico
Jaime Aguilar Ortiz, Universidad Politécnica de Pachuca, Mexico
Eric Simancas-Acevedo, Universidad Politécnica de Pachuca, Mexico
Jazmín Rodriguez-Flores, Universidad Politécnica de Pachuca, Mexico
Josué Román Martínez Mireles, Universidad Politécnica de Pachuca, Mexico
Juan M. Xicoténcatl-Pérez, Universidad Politécnica de Pachuca, Mexico
Marco Antonio Vera Jimenez, Universidad Politécnica de Pachuca, Mexico

Chapter 3
Smart Cities, Transportation, and Regional Development: The Case of the TR32 Regional Development Strategy for 2024-2028 in Turkey 47
Oğuz Özbek, Pamukkale University, Turkey
Ezgi Karaçoban, Evliyaoğlu Textile Inc., Turkey

Chapter 4
The Contribution of Intelligent Systems (IS) on Transport and Planning Studies 83
Metin Mutlu Aydin, Ondokuz Mayıs University, Turkey
Eren Dağlı, Selçuk University, Turkey

Chapter 5
Integrating Cloud Computing for Intelligent Transportation Solutions in Smart Cities: A Short Review 121
Aditya Kumar Panda, School of Computer Engineering, KIIT University (Deemed), Bhubaneswar, India
Amlan Anurag Lenka, School of Computer Engineering, KIIT University (Deemed), Bhubaneswar, India
Ayush Mohapatra, School of Computer Engineering, KIIT University (Deemed), Bhubaneswar, India
Bikram Keshari Rath, School of Computer Engineering, KIIT University (Deemed), Bhubaneswar, India
Ayush Adarsh Parida, School of Computer Engineering, KIIT University (Deemed), Bhubaneswar, India
Hitesh Mohapatra, School of Computer Engineering, KIIT University (Deemed), Bhubaneswar, India

Chapter 6
Analysis on Enhancing Urban Mobility With IoT-Integrated Parking Solutions 143
Supriya Priyadarshi, School of Computer Engineering, KIIT University (Deemed), Bhubaneswar, India
Sanjana Subudhi, School of Computer Engineering, KIIT University (Deemed), Bhubaneswar, India
Shivam Kumar, School of Computer Engineering, KIIT University (Deemed), Bhubaneswar, India
Divya Bhardwaj, School of Computer Engineering, KIIT University (Deemed), Bhubaneswar, India
Hitesh Mohapatra, School of Computer Engineering, KIIT University (Deemed), Bhubaneswar, India

Chapter 7
Development of Analytical Approaches Using Smart Card Data While Investigating Travel Behaviors in Terms of Space and Planning: Izmir Case ... 173
Hakan Uzun, Dokuz Eylul University, Turkey
K. Mert Cubukcu, Dokuz Eylul University, Turkey

Chapter 8
Microsimulation Analysis of Public Transport: The Case of Bagdat Street, Istanbul 201
İrem Merve Ulu, Istanbul Okan University, Turkey
Burcu Ozdemir, Informatics Institute, Istanbul, Turkey

Chapter 9
Road Aesthetics and Its Effect on Road Users 233
Anuj Jaiswal, Maulala Azad National Institute of Technology, Bhopal, India
Arshi Parashar, Maulala Azad National Institute of Technology, Bhopal, India
Vicky Lalramsangi, Maulana Azad National Institute of Technology, Bhopal, India

Chapter 10
Have Rebranding Strategies Improved the Economy and Environment in an Old District of a Metropolis? A Pilot Study in Wanchai, Hong Kong.............. 267
Giovanna Chiu, German Swiss International School, Hong Kong
Chloe Kwun Yee Lee, Hong Kong Baptist University, Hong Kong
Dickson K. W. Chiu, The University of Hong Kong, Hong Kong
Chloe Tsz Yu Tse, German Swiss International School, Hong Kong

Chapter 11
The Circularity in Transportation Completed 291
Eminegül Tekerci, Izmir Institute of Technology, Turkey
Yavuz Duvarcı, Izmir Institute of Technology, Turkey
Volkan Emre Uz, Izmir Institute of Technology, Turkey
Nicel Saygın, Izmir Institute of Technology, Turkey

Chapter 12
The Relationship Between COVID-19 and Urban Features in the Light of Recent Quantitative Studies 329
Tayfun Salihoğlu, Pamukkale University, Turkey
İhsan Tutuk, Gebze Technical University, Turkey

Compilation of References 381

About the Contributors 431

Index 437

Detailed Table of Contents

Preface xiv

Chapter 1
The Role of Transportation in Environmental Sustainability and Climate Change Mitigation 1
Aylin Tepecik, Erciyes University, Turkey

The type of transportation used in urban areas is important in terms of environmental sustainability. Cities trigger greenhouse gas production and thus climate change with the services they offer. Environmental problems have increased with increasing greenhouse gas emissions and traffic congestion. In ensuring environmental sustainability, the importance of green transportation types that reduce the number of motor vehicles and fossil fuel dependency, consume less energy, and reduce vehicle-related pollutant problems are mentioned and play an active role in reducing environmental problems. The importance of this study is to address the transportation sector, which is an important source of greenhouse gases, the relationship between the effects of climate change and the transportation sector, and the examination of green transportation policies in reducing the effects of climate change. The study reveals environmentally friendly transportation types in sustainable transportation policies with green transportation systems.

Chapter 2
The Role of Smart Cities in Creating Sustainable Urban Environments Through Strategies and Innovative Solutions 23
Jorge A. Ruiz-Vanoye, Universidad Politécnica de Pachuca, Mexico
Ocotlán Díaz-Parra, Universidad Politécnica de Pachuca, Mexico
Miguel A. Ruiz-Jaimes, Universidad Politécnica del Estado de Morelos, Mexico
Jaime Aguilar Ortiz, Universidad Politécnica de Pachuca, Mexico
Eric Simancas-Acevedo, Universidad Politécnica de Pachuca, Mexico
Jazmín Rodriguez-Flores, Universidad Politécnica de Pachuca, Mexico
Josué Román Martínez Mireles, Universidad Politécnica de Pachuca, Mexico
Juan M. Xicoténcatl-Pérez, Universidad Politécnica de Pachuca, Mexico
Marco Antonio Vera Jimenez, Universidad Politécnica de Pachuca, Mexico

A smart city is a city that uses the optimum management of water, city technology, energy, mobility, public transportation, distribution of transport routes, road network safety management, the construction of homes / offices / transportation structure, industry / jobs, education, healthcare, public safety, government, civil protection, food, environmental pollution, and solid, hazardous and special waste This chapter will analyse how smart cities can become catalysts for urban sustainability through the implementation of advanced technologies and the integration of innovative strategies. With a focus on carbon footprint reduction, energy efficiency and climate change resilience, it will explore various initiatives and policies that are transforming cities towards a greener and more equitable future.

Chapter 3
Smart Cities, Transportation, and Regional Development: The Case of the TR32 Regional Development Strategy for 2024-2028 in Turkey 47
Oğuz Özbek, Pamukkale University, Turkey
Ezgi Karaçoban, Evliyaoğlu Textile Inc., Turkey

This paper examines the integration of smart cities, transportation policies, and regional development strategies in the TR32 Region (the provinces of Aydın, Denizli and Muğla) of Turkey, focusing on the 2024-2028 Draft Regional Plan prepared by the Southern Aegean Development Agency (GEKA). The research highlights the importance of incorporating smart transportation measures into the regional development framework to enhance the region's competitiveness, sustainability, and quality of life. The draft plan demonstrates a comprehensive approach to integrating smart city principles and technologies into its development strategy, with a focus on digital transformation, innovation, and green growth. The plan effectively addresses the integration of smart technologies in various sectors, such as transportation, manufacturing, agriculture, and tourism. However, the evaluation of the plan's effectiveness in implementing smart transportation strategies is limited by the lack of formal feedback or evaluation of the previous GEKA plan period and the fragmented nature of data sources.

Chapter 4
The Contribution of Intelligent Systems (IS) on Transport and Planning Studies 83
Metin Mutlu Aydin, Ondokuz Mayıs University, Turkey
Eren Dağlı, Selçuk University, Turkey

Intelligent Systems (IS) have emerged as a key tool in revolutionizing transport and urban planning studies by providing innovative solutions. They have supplied transport planners with robust predictive modeling capabilities incorporating artificial intelligence and machine learning. They predict traffic patterns, optimize routes, and anticipate fluctuations in demand. They improve situational awareness, dynamic traffic management, and reduce congestion. These systems are key in optimizing land use, improving accessibility, and promoting sustainable development. In summary, they represent a paradigm shift in transport and planning studies. Intelligent Transport Systems (ITS) are an umbrella under which all transport-related systems are grouped. Therefore, all applications under the umbrella of ITS in cities are carried out to make transport and planning studies more intelligent. In this study, the importance and contribution of ITS in transportation and planning studies are examined and it is aimed to be a guide for future studies.

Chapter 5
Integrating Cloud Computing for Intelligent Transportation Solutions in Smart Cities: A Short Review........ 121
Aditya Kumar Panda, School of Computer Engineering, KIIT University (Deemed), Bhubaneswar, India
Amlan Anurag Lenka, School of Computer Engineering, KIIT University (Deemed), Bhubaneswar, India
Ayush Mohapatra, School of Computer Engineering, KIIT University (Deemed), Bhubaneswar, India
Bikram Keshari Rath, School of Computer Engineering, KIIT University (Deemed), Bhubaneswar, India
Ayush Adarsh Parida, School of Computer Engineering, KIIT University (Deemed), Bhubaneswar, India
Hitesh Mohapatra, School of Computer Engineering, KIIT University (Deemed), Bhubaneswar, India

Urban transportation systems face many challenges, such as traffic congestion, pollution, and inefficient infrastructure. To tackle these problems, cities are increasingly adopting cloud-based smart transportation solutions. These solutions use cloud computing to integrate IoT devices, data analytics, and real-time monitoring, optimizing urban mobility. Cloud platforms allow for efficient data collection, storage, and analysis, helping city authorities make informed decisions. This approach is scalable, flexible, and cost-effective, promoting smarter and more sustainable urban transportation systems. By leveraging cloud computing, cities can improve traffic management, reduce congestion, and enhance decision-making, leading to better efficiency, sustainability, and safety. The proposed model for smart transportation involves using cloud computing, IoT, data analytics, and advanced AI algorithms to create more connected and livable urban environments.

Chapter 6
Analysis on Enhancing Urban Mobility With IoT-Integrated Parking Solutions 143
Supriya Priyadarshi, School of Computer Engineering, KIIT University (Deemed), Bhubaneswar, India
Sanjana Subudhi, School of Computer Engineering, KIIT University (Deemed), Bhubaneswar, India
Shivam Kumar, School of Computer Engineering, KIIT University (Deemed), Bhubaneswar, India
Divya Bhardwaj, School of Computer Engineering, KIIT University (Deemed), Bhubaneswar, India
Hitesh Mohapatra, School of Computer Engineering, KIIT University (Deemed), Bhubaneswar, India

Parking in the conventional way is a very tedious and time-consuming job in the current world due to poor handling. Using Internet of Things (IoT) to ease parking management without any human interference seems to be a convenient technique. IoT is described as the network of intelligent devices such as sensors, actuators, and other technologies which can make informed decisions in accordance with the environment. This technology can be exploited to help in identifying the various empty parking slots, efficiently allocating them, and promoting sustainable urban development. To make parking systems smart, IoT integrated with various common approaches, such as real-time monitoring, predictive analytics, and blockchain models, is being used. Through a thorough analysis of existing literature, this paper highlights the potential and consequential benefits of IoT-enabled Smart Parking Systems. In conclusion, the adoption of IoT in parking management can significantly enhance efficiency, reduce human effort, and contribute to the development of sustainable urban environments.

Chapter 7
Development of Analytical Approaches Using Smart Card Data While Investigating Travel Behaviors in Terms of Space and Planning: Izmir Case ... 173
Hakan Uzun, Dokuz Eylul University, Turkey
K. Mert Cubukcu, Dokuz Eylul University, Turkey

The objective of this study is to ascertain the distinctions in the mobility patterns of elderly individuals and to formulate recommendations that will enhance the accessibility and suitability of public transportation systems for this demographic. To achieve this objective, data from the electronic fare collection system used in public transportation in Izmir city for elderly individuals with smart cards has been utilized. To investigate the variables that differentiate the mobility of elderly individuals, one-way variance analysis (ANOVA) and Student's T-test analysis were employed. The results demonstrate that elderly individuals are not a homogeneous group; gender, age group, and distance from the city center have an impact on mobility. The issue of increasing the mobility of elderly individuals will have an impact on spatial planning and transportation planning.

Chapter 8
Microsimulation Analysis of Public Transport: The Case of Bagdat Street, Istanbul.. 201
İrem Merve Ulu, Istanbul Okan University, Turkey
Burcu Ozdemir, Informatics Institute, Istanbul, Turkey

Metropolitan areas around the world are facing traffic congestion and mobility issues. In this context, improving public transportation systems has become a critical requirement for ensuring the sustainability of cities. This study examines the impacts of public transport strategies implemented on Bagdat Street, one of Istanbul's main corridors, on pedestrian and vehicle networks. Using PTV VISSIM for traffic simulation, various scenarios, including changes in bus frequency, capacity, number of stops, and dedicated lanes, were analyzed. Findings reveal that increasing the frequency and reducing the capacity of the existing public transport buses not only improves traffic flow but also enhances passenger satisfaction. These insights provide valuable guidance for urban transportation planning and policy making, which can be adapted to similar contexts in other urban environments.

Chapter 9
Road Aesthetics and Its Effect on Road Users 233
Anuj Jaiswal, Maulala Azad National Institute of Technology, Bhopal, India
Arshi Parashar, Maulala Azad National Institute of Technology, Bhopal, India
Vicky Lalramsangi, Maulana Azad National Institute of Technology, Bhopal, India

Aesthetics is synonymous to beauty. Anything which adds beauty to the road is a component of road aesthetics. Like safety, utility and comfort; road aesthetics is of equal importance for our road system. Road aesthetics is having a great value to the public and to the visitors who use it every day and in-turn becomes a field of great interest to planners, engineers, architects and designers. It influences road user's perception and behavior and becomes important as effective time spent on roads has increased manifold with increasing globalization and better transport and communication facilities. This paper investigates into the need and the technicalities of an aesthetically positive road environment taking into consideration human characteristics. A possible relationship between aesthetics and behavior is described through interdisciplinary research within arts & aesthetics, environmental psychology and traffic safety.

Chapter 10
Have Rebranding Strategies Improved the Economy and Environment in an Old District of a Metropolis? A Pilot Study in Wanchai, Hong Kong............... 267
Giovanna Chiu, German Swiss International School, Hong Kong
Chloe Kwun Yee Lee, Hong Kong Baptist University, Hong Kong
Dickson K. W. Chiu, The University of Hong Kong, Hong Kong
Chloe Tsz Yu Tse, German Swiss International School, Hong Kong

Wanchai was the first old commercial district in Hong Kong to undergo redevelopment with much economic activity and environmental changes. This study conducted fieldwork, including urban land use surveys, environmental quality surveys, questionnaires, and cross-examined secondary in newspapers, government reports, and online information to find out if these rebranding strategies through building reconstruction improved the economy and environment of Wanchai. Results indicated that though the economy has grown significantly with such development strategies, the quality of the environment has not been improved much. Some suggestions are given accordingly.

Chapter 11
The Circularity in Transportation Completed 291
Eminegül Tekerci, Izmir Institute of Technology, Turkey
Yavuz Duvarcı, Izmir Institute of Technology, Turkey
Volkan Emre Uz, Izmir Institute of Technology, Turkey
Nicel Saygın, Izmir Institute of Technology, Turkey

This study explores the need for radical change in the transport sector, focusing on circularity and its implementation through innovative technologies. The unsustainable nature of current transport systems is discussed, highlighting the urgent need for change due to environmental impacts such as the climate crisis exacerbated by fossil fuels. The research presents a framework for 'Transport Circularity' that integrates solar energy, autonomous vehicles and other intelligent transport systems to create more sustainable, efficient and socially inclusive mobility solutions. Beforehand, drawing on philosophical perspectives, particularly Spinoza's views and Marxist critiques on conventional transportation, the study argues that a systemic overhaul is inevitable. This transition's social and economic implications are also discussed, including its potential to dismantle existing power structures that benefit from the status quo. The study concludes with a vision of a resilient, sustainable and equitable future for urban and regional transport systems, giving a concrete example of Izmir.

Chapter 12
The Relationship Between COVID-19 and Urban Features in the Light of Recent Quantitative Studies 329
Tayfun Salihoğlu, Pamukkale University, Turkey
İhsan Tutuk, Gebze Technical University, Turkey

The COVID-19 pandemic, caused by the novel coronavirus SARS-CoV-2, has profoundly impacted urban environments globally. The virus's dependence on physical proximity for transmission and the necessity for individuals to congregate in urban spaces to meet various needs accelerated its spread. Therefore, as with many other epidemic diseases, investigating the urban factors behind contact and spread has become an important issue for the COVID-19 pandemic as well. This study investigates the relationship between COVID-19 and various urban features by synthesizing recent quantitative research. The review revealed that it is possible to conceptualise urban factors which are influencing COVID-19 spread as sociodemographic characteristics, healthcare services, infrastructure, tourism activities, economic activities, and urban conditions. Understanding these urban determinants is essential for developing targeted strategies to control the spread of COVID-19 and enhance urban resilience against future pandemics.

Compilation of References 381

About the Contributors 431

Index 437

Preface

Transportation and urban planning are foundational elements of modern society. As cities continue to grow at unprecedented rates, the relationship between transportation systems and urban planning strategies becomes increasingly complex and intertwined. In addressing the multifaceted challenges posed by urbanization, sustainability, technological advancement, and climate change, it is clear that a multidisciplinary approach is necessary. "Interdisciplinary Approaches to Transportation and Urban Planning" seeks to address these challenges by bringing together experts from various fields, presenting cutting-edge research, innovative strategies, and practical solutions aimed at creating sustainable, inclusive, and resilient urban environments.

This book explores the intersections between transportation and urban planning, recognizing the synergies required to foster more dynamic and adaptable cities. It emphasizes the growing need for collaboration among urban planners, engineers, environmental scientists, policymakers, and technologists to create future-proof urban mobility systems. These systems are not only tasked with moving people efficiently but must also balance environmental sustainability, economic growth, and social inclusivity. Through the collective expertise presented in this volume, the book offers insights into current trends, challenges, and opportunities while also providing a platform for innovative solutions that respond to the complex demands of contemporary urban life.

In today's rapidly urbanizing world, cities are often considered the engines of economic growth and social progress. However, this rapid development also places significant strain on urban infrastructure, particularly in the areas of transportation and land use planning. Traffic congestion, air pollution, inefficient transportation systems, and unequal access to urban mobility are just a few of the many challenges facing cities globally. These issues are compounded by the increasing urgency of mitigating climate change, managing urban sprawl, and creating sustainable living conditions for all city dwellers.

At the heart of these challenges lies the need for integrated urban planning and transportation systems that can adapt to changing social, environmental, and technological contexts. Traditional, siloed approaches to city development are no longer sufficient; they must be replaced by more holistic, interdisciplinary methods that consider the complex interplay between transportation networks, land use, environmental impact, and human behavior. This book highlights these critical issues, proposing actionable solutions that can shape future policies and development strategies.

The emergence of smart cities, the rise of digital and intelligent systems, and the growing emphasis on sustainability underscore the need for transportation systems that are more than just functional—they must be forward-thinking and resilient to future uncertainties. This book delves into these contemporary developments, offering new perspectives and research methodologies that emphasize the interdisciplinary nature of urban planning and transportation.

This volume is intended for a diverse readership that includes researchers and academics in fields such as urban planning, transportation engineering, geography, environmental studies, sociology, and related disciplines. By providing a thorough examination of the latest research and practical applications in the field, this book will serve as a valuable resource for professionals in both the public and private sectors, including urban planners, policymakers, transportation authorities, infrastructure developers, and urban designers.

In addition, graduate students and advanced undergraduates studying urban studies, transportation planning, sustainable development, and smart cities will find the content highly relevant for their studies and research. The interdisciplinary nature of the book makes it suitable for those who are looking to explore the intersections between different domains, such as technology, policy, and environmental planning.

The practical insights and case studies included in this book are particularly beneficial for international organizations, non-governmental organizations (NGOs), and advocacy groups focusing on urban mobility, transportation equity, climate resilience, and sustainable urban development. The strategies discussed herein offer valuable lessons for cities and regions globally, providing models for replication and adaptation in diverse contexts.

The chapters in this book are carefully selected to offer a broad yet cohesive overview of the critical issues facing transportation and urban planning today. Each chapter provides a unique perspective, grounded in rigorous research and practical case studies, to demonstrate how interdisciplinary approaches can lead to more efficient, equitable, and sustainable cities.

Chapter 1, " The Role of Transportation in Environmental Sustainability and Climate Change Mitigation," authored by Aylin Tepecik explores the pivotal role that transportation systems play in both exacerbating and mitigating climate change. The chapter delves into the impact of motor vehicles on greenhouse gas emissions and environmental degradation, making a compelling case for the adoption of green transportation policies. By examining policies that promote energy efficiency and reduce fossil fuel dependency, Tepecik offers a roadmap for integrating sustainable transportation solutions into broader climate change mitigation strategies.

Chapter 2, "The Role of Smart Cities in Creating Sustainable Urban Environments through Strategies and Innovative Solutions," authored by Jorge A. Ruiz-Vanoye, Ocotlán Díaz-Parra, and Jaime Aguilar-Ortiz present a forward-thinking analysis of how smart cities can serve as catalysts for urban sustainability. This chapter provides an in-depth examination of how the integration of advanced technologies—such as IoT, AI, and big data—into urban planning can enhance energy efficiency, reduce carbon footprints, and foster resilience to climate change. The authors illustrate how smart cities offer not just technological innovation, but also policy frameworks that support sustainable growth and equitable urban environments.

Chapter 3, " Smart Cities, Transportation, and Regional Development: The Case of the TR32 Regional Development Strategy for 2024-2028 in Turkey," authored by Oğuz Özbek and Ezgi Karaçoban focus on the TR32 Region in Turkey to demonstrate how smart transportation policies can be integrated into regional development strategies. Their analysis of the TR32 Regional Development Plan for 2024-2028 highlights the potential for smart city principles to enhance regional competitiveness and sustainability. The chapter provides a valuable case study on the importance of digital transformation and green growth, underscoring the role that transportation policies play in shaping broader economic and social outcomes.

Chapter 4, " The Contribution of Intelligent Systems (IS) on Transport and Planning Studies," authored by Metin Mutlu Aydin and Eren Dağlı delve into the transformative potential of Intelligent Systems (IS) in the realm of transportation and urban planning. The chapter outlines how artificial intelligence, machine learning, and predictive analytics are being used to optimize traffic management, land use, and accessibility. By providing a comprehensive overview of current applications, the authors demonstrate how IS are not just enhancing efficiency but also promoting more sustainable and equitable development patterns.

Chapter 5, " Integrating Cloud Computing for Intelligent Transportation Solutions in Smart Cities: A Short Review," authored by Aditya Kumar Panda and his co-authors present an insightful review of cloud computing as a solution for intelligent transportation systems. This chapter highlights the role of cloud technologies in managing real-time data, optimizing traffic flows, and improving decision-making processes in urban environments. By showcasing the scalability and flexibility of cloud solutions, the authors argue that these technologies are key to developing smarter, more efficient, and sustainable cities.

Chapter 6, " Analysis on Enhancing Urban Mobility with IoT-Integrated Parking Solutions," authored by Supriya Priyadarshi and co-authors explore how IoT-enabled parking systems can enhance urban mobility by reducing congestion and improving efficiency. Through a thorough analysis of existing literature, the chapter demonstrates how IoT technologies can be integrated into urban infrastructure to manage parking in real-time. By optimizing parking allocation and reducing the environmental impact of vehicle emissions, this chapter offers practical solutions for promoting sustainable urban growth.

Chapter 7, " Development of Analytical Approaches Using Smart Card Data while Investigating Travel Behaviors in Terms of Space and Planning: Izmir Case," authored by Hakan Uzun and Mert Çubukçu focus on the Izmir case to explore how smart card data can be utilized to analyze travel behaviors and improve public transportation accessibility. Their analysis of elderly citizens' mobility patterns provides key insights into how public transportation can be tailored to meet the needs of different demographic groups. This chapter offers important lessons for improving spatial and transportation planning through data-driven decision-making.

Chapter 8, " Microsimulation Analysis of Public Transport: The Case of Bagdat Street, Istanbul," authored by İrem Merve Ulu and Burcu Ozdemir use advanced microsimulation techniques to model public transport scenarios on Bagdat Street in Istanbul. This chapter provides valuable insights into how adjustments in public transportation frequency and capacity can reduce congestion and improve service quality. The findings presented here are not only relevant to Istanbul but offer broader implications for urban transportation planning in other high-density urban areas.

Chapter 9, " Road Aesthetics and its Effect on Road Users," authored by Anuj Jaiswal and his co-authors delve into the often-overlooked field of road aesthetics, highlighting its impact on user perception and behavior. This interdisciplinary chapter brings together perspectives from environmental psychology, urban design, and traffic safety to argue that road aesthetics are integral to creating pleasant and functional urban spaces. The authors emphasize the importance of designing roads that are not only safe and efficient but also visually appealing, enhancing the overall urban experience.

Chapter 10, " Have Rebranding Strategies Improved the Economy and Environment in an Old District of a Metropolis?: A Pilot Study in Wanchai, Hong Kong," authored by Giovanna Chiu and colleagues provide a detailed case study of Wanchai, an old commercial district in Hong Kong that has undergone significant redevelopment. The authors assess the impact of rebranding strategies on both the economy and environment, revealing that while economic growth has been substantial, environmental improvements have been limited. This chapter offers important insights into the trade-offs involved in urban renewal projects, providing lessons for policymakers and urban planners seeking to balance economic development with sustainability.

Chapter 11, " The Circularity of Transportation Completed," authored by Yavuz Duvarcı and co-authors propose a radical rethinking of transportation systems, emphasizing the need for circularity in the sector. By integrating concepts like solar energy, autonomous vehicles, and intelligent transport systems, this chapter offers a vision for how transportation can become more sustainable, efficient, and socially inclusive. The authors also explore the philosophical underpinnings of transportation circularity, drawing on critical perspectives to argue for systemic change in the way urban mobility is structured.

Chapter 12, " The Relationship Between COVID-19 and Urban Features in the Light of Recent Quantitative Studies," authored by Tayfun Salihoğlu and İhsan Tutuk analyze the relationship between urban features and the spread of COVID-19, synthesizing recent quantitative studies to explore how various urban factors—such as population density, infrastructure, and healthcare services—contributed to the pandemic's impact. The chapter underscores the importance of considering public health in urban planning, offering recommendations for enhancing urban resilience in the face of future pandemics.

In sum, "Interdisciplinary Approaches to Transportation and Urban Planning" offers a comprehensive exploration of the many ways in which transportation systems and urban planning intersect. The chapters in this book underscore the importance of interdisciplinary collaboration in addressing the complex challenges facing modern cities. From the integration of intelligent systems and cloud computing to the promotion of sustainable transportation policies and urban resilience, the research and case studies presented in this volume provide a roadmap for creating more equitable, efficient, and environmentally sustainable urban environments.

This book contributes to the growing body of literature on urban mobility and planning by offering new perspectives on pressing issues such as climate change, digital transformation, and public health. By bridging the gap between theory and practice, it offers practical solutions that can be implemented in cities around the world, making it an indispensable resource for researchers, practitioners, policymakers, and students alike. Through the collective expertise presented here, this volume advances our understanding of the critical role that transportation and urban planning play in shaping the cities of the future.

Görkem Gülhan
Pamukkale University, Department of Urban and Regional Planning, Turkey

Olcay Polat
Pamukkale University, Department of Industrial Engineering, Turkey

Chapter 1
The Role of Transportation in Environmental Sustainability and Climate Change Mitigation

Aylin Tepecik
https://orcid.org/0000-0002-3886-4301
Erciyes University, Turkey

ABSTRACT

The type of transportation used in urban areas is important in terms of environmental sustainability. Cities trigger greenhouse gas production and thus climate change with the services they offer. Environmental problems have increased with increasing greenhouse gas emissions and traffic congestion. In ensuring environmental sustainability, the importance of green transportation types that reduce the number of motor vehicles and fossil fuel dependency, consume less energy, and reduce vehicle-related pollutant problems are mentioned and play an active role in reducing environmental problems. The importance of this study is to address the transportation sector, which is an important source of greenhouse gases, the relationship between the effects of climate change and the transportation sector, and the examination of green transportation policies in reducing the effects of climate change. The study reveals environmentally friendly transportation types in sustainable transportation policies with green transportation systems.

DOI: 10.4018/979-8-3693-6695-0.ch001

INTRODUCTION

The population living in urban areas in particular is increasing worldwide. While 39% of the world population lived in cities in 1980, this rate increased to 54% in 2015. It is estimated that this rate will increase to 66% by 2050 (United Nations 2014). In parallel with the increase in population, the increasing number of motor vehicles and the dependence of cities on fossil energy sources increase the effects of climate change. Traffic density also increases with the increase in the number of vehicles. This situation causes an increase in the time spent in traffic, an increase in the consumption of petroleum-derived fuels widely used in the transportation sector, and an increase in the share of greenhouse gas emissions, leading to environmental pollution. Cities are the places where climate change occurs and its effects are felt the most. With the increasing human population, consumption needs have started to increase day by day, and in line with these needs, land use patterns, use of natural resources, and energy consumption habits have started to change.

Due to excessive use of fossil fuels, industrialization, misuse of land, agricultural activities, deforestation, production of construction materials, waste, population growth, and urbanization, the amount of greenhouse gas emissions in the atmosphere is increasing (Demirbaş and Aydın 2020). Total greenhouse gas emissions worldwide have increased by approximately 50% between 1990 and 2020 (World Bank, 2022). Fossil fuels, which meet the majority of energy consumption, cause global warming and climate change. When fossil energy sources are used, their negative effects on the environment are greater than renewable energy sources, and they release pollutants and greenhouse gases into the environment (Erdoğan, 2020).

The greenhouse effect is the fact that the gases in the atmosphere are more permeable to short-wave sunlight and less permeable to invisible wavelength light reflected from the earth due to accumulated greenhouse gases, and the parts close to the ground warm more than expected (Öztürk, 2002). The most common greenhouse gases that cause this effect and are formed in the atmosphere as a result of both natural processes and human activities are: water vapor, carbon dioxide, methane, nitrous oxides and ozone (Shah et al. 2024). The greenhouse effect caused by the presence of greenhouse gases occurs in the troposphere layer of the atmosphere, which is responsible for the emission of 30-40% of the toxic gases in the atmosphere. These pollutants that contribute to greenhouse gas emissions contribute to climate change. Rising temperatures triggered and amplified by rising carbon dioxide levels increase ozone formation, and in addition, increased water vapor leads to more chemical reactions and ozone production (Arshad et al. 2024). These pollutants that deteriorate air quality; It affects climate change patterns, changing precipitation, temperature distributions and ecosystem dynamics (Ding et al., 2017).

Cities affect and are affected by climate change, which is a global problem. Greenhouse gas emissions in cities mainly originate from urban activities. The greatest negative effect of transportation, one of the urban activities, on climate change is the use of fossil fuels. On a global scale, cities are responsible for 75% of natural resource consumption and 70% of greenhouse gas emissions (Çelikyay and Bayraktar 2023; Aygün and Torlak 2019). Cities trigger greenhouse gas production and are exposed to the devastating effects of climate change with the increase in their population density. Due to industrial and commercial, transportation service, and heating activities, the energy consumed in urban areas is more than in rural areas, and emission emissions from energy consumption cause environmental problems. Greenhouse gas emissions in cities mainly originate from urban activities such as the construction sector, transportation, and urban waste, and these activities are the main factors affecting climate change. The transportation sector is responsible for a large portion of greenhouse gas emissions and is one of the sectors most dependent on fossil fuels (Robaina and Neves 2021).

Along with the development of technology, the development and diversification of transportation systems provide great convenience in meeting consumer demands, but also bring environmental problems. In the distribution of greenhouse gas emissions by sector, emissions from transportation constitute 21% of total emissions, while 70% of global emissions among transportation types originate from road transportation (Our World in Data 2020; Çalışkan et al. 2017). In 2022, more than 35% of carbon dioxide emissions from road transportation originated from trucks and buses, and approximately 10% from private cars and minibuses (IEA 2022). Motor vehicles have become the most common means of transportation in urban transportation. The continuous increase in carbon emissions as a result of fossil fuel consumption creates sustainability problems (Sardar et al. 2022). The increasing use of motor vehicles in urban transportation, along with the increase in traffic accidents and congestion, has increased environmental problems on a global scale.

The study used literature and document scanning method. The elements affecting environmental sustainability and climate change will be specified, and the aim is to examine the role of transportation in ensuring environmental sustainability and reducing the effects of climate change. The study, which consists of four main sections, addresses the effect of transportation on environmental sustainability and climate change in the first section; the role of transportation in reducing the effects of climate change in the second section. The third section includes sustainable green transportation policies in reducing the effects of climate change prepared by using literature. The fourth section, which follows, addresses the prominent systems and planning approaches in green transportation.

BACKGROUND

Impact of the Transportation Sector on Environmental Sustainability and Climate Change

The concept of sustainability basically focuses on preventing environmental problems and reducing the demand for environmental resources. According to Keleş (1998), the concept of sustainable development is explained as the environmentalist worldview that aims to ensure economic development by using environmental values and natural resources in a rational way that does not lead to waste, by taking into account the rights and benefits of present and future generations. Sustainable development refers to the balance between the environmental, economic and social dimensions of sustainability. One of the most important components of sustainability is environmental sustainability. Environmental sustainability can be defined as the creation and maintenance of conditions that allow people and nature to exist in productive harmony and to meet the social, economic and environmental needs of present and future generations (U.S. Department of Energy, 2020). According to Morelli (2011), environmental sustainability is to meet the resource and service needs of present and future generations without endangering the health of the ecosystems that provide them. The transportation sector, which contributes greatly to economic and social development, has negative effects on the environment, society and economy. It is known that a large part of the negative impacts of the transportation sector on the environment are due to excessive fossil fuel use (Greene and Wegener 1997). Many human activities carried out in cities increase greenhouse gas emissions. The two main sectors that release more carbon dioxide into the environment compared to other sectors are transportation and land use (Yigitcanlar and Kamruzzaman 2014). Today, the transportation sector, whose usage area and frequency are increasing, is also increasing greenhouse gas emissions due to the increasing amount of fossil fuel use.

Figure 1. Carbon dioxide emissions of countries from 1960 to 2021 (Global Carbon Project, 2022)

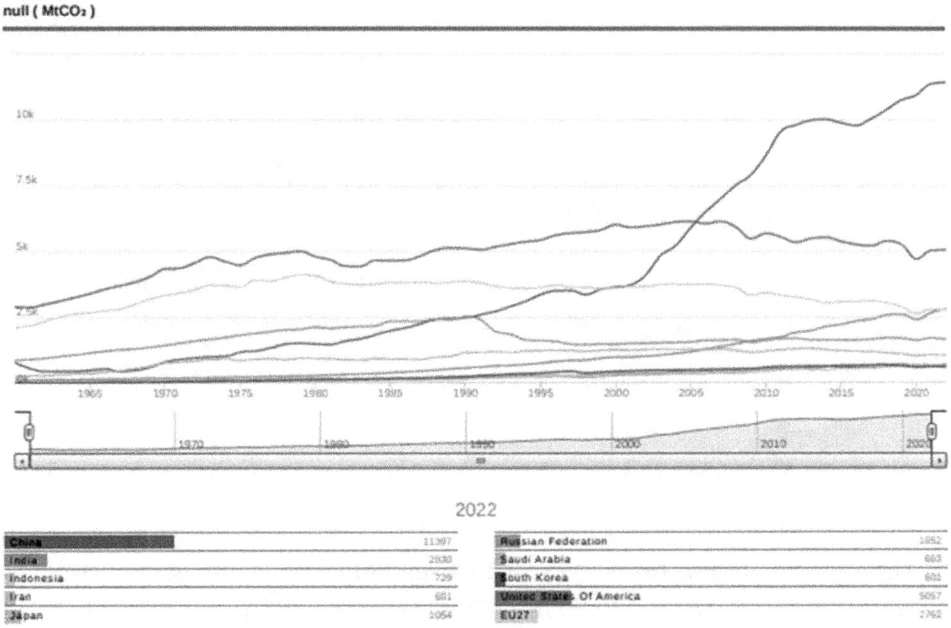

The countries that emitted the most emissions in the world between 1960 and 2022 include China, the United States, India, EU27, Russia and Japan. The top three polluting countries in the world account for 51.1% of total emissions, while the top ten countries account for approximately 70% of greenhouse gas emissions (Global Carbon Project, 2022). China and the United States make the largest contributions to greenhouse gas emissions.

Figure 2. Global greenhouse gas emissions by sector from 1990 to 2023 (Statista, 2023)

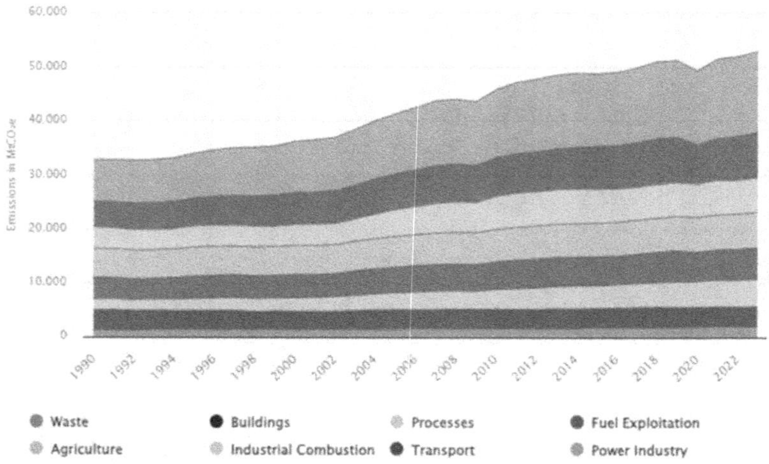

The energy industry was the largest contributor to greenhouse gas emissions in the world between 1990 and 2023. In 2023, the energy industry contributed 38% to carbon dioxide emissions. According to 2023 data, approximately 16% of greenhouse gas emissions are caused by transportation (Statista, 2023). In 2023, transportation sector carbon dioxide emissions increased by more than 21%.

Figure 3. Distribution of carbon dioxide emissions produced by the transportation sector in the world in 2022 (Statista, 2023)

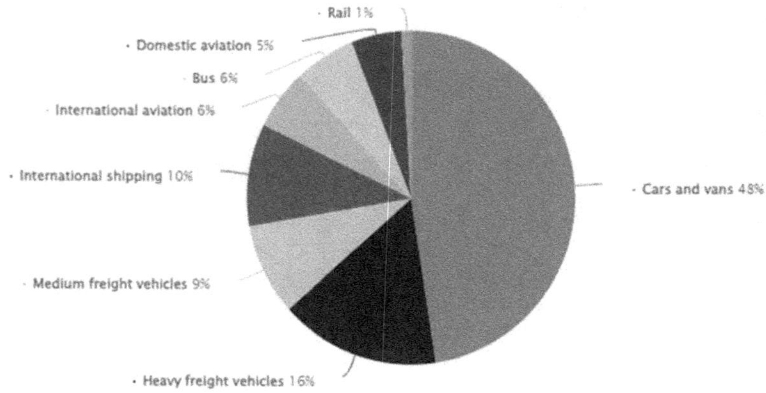

In transportation, especially road transportation is seen as the main source of carbon dioxide. When carbon dioxide emission rates are examined according to transportation networks around the world, road transportation comes first; followed by air, sea, and railway. Cars and minibuses account for approximately 48% of transportation emissions, while medium and heavy-duty vehicles account for approximately one-fourth (Statista, 2023). Climate change, which is a global problem and occurs globally, affects many sectors, including transportation. Due to reasons such as the rapidly increasing number of roads and motor vehicles, their impact on global emissions, and the risk of depletion of fossil-based energy derivatives, the need for efficient use of energy and the trend towards clean and renewable energy sources have become increasingly important (Kaplan and Kaynak 2011). With the need for efficient use of energy, the concept of sustainable transportation that does not harm the environment has gained importance. Over time, the use of vehicles operating with more environmentally friendly fuels has become preferred in passenger and freight transportation. Green transportation is very important in ensuring the sustainability of urban transportation. Green transportation is defined as a transportation service that has less negative impact on public health and the natural environment compared to existing transportation services (Björklund, 2011).

According to Zahedi (2012), green transportation factors are listed as follows:

- People's environmental knowledge and sensitivity
- Managers and employees awareness
- Laws and regulations
- Government priorities and requirements
- Suppliers and manufacturers responsibility for continuous improvement
- Strive for discovering efficiency opportunities in all the above.

In this context, increasing the use of environmentally friendly means of transport is of great importance in terms of sustainable green transport. Transport policies aim to improve air quality, reduce environmental pollution and increase the quality of life. In the transport sector, greenhouse gas emissions can be reduced with measures such as reducing climate change and adaptation strategies such as decarbonization and carbon pricing initiatives, using neutral technologies, and shifting transport to alternative modes of transport such as inland waterways and railways (Melkonyan et al. 2024), indicating a significant paradigm shift with the transition from fossil fuel-based transport systems to low and zero-emission technologies (Yeh et al. 2024).

The Role of Transportation in Mitigating the Impacts of Climate Change

Cities cause climate change through greenhouse gas emissions. Increases in greenhouse gases are the most important cause of climate change. Recently, high population densities in cities, increasing air pollution, decreasing habitats for wild animals, increasing temperatures increasing urban heat island formation, and infrastructure problems increase the effects of climate change. There are two important approaches to minimizing the effects of the factors causing climate change: mitigation and adaptation. Climate change mitigation; tries to reduce greenhouse gas emissions by using alternative energy sources, stops the increase in carbon emissions on earth and encourages more energy saving. Adaptation; covers adaptation policies that aim to prepare societies and economies against the potential effects of climate change (Doğan and Tüzer 2011). In climate change adaptation studies, taking strategic decisions at national, regional and local levels is important in terms of solving possible problems. Among the strategies for adapting to climate change in cities and reducing its effects; increasing the capacity and quality of transportation systems, encouraging non-motorized modes of transportation, and increasing investment in low-carbon or alternative fuel vehicles (Bulkeley et al. 2009). In strategies to reduce emissions from transportation, many cities focus on technological developments such as electrification of transportation (Heidrich et al. 2016). One of the sectors that cities prioritize in greenhouse gas reduction policies is the energy sector. A significant portion of energy consumption in the world is obtained as a result of burning fossil fuels. Greenhouse gas emissions of cities are mainly caused by transportation, construction, waste, and agricultural activities (Gergin 2024). The transportation sector, which has an important role in sustainable development, contributes to the economy on a national and global scale, as well as causing an increase in greenhouse gases caused by fossil fuels. The increase in the emissions of these gases causes global warming and climate change. According to Bıyık and Civelekoğlu (2018), the only way to prevent climate change is to reduce the share of fossil fuels used in vehicles in energy production and to turn to renewable energy sources.

In solving environmental problems originating from transportation; the tariff and taxation of environmental costs, improvement of environmental norms regarding emissions, and penalization of violations in transportation come to the forefront (Şekercioğlu and İncekara 2017). In this direction, many stakeholders have focused on the concepts of sustainability and green transportation in order to reduce the negative effects of transportation systems on the environment. One of the most important sources of fossil fuel-based emissions that cause air pollution and climate change is motor vehicles (Bıyık and Civelekoğlu 2018).

With the widespread use of motor vehicles in urban transportation, air pollution is also increasing. The use of motor vehicles, which constitutes an environmentally unsustainable structure, brings with it problems such as air pollution, traffic congestion, and fossil fuel dependency. The widespread use of bicycles and electric vehicles to prevent air pollution caused by motor vehicles is an important step in this direction (Akyüz 2021). In this context, studies carried out to reduce greenhouse gases will reduce emissions that cause air pollution. Measures to reduce greenhouse gas emissions in transportation are listed as the use of alternative fuels, encouraging vehicles that save fuel, reducing vehicle use, reducing the areas of vehicles used for transportation, developing smart transportation systems, integrated transportation systems, and reducing travel (Peng et al. 2010). According to Boarnet (2010), the effectiveness of greenhouse gas emission pricing policies and regulations that encourage land use are effective policy approaches.

Countries' approaches to reducing and preventing greenhouse gases also guide their economic policies. Some countries' efforts to reduce these negative effects are listed as energy-efficient cars, fuel taxes, improving traffic flow, financial measures, and voluntary agreements (Demirtürk 2021). The rate of fossil fuel use in road transportation is quite high within transportation networks. The measures to be taken by the leading countries in the world in greenhouse gas emissions to reduce this increase are as follows; reducing fuel consumption, alternative fuel preference, life cycle analyses, designs for fuel saving, sustainable road construction, vehicle technology development, traffic management systems, smart transportation systems, administrative sanctions, and reducing mobility (Demirtürk 2021). Carbon pricing is used to reduce greenhouse gas emission rates in preventing climate change. Carbon pricing is an application that causes a change in the behavior of consumers with high carbon footprints and encourages consumers to consume less energy (Karakaya et al. 2023).

Sustainable Green Transportation Policies in Mitigating the Impacts of Climate Change

Sustainable urban transportation is a system that is safe, environmentally friendly, energy efficient, low cost, accessible and fuel efficient (Cirit 2014). A sustainable transportation system requires the use of integrated transportation systems instead of individual vehicle use. In reducing the impact of transportation systems on climate change, it is aimed to reduce dependency on fossil fuels, take measures to reduce greenhouse gas emissions, and spread climate-friendly technologies. In this context, it is aimed to reduce the negative impacts of the transportation sector on the environment and to provide sustainable transportation.

The environmental, social and economic dimensions of sustainable development have been accepted at national and international levels, and the role of the transportation sector was explained in the UN Rio Earth Summit in 1992 and in the action plan titled Agenda 21. The scope of sustainability-related studies was expanded at the UN Sustainable Development Conference organized by the UN in 2012 (Rio+20), and it was emphasized that transportation and mobility increase economic growth and accessibility within the framework of the green economy and green sustainability concept. The High-Level Advisory Group on Sustainable Transportation organized by the UN was established in 2014, and the Housing and Sustainable Urban Development Conference organized by the UN was held in 2016. The importance of green transportation is emphasized in the Sendai Disaster Risk Reduction Framework adopted at the UN World Conference on Disaster Risk Reduction held in 2015, and in the New Urban Agenda document adopted at the UN Habitat III Summit held in 2016, and local climate actions include mitigation and adaptation issues.

Agenda 21, which sets forth the principles for coping with environmental and development problems in the 21st century and achieving sustainable development goals, emphasizes the relationship between the transportation sector and the energy sector, and reveals the need for countries to plan their national goals with sustainable development principles. Agenda 2030 (UN Sustainable Development Goals) was adopted in New York in 2015. In order to create a city and environment compatible with sustainable development goals, 17 goals were established to be implemented by 2030. In line with the 2030 Sustainable Development Goals, in order to ensure green transportation; accessible and clean energy was aimed in the 7th goal, and sustainable cities and living spaces were aimed in the 11th goal, and the implementation of other goals towards this end has gained importance.

In national and supranational transportation policy documents, the development and use of sustainable and clean transportation modes are targeted in various parts of the world. Babalık-Sutcliffe (2009) stated that in many cities of the world, there has been a policy change from mobility models that emphasize public transportation and non-motorized modes to an approach that implements innovative programs that increase awareness about clean transportation. Green transportation, also known as sustainable transportation, is a type of transportation that does not negatively affect the environment, unlike fossil fuel transportation modes, and is based on renewable energy sources (DBSA 2021). The reduction of negative effects caused by pollutants, especially greenhouse gases, in urban transportation can be achieved with sustainable green transportation policies (Kabakuş and Eyüboğlu 2023). The common points of sustainable transportation policies are to expand transportation with micro mobility vehicles, encourage public transportation, reduce individual motor vehicle use, and reduce environmental impacts by using vehicles such as bicycles and e-scooters (Özden and Kun 2024).

By examining academic studies on the subject, inferences regarding sustainable green transportation policies are presented in Table 1.

Table 1. Green policy proposals

No	Policy Recommendations	References
1	Infrastructure provision and management, technology, regulation, information and awareness, pricing and taxation	May and Crass (2007)
2	The environmental efficiency model that measures the energy efficiency of transportation systems is proposed, and it is suggested that emissions trading systems should be established between developed and developing countries.	(Yoshino et al., 2010)
3	Developing and using green technologies that aim to reduce the impact of transportation and land use on the environment, encourage sustainable behavioral changes, and provide multimodal transportation infrastructures.	Yigitcanlar and Kamruzzaman (2014)
4	Sustainable low-carbon transportation plans are proposed to improve transportation structure and reduce demand.	Dhar and Shukla (2015)
5	Promoting low-carbon and environmentally friendly green transportation	(Li, 2016)
6	Use of low-carbon public transport in the road network of historical areas, taking scientific and strict traffic management measures to improve the traffic problem in these areas	Wang and Sun (2019)
7	All types of transport that are greener for all stakeholders, promote sustainability by limiting the loss of natural areas, and take into account social and environmental costs.	(Bibri et al., 2020)
8	Three strategies (prevention-modification-improvement) are proposed to overcome the obstacles to the implementation of green transport. It is suggested to adopt innovative technologies or management approaches to green public transport systems.	(Shah et al., 2021)
9	Ensuring green supply chains, creating sustainable technology centers for green transportation systems, managing urban waste, and providing cleaner energy conversion areas by recycling waste.	(Rehman et al., 2023)
10	Ensuring public-private partnerships in reducing environmental damage in the transportation sector, integrating renewable energy into the transportation sector, use of biodiesel, use of fuel-efficient vehicle systems	(Alnour et al., 2024)
11	Investing in green transport infrastructure, raising public awareness and implementing effective policy interventions to reduce air pollution	(Sari et al., 2024)

Source: Drawn up by the author

Considering the policies developed in sustainable green transportation in reducing the effects of climate change; policies such as easing traffic congestion, reducing energy consumption in transportation systems, spreading and promoting green and environmentally friendly sustainable transportation systems provide positive contributions to solutions for improving air quality and reducing greenhouse gas emissions. In policies addressed in the context of the effects of the transportation sector causing climate change, studies and practices on green energy issues should be developed and strengthened. This deficiency in current policies should be addressed by conducting studies on green energy investments to ensure energy efficiency, encouraging local governments to develop policies to increase green areas and reduce carbon emissions, and investing in renewable energy sources.

Green Transportation Systems and Planning Approaches

The contribution of green transportation to sustainability is that transportation activities are carried out with the right timing, the right infrastructure and low cost by eliminating the elements that will harm the environment (Kabakuş and Eyüboğlu 2023). The prominent systems within the scope of green transportation systems are; bicycle and pedestrian transportation, electric vehicles, micromobility and public transportation systems.

Bicycle and Pedestrian Transportation

Pedestrian and cycling transportation are among the non-motorized transportation types. The increase in the use of motor vehicles in urban areas leads to environmental, social and economic problems. Automobile-focused urban areas have caused the formation of environmentally unsustainable living spaces (Erçetin 2023). Pedestrian and cycling transportation types are included in the scope of green transportation in ensuring environmental sustainability. The Sustainable Urban Mobility Plan prepared by the European Commission develops policies to increase bicycle and pedestrian transportation. SUMPs are strategic plans that focus on the mobility of individuals, aim to change travel preferences as individual mobility, and propose regulations to ensure the sustainability of the environment and urban mobility, unlike the traditional transportation approach (Ayaz 2023). Transportation types that include the practice of cycling and walking and support these activities with digital tools are becoming increasingly widespread. Recently, the use of transportation vehicles such as electric scooters, electric bicycles, and e-micromobility as micromobility vehicles has become widespread in many cities (Önder and Akdemir 2022).

Micromobility

Micromobility is an urban transportation mode that reduces the use of private vehicles and facilitates the transition to low-carbon transportation modes (Abduljabbar et al. 2021). Today, micromobility solutions are increasing due to both economic and environmental benefits. Micromobility solutions include mini vehicles such as bicycles, skateboards, and electric scooters. E-scooters are one of the most prominent micromobility vehicles recently. It is defined as a cost-effective, sustainable, and innovative transportation type that provides users with short-distance travel options (Shaheen et al. 2020). The basic strategies and goals of cycling in urban transportation in developing and emerging countries are; reducing the use of motor vehicles, popularizing cycling, integrating cycling with other public transportation vehicles, creating hierarchies of bicycle paths, increasing awareness of bicycle transportation, increasing the safety of bicycle paths, increasing bicycle parking areas, and ensuring a safe infrastructure for bicycle paths (Ardıçoğlu 2023).

Electric Vehicles

As the world population increases, the rate of vehicle use is also increasing. The use of fossil fuels is seen as the main source of carbon dioxide. It is known that a significant part of the negative impacts of the transportation sector on the environment are caused by the use of fossil fuels. Electric vehicles, which do not emit greenhouse gas emissions, have been considered as a solution to prevent environmental concerns. Electric vehicles have great potential to solve climate change problems by limiting carbon dioxide emissions (Shah et al. 2021). Electric vehicle sales have increased in recent years. In constantly changing and developing electric vehicles, batteries are used instead of gasoline tanks and electric motors are used instead of internal combustion engines. In the last decade, electric vehicles have become more popular as a green transportation mode thanks to their high energy efficiency and zero exhaust emissions (Ouyang and Xu 2022). Electric vehicles; It provides more economical, quieter and more environmentally friendly transportation due to its low fuel consumption and lack of oil dependency. While these vehicles provide the advantage of being highly efficient, low-cost and able to be charged overnight; They have disadvantages such as deficiencies in battery technology, being expensive compared to internal combustion engines, and having limited range (Kocagöz and İğde 2022). It is predicted that by 2050, sales of electric vehicles will increase in the world and will be the transportation choice of the future.

Public Transportation Systems

Providing easy access to public transportation systems and using environmentally friendly fuel types are among the basic features of a sustainable city (Cirit 2014). Access to public transportation services in sustainable urban mobility is facilitated by information and communication technologies. Solutions for smart transportation targets in cities include various smart transportation applications such as smart stop information, smart parking systems, smart bicycle systems and paths, pedestrian paths, and traffic density maps (Yavuz 2024). These applications that encourage the effective use of public transportation should be made widespread. The basic strategies and goals in urban transportation in developing and emerging countries; integrating the green wave system into the existing traffic system, establishing a safe transportation infrastructure, providing safe vehicle parking and stop areas, ensuring the integration of transportation types with smart applications, and taking shared road arrangements in the city as a basis. When the compact city and eco-compact city, which are among the planning approaches that prioritize sustainable development, are examined, the common points of these approaches are; reducing energy consumption and greenhouse gas emissions, creating walkable access distances, and structuring alternative models in public transportation systems.

CONCLUSION AND EVALUATION

Due to greenhouse gas emissions, climate change is negatively affected and one of the most fundamental sources of this is transportation. The transportation sector is responsible for approximately ¼ of greenhouse gas emissions globally. In this context, adaptation to climate change and reduction policies are important in terms of solving problems to reduce emissions originating from transportation. One of the most important actions for reducing greenhouse gas emissions of cities is sustainable transportation or green transportation. Methods that will reduce the use of fossil fuels in ensuring environmental sustainability have become increasingly important in recent years. The targets of developing sustainable and clean transportation modes in transportation and increasing their use are implemented in many cities of the world. In this context, the study attempts to explain the targets and measures to reduce the negative effects of the transportation sector on the environment. In these studies, the common points of sustainable/green transportation policies; expanding transportation with vehicles such as bicycles, e-scooters, electric bicycles, which are known as low-carbon, increase accessibility, take up little space in transportation areas, and provide mobility at a micro scale, encouraging public transportation vehicles, and prioritizing non-motorized transportation types such as reducing individual vehicle use were discussed. Micro mobility vehicles that help alleviate traffic should be integrated with urban transportation systems. Various measures are being taken to reduce the share of fossil fuels in energy production, to encourage the use of renewable energy sources and alternative fuel vehicles. In addition, deterrent measures are being implemented against individual vehicle use through measures such as increasing fuel prices, fuel taxes, traffic congestion, and extra charges. It is necessary to support individual vehicle use with improvements in alternative transportation modes to discourage it and increase the quality of life.

In this direction, local government units play a major role in the development of transportation infrastructure and quality. In urban planning approaches, green transportation systems that enable compact cities and eco-compact cities should be used. Bicycle transportation, micromobility, electric vehicles, and public transportation systems should be integrated with the entire transportation system. These green transportation systems should be integrated with the transportation infrastructure on a national, regional and urban scale. In this way, the amount of carbon footprint, especially from road transportation, can be reduced in the transportation sector. In order to reduce greenhouse gas emissions and carbon footprint, it is necessary to prefer low-emission fuels as another option, use electric vehicles, popularize public transportation provided by electric buses, and develop shared road system applications. In compositions where public vehicles meet pedestrians, the use of common platforms has the potential to create problems for safety. Considering this situation, vehicle-bicycle paths should be specified with separate platforms, sidewalks should be arranged with wide distances, some streets should be completely turned into pedestrian paths, vehicles that allow for micro mobility compatible with the environment should be popularized, and road design guides should be provided for the use of these vehicles.

REFERENCES

Abduljabbar, R. L., Liyanage, S., & Dia, H. (2021). The Role of Micro-mobility in Shaping Sustainable Cities: A Systematic Literature Review''. *Transportation Research Part D, Transport and Environment*, 92, 102734. DOI: 10.1016/j.trd.2021.102734

Afifa, K. A., Hussain, N., Ashraf, M. H., & Saleem, M. Z. (2024). Air Pollution and Climate Change as Grand Challenges to Sustainability''. *The Science of the Total Environment*, 928, 172370. Advance online publication. DOI: 10.1016/j.scitotenv.2024.172370 PMID: 38604367

Akyüz, E. (2021). *''Urban Transportation: Pros and Cons of Elektric Vehicles''*. Hagia Sophia 3. İnternational Conference on Multidisciplinary Scienific Studies, İstanbul, 685-696.

Alnour, M., Awan, A., & Hossain, M. E. (2024). Towards a green transportation system in Mexico: The Role of Renewable Energy and Transport Public-Private Partnership to Curb Emissions. *Journal of Cleaner Production*, 442, 140984. Advance online publication. DOI: 10.1016/j.jclepro.2024.140984

Ardıçoğlu, R. (2023). Bicycle Network Planning in Urban Transportation: The Case of Elazığ. *Urban Academy*, 16(1), 194–223.

Arshad, K., Hussain, N., Ashraf, M. H., & Saleem, M. Z. (2024). Air pollution and climate change as grand challenges to sustainability. *The Science of the Total Environment*, •••, 172370. PMID: 38604367

Ayaz, Ç. E. (2023). A Strategic Overview on Urban Mobility Management: A Qualitative Analysis on Istanbul Sustainable Urban Mobility Plan. *Eskişehir Osmangazi University Journal of Social Sciences*, 24(1), 41–55.

Aygün, A., & Torlak, S. E. (2020). An Examination on Climate Change Resilience on Denizli''. *Nigde Omer Halisdemir University Journal of Engineering Sciences*, 9(1), 648–664.

Babalık-Sutcliffe, E. (2009). "Clean Transport: Innovative Solutions to the Creation of a More Sustainable Urban Transport System". Prospects for Research in Transport and Logistics on a Regional: Global Perspective (pp. 89-94). İstanbul: Dogus University Publications.

Bibri, S. E., Krogstie, J., & Kärrholm, M. (2020). "Compact city planning and development: Emerging practices and strategies for achieving the goals of sustainability". *Developments in the Built Environment*, 4(20), 1–20. DOI: 10.1016/j.dibe.2020.100021

Bıyık, Y., & Civelekoğlu, G. (2018). Investigation of Carbon Footprint Change Originated from Transportation Sector''. *Bilge International Journal of Science and Technology Research*, 2(2), 157–166.

Björklund, M. (2011). Influence From The Business Environment on Environmental Purchasing -Drivers and Hinders of Purchasing Green Transportation Services. *Journal of Purchasing and Supply Management*, 17(1), 11–22. DOI: 10.1016/j.pursup.2010.04.002

Boarnet, M. G. (2010). Planning, climate change, and transportation: Thoughts on policy analysis. *Transportation Research Part A, Policy and Practice*, 44(8), 587–595. DOI: 10.1016/j.tra.2010.03.001

Bulkeley, H., Schroeder, H., Janda, K., Zhao, J., Armstrong, A., Chu, S. Y., & Ghosh, S. (2009). ''Cities and Climate Change: The role of İnstitutions, Governance and Urban Planning''.*5th Urban Research Symposium: Cities and Climate Change*, Marseille, France, 1-92.

Çalışkan, D. Z., Kurt, Ü., & Timur, M. C. (2017). ''Econometric Analysis of Climate Change and Transportation Sector Relation: The Case of Turkey''.*International Congress of Energy, Economy and Policy*, İstanbul.

Çelikyay, H. H., & Bayraktar, H. K. (2023). *"Climate Friendly Cities and Resilience Strategies in Climate Policy Papers"*. HUMANITAS –. *The International Journal of Social Sciences (Islamabad)*, 11, 87–107.

DBSA. (2021). ''How Green Transport Can be a Catalyst for Economic Development and Growth *[online]*''. (18.07.2024), https://www.dbsa.org/article/how-green-transport-can-be-catalyst-economic-development-and-growth

Demirbaş, M., & Aydın, R. (2020). The Biggest Threat of the 21st Century: Global Climate Change. *Ecological Life Sciences*, 15(4), 163–179.

Dhar, S., & Shukla, P. R. (2015). Low Carbon Scenarios for Transport in India: Co-benefits Analysis. *Energy Policy*, 81, 186–198. DOI: 10.1016/j.enpol.2014.11.026

Ding, A., Huang, X., & Fu, C. (2017). Air Pollution and Weather Interaction in East Asia''. *Environmental Sciences (Ruse)*. Advance online publication. DOI: 10.1093/acrefore/9780199389414.013.536

Doğan, S., & Tüzer, M. (2011). Combating Global Climate Change: General Approaches and International Efforts''. *Journal of Sociology Conferences*, 44, 157–194.

Erçetin, C. (2023). Environmental Sustainability and Cycling as a Transport Mode: Best Pracices". *Environment, City and Climate Journal.*, 2(3), 238–269.

Erdoğan, S. (2020). *''Energy, Environment and Greenhouse Gas''*. Çankırı Karatekin University Faculty of Economics and Administrative Sciences Journal, 10(1), 277-303.

Gergin, E. D. (2024). Cities Resistant to Climate Change: Examples of Good Local Governance Practice in The World and In Turkey''. *Erzurum Technical University Social Sciences Institute Journal*, 18, 94–111.

Global Carbon Project. (2022). *''Carbon dioxide emissions 2022 [online]''*. (15.08.2024). https://www.globalcarbonproject.org/

Heidrich, O., Reckien, D., Olazabal, M., Foley, A., Salvia, M., de Gregorio Hurtado, S., Orru, H., Flacke, J., Geneletti, D., Pietrapertosa, F., Hamann, J. P., Tiwary, A., Feliu, E., & Dawson, R. J. (2016). National climate policies across Europe and their impacts on cities strategies''. *Journal of Environmental Management*, 168, 36–45. DOI: 10.1016/j.jenvman.2015.11.043 PMID: 26696604

Holden, E., Linnerud, K., & Banister, D. (2013). Sustainable passenger transport: Back to Brundtland''. *Transportation Research Part A, Policy and Practice*, 54, 67–77. DOI: 10.1016/j.tra.2013.07.012

IEA. (2021). *''Global Energy Review 2021 [online]''*. (07.07.2024), https://www.iea.org/reports/global-energy-review-2021

IEA. (2022). *''Uses of Energy' [online]'*. (10.07.2024), https://www.iea.org/energy-system

Kabakuş, N., & Eyüboğlu, M. (2023). *''Use of Environmentally Friendly Transportation Vehicles''*. In 3 th International Congress of Engineering and Natural Sciences, Ankara 258-270.

Kaplan, H., & Kaynak, Z. (2011). *''Green transportation types and regulations in urban transportation: an analysis in terms of authority and responsibilities of local governments''*. 2nd Public Transportation Week Transist, IV. Transportation Symposium and Exhibition, Haliç Congress Center, Istanbul, 01-02 December, 160-162.

Karakaya, E., Akkoyun, G., & Hiçyılmaz, B. (2023). *''Carbon Pricing for Greenhouse Gas Emission Reduction: Carbon Tax or Emission Trading?''*. Journal of Research in Economics. *Politics & Finance*, 8(4), 813–841.

Keleş, R. (1998). *Dictionary of Urban Science Terms* (2nd ed.). İmge Bookstore Publications.

Kocagöz, E., & İğde, Ç. S. (2022). Which Factors Affect the Intention to Buy an Electric Vehicle? A Consumer Study''. *Journal of Social Sciences of Kahramanmaraş Sütçü İmam University*, 19, 104–120.

Li, H. R. (2016). Study on Green Transportation System of International Metropolises''. *Procedia Engineering*, 137, 762–771. DOI: 10.1016/j.proeng.2016.01.314

May, T., & Crass, M. (2007). Sustainability in Transport: Implications for Policy Makers. *Transportation Research Record: Journal of the Transportation Research Board*, 2017(1), 1–9. DOI: 10.3141/2017-01

Morelli, J. (2011). Environmental Sustainability: A Definition for Environmental Professionals. *Journal of Environmental Sustainability*, 1(1), 1–10. DOI: 10.14448/jes.01.0002

Önder, H. G., ve Akdemir, F. (2022). ''*Rebuilding Sustainable Transportation Infrastructure in the Period of Pandemic: Micromobility Trends and Turkey*''. İdealkent, 13(2), 748-770.

Our World in Data. (2020). ''Cars, planes, trains: where do CO_2 emissions from transport come from *[online]*''. (12.07.2024), https://ourworldindata.org/co2-emissions-from-transport

Ouyang, X., & Xu, M. (2022). Promoting Green Transportation under the Belt and Road Initiative: Locating Charging Stations Considering Electric Vehicle Users' Travel Behavior''. *Transport Policy*, 116, 58–80. DOI: 10.1016/j.tranpol.2021.11.023

Özden, A., & Kun, S. B. K. (2024). *Evaluation of the Bicycle and E-Scooter Infrastructure in Turkey in terms of Urban Transportation*. Pamukkale University Journal of Engineering Sciences., DOI: 10.5505/pajes.2024.09633

Öztürk, K. (2002). Global Climatic Changes and Their Probable Effect upon Turkey''. *Gazi University Gazi Faculty of Education Journal*, 22(1), 47–65.

Peng, Z., Shen, S., Lu, Q., & Perch, S. (2010). *''Climate Change and Transportation''*, M. Kutz (eds.), Handbook of Transportaion Engineering Volume II, New York: McGraw Hill, pp. (35.1-35.24).

Rehman, F. U., Islam, M. M., & Miao, Q. (2023). Environmental Sustainability via Green Transportation: A Case of the Top 10 Energy Transition Nations. *Transport Policy*, 137, 32–44. DOI: 10.1016/j.tranpol.2023.04.013

Robaina, M., & Neves, A. (2021). Complete decomposition analysis of CO2 emissions intensity in the transport sector in Europe''. *Research in Transportation Economics*, 90(101074), 101074. DOI: 10.1016/j.retrec.2021.101074

Sardar, M. S., Asghar, N., & Rehman, H. U. (2022). *''Moderation of competitiveness in determining environmental sustainability: Economic growth and transport sector carbon emissions in global perspective''.Environment, Development and Sustainability*, 26(1), 1–23. DOI: 10.1007/s10668-022-02771-z

Sari, A. R., Sarie, F., Suparni, S., Nadi, M. A. B., & Karunia, M. N. (2024). The Effect of Environmentally Friendly Transportation Implementation and Supporting Policies on Air Pollution Reduction in Central Kalimantan. *West Science Interdisciplinary Studies*, 2(3), 605–614. DOI: 10.58812/wsis.v2i03.739

Şekercioğlu, S., & İncekara, B. (2017). *''Transforming Transport Policies: Planning for Greenhouse Gas Reduction''.* 12th Transportation Congress (Transportation Policies), 181-190.

Shah, I. H., Manzoor, M. A., Jinhui, W., Li, X., Hameed, M. K., Rehaman, A., Li, P., Zhang, Y., Niu, Q., & Chang, L. (2024). Comprehensive Review: Effects of Climate Change and Greenhouse Gases Emission Relevance to Environmental Stress on Horticultural Crops and Management. *Journal of Environmental Management*, 351, 119978. DOI: 10.1016/j.jenvman.2023.119978 PMID: 38169258

Shah, K. J., Pan, S. Y., Lee, I., Kim, H., You, Z., Zheng, J. M., & Chiang, P. C. (2021). Green Transportation for Sustainability: Review of Current Barriers, Strategies, and İnnovative Technologies''. *Journal of Cleaner Production*, 326, 129392. DOI: 10.1016/j.jclepro.2021.129392

Shaheen, S., Cohen, A., Chan, N., & Bansal, A. (2020). *''Sharing Strategies: Carsharing, Shared Micromobility (bikesharing and scooter sharing), Transportation Network Companies, Microtransit, and Other İnnovative Mobility Modes''*. Transportation, Land Use, and Environmental Planning, 237–262.

Statista, (2023). *''Global transportation sector CO_2 emissions 2000-2022 [online]''*,(19.08.2024), https://www.statista.com/statistics/1387814/carbon-dioxide-emissions-transport-subsector-worldwide/

Tuğaç, Ç. (2022). *''Green Transportation in the Context of Climate Change and Urban Resilience''*. İdealkent, 13(36), 545-575.

United Nations. (2014). *"Department of Economic and Social Affairs, Population Division [online]''*,(17.07.2024), https://www.urbanet.info/world-urban-population/

U.S. Department of Energy. (2020). *''Environmental Sustainability''*, (15.06.2024), https://www.energy.gov/lm/listings/environmental-sustainability

Wang, Q., & Sun, H. (2019). Traffic Structure Optimization İnhistoric Districts Based on Green Transportation and Sustainable Development Concept. *Advances in Civil Engineering*, 2019(1), 9196263. Advance online publication. DOI: 10.1155/2019/9196263

World Bank. (2022). ''*The World Bank Open Data*'', (15.07.2024), https://data .world bank.org/

Yavuz, H. R. (2024). *"Istanbul Example in Smart Cities Regarding Transportation"*. Premium e-. *Journal of Social Sciences*, 8(38), 91–96.

Yeh, S., Paltsev, S., Reilly, J. M., Daniels, D., & Linares, P. (2024). *''Designing Resilience for Multi-System Dynamics of Future Transportation''*. MIT Joint Program on the Science and Policy of Global Change, 371.

Yigitcanlar, T., & Kamruzzaman, M. (2014). Investigating the İnterplay Between Transport, Land Use and The Environment: A Review of the Literature''. *International Journal of Environmental Science and Technology*, 11(8), 2121–2132. DOI: 10.1007/s13762-014-0691-z

Yoshino, D., Fujiwara, A., & Zhang, J. (2010). Environmental Efficiency Model based on data Envelopment Analysis and its Application to Environmentally Sustainable Transport Policies. *Transportation Research Record: Journal of the Transportation Research Board*, 2163(1), 112–123. DOI: 10.3141/2163-13

Zahedi, S. (2012). Green Transportation a More Eco-efficient Option''. *Journal of Social and Development Sciences*, 3(7), 223–228. DOI: 10.22610/jsds.v3i7.706

Chapter 2
The Role of Smart Cities in Creating Sustainable Urban Environments Through Strategies and Innovative Solutions

Jorge A. Ruiz-Vanoye
https://orcid.org/0000-0003-4928-5716
Universidad Politécnica de Pachuca, Mexico

Ocotlán Díaz-Parra
Universidad Politécnica de Pachuca, Mexico

Miguel A. Ruiz-Jaimes
https://orcid.org/0000-0002-2585-9896
Universidad Politécnica del Estado de Morelos, Mexico

Jaime Aguilar Ortiz
Universidad Politécnica de Pachuca, Mexico

Eric Simancas-Acevedo
https://orcid.org/0000-0001-7823-709X
Universidad Politécnica de Pachuca, Mexico

Jazmín Rodriguez-Flores
https://orcid.org/0000-0002-3313-0325
Universidad Politécnica de Pachuca, Mexico

Josué Román Martínez Mireles
https://orcid.org/0000-0002-3167-5209
Universidad Politécnica de Pachuca, Mexico

Juan M. Xicoténcatl-Pérez
https://orcid.org/0009-0008-8554-4192
Universidad Politécnica de Pachuca, Mexico

Marco Antonio Vera Jimenez
https://orcid.org/0009-0009-3712

DOI: 10.4018/979-8-3693-6695-0.ch002

-1440
Universidad Politécnica de Pachuca, Mexico

ABSTRACT

A smart city is a city that uses the optimum management of water, city technology, energy, mobility, public transportation, distribution of transport routes, road network safety management, the construction of homes / offices / transportation structure, industry / jobs, education, healthcare, public safety, government, civil protection, food, environmental pollution, and solid, hazardous and special waste This chapter will analyse how smart cities can become catalysts for urban sustainability through the implementation of advanced technologies and the integration of innovative strategies. With a focus on carbon footprint reduction, energy efficiency and climate change resilience, it will explore various initiatives and policies that are transforming cities towards a greener and more equitable future.

1. INTRODUCTION

A smart city is defined as an urban area that effectively manages resources such as water, technology, energy, transportation, and public services. This includes optimizing mobility and public transit systems, planning transport routes, ensuring the safety of road networks, and overseeing the construction of residential and commercial buildings as well as transportation infrastructure. Additionally, it encompasses the management of industries and job creation, educational facilities, healthcare services, public safety measures, governmental functions, civil protection efforts, food supply chains, environmental pollution control, and the handling of solid, hazardous, and special waste (Ruiz-Vanoye et al., 2023).

A smart city is defined as an urban area that leverages digital technologies and interconnected data to enhance the living standards of its residents, maximize the efficiency of urban resources and services, and foster sustainable development (Ruiz-Vanoye et al., 2023). Some Key Aspects of Sustainability in Smart Cities: Some key aspects of sustainability in smart cities include energy efficiency, sustainable mobility is fundamental, developing efficient public transport infrastructures, smart water management is using advanced technologies to monitor and efficiently distribute the resource, detecting leaks and optimizing consumption, waste management benefits from smart systems that facilitate collection and recycling, environmental control, with continuous monitoring of air quality through sensors and the implementation of policies to reduce pollution, green infrastructure, citizen participation is strengthened through the use of digital platforms that allow inhabitants to get involved in decision making, promoting inclusive and participatory governance. These combined elements ensure that smart cities are not only technologically advanced, but also sustainable, resilient and equitable, aligning with global sustainable development goals.

Moreover, the role of data analytics in enhancing urban services cannot be overstated. The work of underscores the cost-effectiveness of Internet of Things (IoT) technologies in smart cities, particularly in optimizing healthcare delivery and improving public services (Febrer et al., 2021). As cities increasingly adopt smart technologies, the ability to analyze and utilize data effectively becomes paramount for driving innovation and sustainability.

However, the integration of these technologies also raises concerns regarding privacy and security, particularly with the rapid generation of personal data by connected devices. This challenge is echoed in the literature, where scholars emphasize the need for robust governance frameworks that can adapt to the evolving landscape of smart city initiatives (Mohammadzadeh, 2023). Ensuring that policies keep pace with technological advancements is crucial for fostering trust among citizens and stakeholders.

Smart cities represent a crucial intersection between technology and sustainability. This chapter will provide an overview of how smart cities can drive the transition to urban sustainability, addressing both the benefits and associated challenges.

2. KEY TECHNOLOGIES FOR SUSTAINABILITY IN SMART CITIES

The rapid increase in population necessitates urban development that aligns with the evolving needs of communities, presenting various challenges related to environmental sustainability, resource management, and the quality of life for residents. In response to these complex issues, the concept of smart cities has emerged as a promising solution. Technologies such as the Internet of Things (IoT), Big Data, renewable energy sources, and artificial intelligence can be utilized to transform urban areas into more efficient, resilient, and livable environments for their inhabitants. These technologies enable cities to gather real-time data on multiple facets of urban life, analyze patterns, and make informed decisions that optimize resource utilization while minimizing adverse environmental effects. This approach contributes to reducing carbon emissions, enhancing air and water quality, and promoting overall sustainability for both present and future generations.

In the field of urban development, sustainability plays a crucial role in ensuring the long-term viability and resilience of cities. By incorporating sustainable practices into urban planning and design, cities can reduce their environmental impact, promote social equity and increase economic prosperity. Sustainability in urban development involves a comprehensive approach that considers factors such as energy efficiency, waste management, transportation systems, green spaces and water conservation. In addition, sustainability in urban development can lead to cost savings, improved public health outcomes, and greater community engagement and empowerment. As cities continue to grow and face challenges related to climate change and resource depletion, prioritizing sustainability in urban development is essential to building resilient and thriving urban spaces.

Internet of Things (IoT) in Smart Cities

The integration of the Internet of Things (IoT) in the development of smart cities presents a dynamic and multifaceted approach to urban development.

The Internet of Things (IoT) is crucial for promoting urban sustainability as it facilitates a network of connected devices that gather and exchange data to enhance resource management efficiency. In smart cities, IoT devices are employed to oversee and regulate different systems, including transportation, energy, and waste management. By collecting data in real time, these devices allow for informed decision-making and the optimization of urban operations. For example, IoT sensors can track energy consumption patterns to identify areas for improvement and reduce the overall carbon footprint. By incorporating IoT solutions into urban planning and infrastructure development, cities can create safer and healthier environments for their residents while minimizing their ecological footprint (Bibri S. E., 2018).

The development of smart cities represents a multifaceted approach that integrates technology, governance, and community engagement to address urban challenges. The successful implementation of smart city initiatives hinges on the effective use of technology, active citizen participation, and robust governance structures that collectively contribute to sustainable urban development. This integration of technology not only facilitates efficient resource management but also fosters sustainable economic growth and improved quality of life for residents (Shieh, 2022; Dimitrova, 2024).

The integration of big data analytics and context-aware computing into urban planning is pivotal for advancing the sustainability of smart cities. discusses the untapped potential of these technologies in shaping smart sustainable cities, emphasizing that big data can significantly enhance urban decision-making processes and operational efficiencies (Bibri, 2019).

Big Data Analytics for Sustainable Urban Development

By applying these tools, cities can optimize energy consumption, mitigate ecological degradation, and improve the overall quality of life for residents. In addition, the application of big data for traffic management and urban planning in smart cities has shown promising results in improving overall sustainability. By analyzing large data sets collected from various sources, such as IoT devices, cameras and sensors, urban planners can obtain valuable information on traffic patterns, congestion hotspots and transportation demands, enabling the optimization of traffic flow, the development of efficient public transportation systems and the establishment of pedestrian-friendly urban spaces.

The integration of Big Data analytics into smart city actions presents immense opportunities to improve urban management efficiency, sustainability, and residents' quality of life. However, this transformative potential must be accompanied by a strong focus on ethical and privacy considerations to ensure responsible deployment. One of the critical contributions of technology to urban infrastructure is its role in promoting sustainability. emphasize that innovative technologies can significantly enhance the performance of sustainable urban infrastructures through a set of indicators that assess their contribution to sustainable development (Schinkel et al., 2022).

Renewable Energy Technologies in Smart Cities

The integration of technologies for renewable energy production in the smart city environment detonates a fundamental opportunity for sustainable urban development.

The development of smart cities is increasingly intertwined with the integration of Big Data and the Green Internet of Things (G-IoT), which collectively aim to enhance urban sustainability and efficiency. Butt and Tariq highlight that utilizing Big Data analytics in conjunction with G-IoT technologies can significantly improve resource management, energy efficiency, and overall urban livability Wu et al. (2018). The synergy between these technologies allows for the collection and analysis of vast amounts of data generated by IoT devices, facilitating informed decision-making that can lead to more sustainable urban environments (Eckhoff & Wagner, 2018).

Urban areas are increasingly adopting sustainable energy solutions, with a focus on integrating solar energy systems into existing infrastructure, as highlighted in the comprehensive review by (Etukudoh E.A., et al., 2024), the unique challenges of urban environments, such as limited space and shading, drive the need for innovative design solutions to optimize energy capture. Seamless integration of solar panels into buildings, windows, and urban infrastructure shows the potential to transform cities into energy efficient hubs.

In the realm of sustainable urban development, innovations in energy storage from renewable energy technologies play a key role in shaping the trajectory towards a more environmentally friendly future.

These technological advances create immense potential for the development of sustainable smart cities, but their effectiveness ultimately depends on informed decision-making and strategic planning. Local governments and urban planners must prioritize investments in smart infrastructure that not only improve efficiency but also promote environmental responsibility, but for the future of smart cities, it is imperative to consider the social and ethical implications of these technologies. Community engagement and collaboration are essential to ensure that smart city initiatives prioritize the well-being of all residents and create inclusive and equitable urban environments.

Sustainable Smart Cities Case Studies

The swift expansion of urban areas poses considerable challenges related to sustainability and the overall quality of life. In response, smart cities have been proposed as a viable solution, utilizing digital technologies to enhance the efficiency of resource management and urban services. This research is grounded in a qualitative analysis of multiple case studies, drawing on secondary data sourced from reports, academic literature, and particular smart city initiatives.

Some Case Studies are:

- **In Bangalore, India, Cisco's Smart City project** focuses on smart traffic management, air quality monitoring and waste management, with the aim of improving the quality of life in one of the country's most dynamic cities. This project has implemented advanced technologies such as IoT sensors and real-time traffic management systems, enabling more efficient monitoring and control of urban conditions. The implementation of these systems has had a remarkable impact on urban mobility by significantly reducing travel times and traffic congestion, which has not only improved transport efficiency in the city, but also contributed to the reduction of greenhouse gas emissions from urban traffic. In addition, the constant monitoring of air quality by these sensors has enabled authorities to make more informed and proactive decisions to mitigate air pollution, thereby improving public health and the well-being of residents. Cisco's project in Bangalore exemplifies how the integration of smart technologies can address critical urbanization issues, promoting a more sustainable and livable environment in rapidly expanding cities like Bangalore.

- **The Multimedia Super Corridor in Malaysia** is an ambitious initiative that seeks to transform a region into a technology and multimedia hub, promoting energy efficiency and the use of renewable energy sources. This project has implemented an advanced Information and Communication Technology (ICT) infrastructure together with the use of renewable energy, which has significantly reduced the dependence on fossil fuels for power generation. As a result, a considerable decrease in carbon emissions associated with energy production has been achieved, thus contributing to energy sustainability and economic development in the region. The combination of state-of-the-art technology infrastructure with a focus on clean energy not only drives economic growth, but also positions the Multimedia Super Corridor as a model of sustainable development for other regions seeking to integrate technology with environmental sustainability. This project demonstrates how the development of a technology hub can go hand in hand with responsible practices that benefit both the economy and the environment.
- **The Eko Atlantic Project in Nigeria** is an ambitious sustainable urban development project being implemented on land reclaimed from the ocean, with the aim of creating a modern and resilient city in the face of environmental challenges. This project uses smart technologies for water, energy, and waste management, integrating advanced desalination and smart waste management systems. The implementation of these technologies has minimized methane emissions from landfills and optimized energy use in water treatment systems, contributing significantly to the environmental sustainability of the region. In addition to providing robust and adaptable urban infrastructure, the Eko Atlantic Project sets a new standard in urban development, demonstrating how the combination of smart technologies and sustainable practices can create cities that are not only responsive to today's needs, but also ready to meet future challenges. This approach not only ensures a more efficient use of natural resources, but also strengthens the resilience of infrastructure in the face of climate hazards, positioning Eko Atlantic as a model for sustainable urbanization in Africa and beyond.
- **Konza Technology City in Kenya** is a planned city that stands out for its advanced infrastructure for technology and innovation. This ambitious project has implemented smart technologies for resource and transport management, resulting in significant efficiency gains in urban management and a boost to innovation. Smart resource and transport management has improved energy efficiency by reducing energy consumption in transport systems and buildings through real-time monitoring and

control. This approach not only optimizes the use of energy resources, but also contributes to environmental sustainability by reducing carbon emissions and other pollutants associated with energy consumption. Konza Technology City is presented as a model of efficient and innovative urbanization, where the integration of smart technologies plays a crucial role in creating a sustainable urban environment conducive to technological and economic development.

- **The Petronia City Project in Ghana** is a large-scale urban development project that aims to create a smart and sustainable city. This ambitious project has implemented an integrated smart infrastructure designed to improve urban living standards and promote sustainable development. The smart infrastructure has enabled a more efficient use of energy in the city, optimizing consumption in various urban systems and reducing environmental impact. By harnessing advanced technologies for resource and service management, Petronia City is positioned as a model of modern urbanization in West Africa, offering an improved living environment aligned with sustainability principles. This approach not only contributes to more efficient urban development, but also sets a new standard for future smart cities in the region.
- **The Smart Nation Program in Singapore** is a comprehensive initiative covering urban mobility management, energy efficiency, water management and citizen engagement. This program has implemented advanced technologies such as big data, IoT and citizen engagement platforms to transform the way the city manages its public resources and services. Smart technologies have significantly improved efficiency in energy management, reducing energy consumption and associated costs. This has not only optimized the use of resources, but also contributed to improving the quality of life of citizens and the efficiency of public services. Through this approach, Singapore has established itself as a world leader in implementing technological solutions to create a more sustainable, efficient, and inclusive urban environment.
- **The Eco Smart City Project in Vietnam** is an initiative focused on building sustainable infrastructure and implementing smart technologies to improve the quality of life of its inhabitants. This project has incorporated green buildings and smart energy management systems, leading to a significant reduction in energy consumption. Green buildings, designed to maximize energy efficiency, together with smart systems that monitor and optimize energy use, have contributed to improving urban sustainability. This approach not only decreases the city's carbon footprint, but also sets a new standard for sustainable urban development in

Vietnam, demonstrating how the combination of advanced technologies and green design can transform cities into more efficient and environmentally friendly spaces.

- **The Digital and Knowledge-Based City in Mexico**, known as Ciudad Creativa Digital in Guadalajara, is a project focused on promoting the creative and technology industry in the region. This initiative has implemented smart urban management solutions that have optimized energy use, improved efficiency and reduced operating costs. By integrating advanced technologies into urban resource management, the Digital Creative City not only fosters innovation and economic development, but also establishes a more sustainable and efficient environment for the creative and technology industries. This project highlights how the implementation of smart technologies can transform the urban landscape, making the city an engine of economic growth and a benchmark for sustainability and operational efficiency in Mexico.
- **The Skolkovo Project in Russia** represents a bold initiative aimed at establishing a high-tech innovation hub. This endeavor has introduced a sophisticated Information and Communication Technology (ICT) infrastructure along with smart building technologies, enhancing energy efficiency within the urban landscape. The incorporation of these technologies not only fosters economic growth and innovation but also aids in developing a more sustainable and efficient urban environment, aligning with Skolkovo's objectives to become a focal point for technological advancement in Russia.
- **The Living Lab in the Netherlands** is a project located in Amsterdam that focuses on fostering urban innovation through a living laboratory, where smart solutions are tested and applied. This approach has enabled the development of strategies that improve energy efficiency in the city, promoting both innovation in urban management and citizen participation. The continuous experimentation and implementation of smart solutions has turned the Living Lab into a model of how cities can evolve in a sustainable way, involving citizens in the innovation process and improving the quality of urban life.
- **The Chamber Campaign in Colombia** is a campaign focused on improving urban services through the implementation of smart technologies. This initiative has developed smart platforms for the delivery of public services, resulting in increased energy efficiency and reduced consumption. The implementation of these technologies has not only improved the quality and efficiency of public services, but also increased citizen satisfaction by providing faster and more effective services. This

approach demonstrates how the integration of technological solutions can transform urban management, making services more sustainable and aligned with community needs.

- **The Large Outdoor Escalator System in Colombia** is an urban mobility project designed to improve accessibility in mountainous areas. This smart escalator system has significantly improved urban mobility, optimizing the use of energy in transporting people. By facilitating travel in hard-to-reach areas, this project not only improves the quality of life of residents, but also contributes to a more efficient use of energy resources in transport. The integration of smart technologies in this mobility system is an example of how innovative urban solutions can improve both accessibility and sustainability in urban environments.
- **The Integrated Transportation Network in Brazil** is a project that seeks to integrate various modes of urban transport through the use of smart technologies. Through the implementation of intelligent transport management systems, this project has succeeded in improving the efficiency of the transport system and reducing urban congestion. The integration of different transport modes has optimized the efficiency of the system as a whole, resulting in a significant decrease in energy consumption. This approach demonstrates how the coordination and intelligent management of transport resources can transform urban mobility, making it more fluid, efficient and sustainable.
- **The Sustainable Barcelona Map Project in Spain** is an initiative aimed at creating a comprehensive map of sustainable practices and resources in Barcelona, using geographic information systems (GIS) and open data platforms. This project has increased awareness and participation in sustainable practices, promoting the use of smart technologies to improve energy efficiency in the city. By providing access to detailed information on sustainable resources, the project has encouraged greater adoption of green practices, contributing to the reduction of energy consumption, and strengthening Barcelona's commitment to urban sustainability.
- **Open Data - GXBus in Uruguay** is an initiative focused on the use of open data to improve urban transport systems. Through open data platforms and intelligent transport management systems, route optimization and greater transport efficiency have been achieved, significantly reducing fuel consumption. This project not only improves transport efficiency, but also encourages citizen participation by providing access to crucial information that can be used to improve urban mobility.

- **The Built Smart Program in the United States** is a program aimed at improving the efficiency and sustainability of buildings through the use of smart technologies. The implementation of smart building management systems has led to greater operational efficiency, reducing energy consumption and operating costs. This program not only contributes to environmental sustainability, but also provides a model for future sustainable building initiatives.
- **The New York City 311** is a smart city initiative in the United States that provides a centralized platform for citizens to report problems and access city services. The centralized platform, coupled with big data analytics, has enabled more efficient management of public services, reduced waste of resources and improving energy efficiency in public administration. This has resulted in more effective service delivery and increased citizen participation.
- **The Wise City Initiative in China** is an initiative aimed at developing smart urban infrastructure in several Chinese cities. The use of smart infrastructure and big data has improved urban management and sustainability, increasing energy efficiency in city administration. This initiative reinforces China's commitment to sustainable urban development through technology.
- **The Tianjin Eco-City in China** is a joint project between China and Singapore that focuses on creating a sustainable city. Smart water, energy and waste management has improved the city's energy efficiency and climate resilience. This project serves as a model of sustainable urbanization, combining advanced technology with efficient resource management to create a resilient and ecologically responsible urban environment.
- **Guimarães Green City in Portugal** has implemented green initiatives and smart technologies for waste management and sustainable mobility. The implementation of smart waste management and sustainable mobility systems has improved environmental quality and reduced energy consumption in the city. This approach has not only improved urban sustainability, but also raised the quality of life of residents, establishing Guimarães as an example of a green city in Europe.

The implementation of smart city initiatives requires a collaborative approach that involves multiple stakeholders, including government agencies, private sector entities, and citizens. emphasize the significance of innovative city pilot policies in stimulating regional innovation and fostering collaboration among various stakeholders (Muniba & Yu, 2023).

Challenges and Opportunities in the Implementation of Smart Solutions

Incorporating smart technologies into urban planning involves numerous challenges that need to be tackled to guarantee the effectiveness of smart city initiatives. A critical concern is privacy and data security. The gathering and examination of large volumes of data can make cities vulnerable to considerable cybersecurity threats and privacy violations, potentially eroding public confidence in smart city endeavors. Implementing robust data protection measures is essential to safeguard citizens' information and maintain their confidence in these technologies (Adenekan, Ezeigweneme, & Chukwurah, 2024) emphasize that without addressing these concerns, the potential benefits of smart city initiatives may be overshadowed by public apprehension regarding data misuse (Adenekan, Ezeigweneme, & Chukwurah, 2024).

Ensuring last-mile connectivity is another major challenge, as effective smart mobility solutions require seamless door-to-door service regardless of the transportation mode. Additionally, the rapid generation of personal data by connected devices poses serious privacy and security concerns, making data sharing among various stakeholders complexes. Governance is also a critical challenge, as the inclusion of tech companies and service providers requires updated policies and regulations that are currently lagging. Finally, fostering adoption among users and stakeholders, managing dynamic routing and transportation demands, and ensuring interoperability between legacy systems and new technologies are all significant hurdles that need to be addressed.

The integration of smart technologies into urban planning is essential for creating efficient and sustainable urban environments. discuss how smart innovation ecosystems can meet citizens' needs by deploying sensor-based solutions and IoT technologies to improve local quality of life (Agbali et al., 2019).

Moreover, the principles of ecodevelopment, as outlined by, emphasize the need for high-tech solutions that not only reduce environmental impact but also promote the health and well-being of urban inhabitants (Korol et al., 2018).

Copaja Alegre and Esponda Alva (2019) articulate that the smart city model seeks to tackle contemporary urban challenges through interconnected intelligent systems that leverage new information and communication technologies (ICT).

3. SUGGESTIONS FOR URBAN PLANNERS AND POLICY MAKERS

This section provides key recommendations for urban planners and policy makers to guide the development of sustainable and resilient smart cities. By focusing on the integration of advanced technologies, promoting community engagement, and implementing robust environmental policies, these strategies aim to address the challenges of urbanization while fostering a more equitable and sustainable urban future.

1. Recommendation: Establish legal and regulatory frameworks that facilitate private investment in smart city projects, while ensuring that the benefits are distributed equitably among all citizens. Encourage public-private partnerships for the implementation of smart technologies and the financing of sustainable projects. Collaborations can accelerate the adoption of innovative solutions and improve efficiency in the implementation of urban projects. This table provides a summarized and organized view of examples of public-private partnerships in different cities and their impacts on sustainability and urban efficiency.

Table 1. Projects per city, part 1

City	Project Description	Impact
Barcelona, Spain	Public-private partnership to develop smart city strategy, including energy and transport management.	Modernizing urban infrastructure, improving energy efficiency and the quality of public services, and encouraging citizen participation.
Songdo, South Korea	Development of a smart city from scratch, in collaboration with companies such as Cisco and POSCO Engineering & Construction.	Creation of highly efficient and sustainable urban infrastructure with real-time monitoring of resources, demonstrating the potential of large-scale partnerships.
Amsterdam, The Netherlands	Amsterdam Smart City project that brings together business, government, and research organizations to develop sustainable urban solutions.	Implementation of pilot projects in electric mobility and energy efficiency, promoting an open innovation approach that can be replicated in other cities.
New York, United States	LinkNYC project to install smart kiosks offering free Wi-Fi, phone calls and digital services.	Improved urban connectivity and provision of essential services, especially in areas with less access, financed through advertising revenues.
Copenhagen, Denmark	Copenhagen Connecting project in collaboration with companies such as Cisco to use IoT and big data in traffic management.	Progress towards carbon neutrality by 2025, optimizing transport flow and reducing congestion and pollution in the city.

2. Recommendation: Incorporate clear targets for green infrastructure into urban development plans and ensure that these are implemented as part of a comprehensive approach to sustainability. Integrate green infrastructure into urban design to improve air quality, reduce the heat island effect, and promote biodiversity. This includes the creation of parks, green roofs, and green corridors.

Table 2. Projects per city, part 2

City	Project Description	Impact
Singapore	Singapore has integrated parks, green roofs and vertical gardens into its urban planning.	Significant improvement in air quality, reduction of the heat island effect, and increased biodiversity in a highly urbanized environment.
Copenhagen, Denmark	Copenhagen has developed a network of green corridors and green roofs as part of its strategy to become a carbon neutral city by 2025.	Improving air quality and reducing urban temperatures, as well as increasing accessible recreational areas and improving stormwater management.
Toronto, Canada	Toronto was the first city in North America to implement a bylaw requiring the installation of green roofs on new buildings.	Reducing stormwater runoff, improving the energy efficiency of buildings, and expanding green spaces in the urban environment.
Melbourne, Australia	Melbourne has implemented a strategy to increase tree cover in the city and create more green spaces.	Reduction of the heat island effect, improvement of air quality, and strengthening of urban biodiversity, along with an increase in the quality of life of citizens.
New York, United States	Initiative to plant one million trees in New York City as part of an effort to expand green infrastructure.	Improving air quality, reducing pollution and the heat island effect, and creating habitats for urban wildlife.

3. Recommendation: Develop policies that prioritize investment in public transport and smart mobility technologies, and create incentives for the adoption of electric vehicles and other sustainable modes of transport. Promote the adoption of smart transport systems that optimize urban mobility, reduce congestion, and minimize carbon emissions. This includes the expansion of efficient public transport networks, electric vehicles, and real-time traffic management systems.

Table 3. Projects per city, part 3

City	Project Description	Impact
Copenhagen, Denmark	Integration of a public bicycle system with electric vehicle infrastructure and charging stations.	Significant reduction of carbon emissions and vehicle congestion, promoting more sustainable and healthier mobility.
London, United Kingdom	Implementation of congestion charging zones and low emission zones to reduce traffic and pollution.	Reducing congestion and air pollution in central London by encouraging the use of public transport and electric vehicles.
Amsterdam, Netherlands	Expansion of electric vehicle infrastructure and promotion of bicycle mobility as a primary means of transport.	Amsterdam has achieved a high use of bicycles and electric vehicles, reducing car dependency and greenhouse gas emissions.
Barcelona, Spain	Implementation of rapid buses and a smart traffic light system that prioritizes public transport.	Improved punctuality and efficiency of public transport, reduced journey times and lower CO2 emissions due to traffic optimizations.
Seoul, South Korea	Use of sensors and cameras to monitor and manage traffic in real time, and expansion of public transport.	Reducing congestion, improving road safety, and optimizing the use of transport infrastructure, with a reduction in carbon emissions.

4. Recommendation: Develop and implement citizen participation tools that facilitate two-way communication between citizens and authorities, ensuring that urban decisions reflect the needs and desires of the community. Engage citizens in urban planning and management through digital platforms for participation and transparency in decision-making. This helps to create a sense of shared ownership and responsibility towards sustainability initiatives.

Table 4. Projects per city, part 4

City	Project Description	Impact
Reykjavik, Iceland	Better Reykjavik is an online platform that allows citizens to propose ideas and vote on urban projects, encouraging transparency and participation in decision-making.	It has generated greater citizen engagement, with many citizens' ideas implemented, strengthening transparency and trust in public administration.
Madrid, Spain	Decide Madrid is a citizen participation platform that allows residents to propose, debate and vote on urban initiatives and participatory budgets.	It has empowered citizens, allowing them to directly influence urban decisions and resource allocation, which has increased shared responsibility.
Paris, France	Budget Participatif is a participatory budget program where citizens vote on how part of the city's budget should be spent.	It has promoted transparency and direct participation in the allocation of public funds, resulting in community projects that better reflect the needs of the population.

continued on following page

Table 4. Continued

City	Project Description	Impact
Helsinki, Finlandia	Helsinki's Open Ahjo is a system that allows citizens to access local government documents and decision-making processes in real time, and to participate in online discussions.	It has improved government transparency and facilitated greater citizen participation in policy decisions, increasing trust in local government.
New York, United States	NYC 311 is a platform that allows citizens to report problems, make suggestions and access information about city services, facilitating communication with authorities.	It has improved efficiency in solving urban problems and strengthened communication between citizens and authorities, creating a sense of community collaboration.

5. Recommendation: Create policy frameworks that incentivize waste reduction, promote recycling and support innovation in waste management technologies, with a focus on the circular economy. Establish robust waste management policies that include source separation, recycling, and hazardous waste treatment. Smart cities can use advanced technologies to optimize waste collection and management.

Table 5. Projects per city, part 5

City	Project Description	Impact
San Francisco, United States	Zero Waste by 2020 is a waste management policy that includes source separation, mandatory recycling, and composting, with the aim of achieving zero waste by 2020.	San Francisco has successfully diverted more than 80% of its waste from landfills, making it a global leader in reducing waste and promoting the circular economy.
Stockholm, Sweden	Implementation of an automated underground system that transports waste from collection points to processing stations.	It has improved waste collection efficiency, reduced waste truck traffic, emissions and operating costs, while increased recycling.
Ljubljana, Slovenia	Ljubljana became the first European capital to adopt a zero-waste strategy, with policies including source separation and recycling.	The city has managed to significantly reduce waste sent to landfill, increasing recycling rates and promoting a culture of sustainability and circular economy.
Tokyo, Japan	Tokyo has implemented a complex system of separating waste into various categories, including plastics, metals, and organics, with strict recycling policies.	It has improved recycling rates and waste treatment efficiency, reducing the amount of waste going to landfill and promoting the reuse of materials.
Milan, Italy	Milan has implemented a selective collection system for organic waste, which is converted into compost or biogas, promoting urban sustainability.	It has achieved one of the highest organic waste recycling rates in Europe, significantly reducing the amount of waste sent to landfill and supporting the circular economy.

6. Recommendation: Adopt climate action plans at the municipal level that include adaptation and mitigation measures, supported by scientific and technological data to enable informed decision-making. Incorporate climate resilience strategies into urban planning, including preparedness for extreme weather events and mitigation of climate change impacts. This includes resilient infrastructure planning and natural risk management.

Table 6. Projects per city, part 6

City	Project Description	Impact
New York, United States	PlaNYC (OneNYC): A comprehensive plan that includes climate change adaptation measures, such as building resilient infrastructure and creating green spaces for water absorption.	It has enhanced the city's ability to manage extreme weather events like hurricanes and floods, thereby lowering related risks and safeguarding at-risk communities.
Rotterdam, The Netherlands	Rotterdam Climate Proof is a strategy to make the city resilient to climate change by 2025, including the construction of blue-green infrastructure such as water squares and green roofs.	It has reduced flood risk and improved water management, making Rotterdam a global model of urban climate resilience.
Copenhagen, Denmark	Cloudburst Management Plan: Plan to manage large amounts of rainfall through green and blue infrastructure, including flood parks and streets with advanced drainage systems.	It has mitigated the effects of flash floods, protecting both infrastructure and the population, and improved the city's ability to adapt to climate change.
Tokyo, Japan	Tokyo Metropolitan Resilience Project: An initiative aimed at developing infrastructure that can withstand climate impacts and establishing early warning systems for natural disasters.	It has strengthened Tokyo's ability to respond to earthquakes, typhoons, and other natural disasters, minimizing the impact on infrastructure and the lives of citizens.
Mexico City, Mexico	CDMX Resilience Strategy: A strategy that addresses the city's vulnerability to earthquakes, floods and other hazards, with adaptation measures and resilient infrastructure projects.	It has increased the city's capacity to manage extreme weather events and natural disasters, protecting vulnerable communities and ensuring the continuity of basic services.

7. Recommendation: Develop policies that facilitate investment in renewable energy infrastructure and promote the adoption of clean technologies at the community level, such as solar roofs and energy storage systems. Promote the integration of renewable energy sources into the urban energy grid, including solar, wind and other clean technologies. Smart cities should be leaders in the adoption of renewable energy to reduce their dependence on fossil fuels.

Table 7. Projects per city, part 7

City	Project Description	Impact
Freiburg, Germany	Green City Freiburg: Freiburg is known as the green city for its focus on solar energy, with solar roofs on public and private buildings, and the integration of solar energy into the urban grid.	It has significantly reduced dependence on fossil fuels, positioning itself as a model in the adoption of clean and sustainable technologies at EU level.
Copenhagen, Denmark	Copenhagen Energy Plan: Includes the integration of wind and solar energy into the city's electricity grid, with the goal of becoming carbon neutral by 2025.	Copenhagen has achieved a high penetration of renewable energy in its grid, drastically reducing carbon emissions and moving closer to its carbon neutrality target.
San Diego, United States	San Diego Climate Action Plan: Plan that includes a commitment to achieve 100% renewable energy by 2035, with a focus on solar and wind energy, and the promotion of solar roofs on homes.	San Diego has made progress towards its goal of 100% renewable energy, with a significant increase in the adoption of solar roofs and reduced reliance on fossil energy.
Sydney, Australia	Sustainable Sydney 2030: Initiative promoting the adoption of solar energy in public buildings and the installation of energy storage systems throughout the city.	Sydney has improved the city's energy resilience, with an increase in installed solar capacity and reduced carbon footprint, contributing to urban sustainability.
Barcelona, Spain	Barcelona Solar Project: A project that promotes the installation of solar roofs on residential and commercial buildings, as well as the creation of local energy cooperatives.	It has increased local solar energy production and reduced dependence on external energy sources, improving sustainability and community participation in renewable energy.

8. Recommendation: Invest in educational programs that include urban sustainability as part of the curriculum and provide workshops and training for public officials and citizens on the use and benefits of smart technologies. Drive education and training programs for city residents and professionals on sustainability issues and smart technologies. This ensures that the transition to a smart and sustainable city is understood and supported by all.

Table 8. Projects per city, part 8

City	Project Description	Impact
Vancouver, Canada	Greenest City Action Plan (GCAP): Includes educational programs in schools and for the community on urban sustainability, recycling, and energy efficiency.	It has increased awareness and participation in sustainability issues among residents, promoting sustainable practices at community and school level.
Copenhagen, Denmark	Copenhagen School of Design and Technology: Offers specific programs in sustainability and smart technologies, training students and professionals in sustainable urban design.	It has trained a new generation of professionals with skills in urban sustainability, who are actively contributing to the development of the smart and green city.
Melbourne, Australia	Sustainable Melbourne Fund: Provides training to businesses and citizens on how to implement sustainable practices and leverage clean technologies to improve efficiency.	It has facilitated the adoption of sustainable technologies in local businesses and increased community participation in green initiatives, improving urban sustainability.
Singapore	Eco-Smart Learning Hub: An initiative that integrates urban sustainability and the use of smart technologies into the curriculum of schools and training programs for civil servants.	It has ensured that citizens and civil servants alike are well informed about sustainability and smart technologies, supporting Singapore's transition to a smart city.
New York, United States	NYC CoolRoofs: A program that trains residents to install white roofs that reflect sunlight, reducing the need for air conditioning and improving energy efficiency.	It has created local employment and improved energy efficiency in buildings, reducing carbon emissions and raising environmental awareness among participants.

9. Recommendation: Create mechanisms for periodic review of urban policies, based on performance indicators, and adapt strategies according to changing environmental, technological, and social conditions. Implement monitoring and evaluation systems that allow for reviewing and adjusting urban policies based on tangible data and results. This ensures that implemented strategies remain effective and aligned with long-term sustainability objectives.

Table 9. Projects per city, part 9

City	Project Description	Impact
London, United Kingdom	London Environment Strategy (LES): A strategy that includes an ongoing monitoring system to assess progress towards air quality, energy efficiency and waste reduction targets.	It has enabled policies and projects to be adjusted based on the data collected, ensuring that London remains on track towards its long-term sustainability goals.
Seoul, South Korea	Seoul Smart City Plan: Implementation of a real-time monitoring system to evaluate the efficiency of mobility, energy and waste management policies.	It has facilitated the rapid adaptation of results-based policies, improving the effectiveness of sustainability strategies and the quality of life of citizens.
Stockholm, Sweden	Stockholm's Vision 2040: Using performance indicators and an annual review system to adjust policies related to transport, energy and green infrastructure.	It has ensured that urban policies adapt to new technologies and social conditions, keeping Stockholm on the road to urban sustainability and resilience.
New York, United States	OneNYC: Sustainability and resilience strategy that includes a monitoring system and an annual report that assesses progress and allows for adjustments in housing, transport, and energy policies.	It has allowed for the modification of strategies and the allocation of resources based on the results obtained, improving the effectiveness of urban policies over time.
Copenhagen, Denmark	Copenhagen Climate Plan: A plan that uses a system of performance indicators to review and adjust climate and energy policies on an ongoing basis.	It has enabled Copenhagen to remain a leader in urban sustainability, adjusting policies to meet carbon neutrality targets for 2025.

CONCLUSIONS

Smart cities are essential for fostering sustainable urban environments by utilizing cutting-edge technologies and innovative approaches. By incorporating renewable energy sources, enhancing urban mobility, encouraging citizen engagement, and establishing effective waste management policies, these cities not only boost operational efficiency but also lower their carbon emissions and enhance resilience to climate change. Furthermore, education and training focused on sustainability empower both citizens and professionals to engage actively in the shift towards a smart and sustainable urban future. Continuous assessment and policy adaptation, supported by data-driven monitoring systems, enable urban strategies to remain aligned with long-term sustainability goals.

As cities around the world face increasingly complex environmental, social and economic challenges, the adoption of urban intelligence-based approaches becomes an imperative. Supported by appropriate public policies, investments in sustainable infrastructure and a focus on equity and inclusion, smart cities can lead the transformation to a greener, more efficient and equitable urban future. This approach not only benefits current generations, but also the future of cities.

REFERENCES

Adenekan, O. A., Ezeigweneme, C., & Chukwurah, E. G. (2024). The evolution of smart cities: Integrating technology, governance, and sustainable development. *International Journal of Applied Research in Social Sciences*, 6(5), 891–902. DOI: 10.51594/ijarss.v6i5.1131

Agbali, M., Trillo, C., Ibrahim, I., Arayıcı, Y., & Fernando, T. (2019). Are smart innovation ecosystems really seeking to meet citizens' needs? insights from the stakeholders' vision on smart city strategy implementation. *Smart Cities*, 2(2), 307–327. DOI: 10.3390/smartcities2020019

Bibri, S. (2019). The anatomy of the data-driven smart sustainable city: Instrumentation, datafication, computerization and related applications. *Journal of Big Data*, 6(1), 59. Advance online publication. DOI: 10.1186/s40537-019-0221-4

Bibri, S., Alahi, A., Sharifi, A., & Krogstie, J. (2023). Environmentally sustainable smart cities and their converging ai, iot, and big data technologies and solutions: An integrated approach to an extensive literature review. *Energy Informatics*, 6(1), 9. Advance online publication. DOI: 10.1186/s42162-023-00259-2 PMID: 37032812

Copaja-Alegre, M., & Esponda-Alva, C. (2019). Tecnología e innovación hacia la ciudad inteligente. avances, perspectivas y desafíos. *Bitácora Urbano Territorial*, 29(2), 59–70. DOI: 10.15446/bitacora.v29n2.68333

Dimitrova, V. (2024). Education in the era of smart cities: transformation and opportunities. *Environment Technology Resources Proceedings of the International Scientific and Practical Conference*, 2, 352-357. DOI: 10.17770/etr2024vol2.8084

Eckhoff, D., & Wagner, I. (2018). Privacy in the smart city-applications, technologies, challenges, and solutions. *IEEE Communications Surveys and Tutorials*, 20(1), 489–516. DOI: 10.1109/COMST.2017.2748998

Etukudoh, E. A., Queen, Z., Nwokediegwu, S., Umoh, A. A., Ibekwe, K. I., & Adefemi, A. (2024). Solar power integration in urban areas: A review of design innovations and efficiency enhancements. *World Journal of Advanced Research and Reviews*, 21(1), 1383–1394. DOI: 10.30574/wjarr.2024.21.1.0168

Febrer, N., Folkvord, F., & Lupiañez-Villanueva, F. (2021). Cost-effectiveness assessment of internet of things in smart cities. *Frontiers in Digital Health*, 3, 662874. Advance online publication. DOI: 10.3389/fdgth.2021.662874 PMID: 34713138

Korol, O., Shushunova, N., Lopatkin, D., Zanin, A., & Шушунова, T. (2018). Application of high-tech solutions in ecodevelopment. Matec Web of Conferences, 251, 06002. DOI: 10.1051/matecconf/201825106002

Mohammadzadeh, Z. (2023). Smart city healthcare delivery innovations: a systematic review of essential technologies and indicators for developing nations. DOI: 10.21203/rs.3.rs-3297004/v1

Muniba, M., & Yu, B. (2023). Does innovative city pilot policy stimulate the chinese regional innovation: An application of did model. *International Journal of Environmental Research and Public Health*, 20(2), 1245. DOI: 10.3390/ijerph20021245 PMID: 36673993

Ruiz-Vanoye, J. A., Diaz-Parra, O., Marroquín-Gutiérrez, F., Rodríguez-Lara, B. M., & Aguilar-Ortiz, J. (2023). Smart and Sustainable Cities. In Ruiz-Vanoye, J. A. (Ed.), *Management, Technology, and Economic Growth in Smart and Sustainable Cities*. IGI Global., DOI: 10.4018/979-8-3693-0373-3.ch001

Schinkel, U., Becker, N., Trapp, M., & Speck, M. (2022). Assessing the contribution of innovative technologies to sustainable development for planning and decision-making processes: A set of indicators to describe the performance of sustainable urban infrastructures (isi). *Sustainability (Basel)*, 14(4), 1966. DOI: 10.3390/su14041966

Shieh, A. (2022). Smart cities for technological and social innovation: Case studies, current trends, and future steps. *Urban Policy and Research*, 40(3), 281–284. DOI: 10.1080/08111146.2022.2104441

Wu, S., Dong-qiang, G., Wu, Y., & Wu, Y. (2018). Future development of taiwan's smart cities from an information security perspective. *Sustainability (Basel)*, 10(12), 4520. DOI: 10.3390/su10124520

Chapter 3
Smart Cities, Transportation, and Regional Development:
The Case of the TR32 Regional Development Strategy for 2024-2028 in Turkey

Oğuz Özbek
https://orcid.org/0000-0002-6742-4006
Pamukkale University, Turkey

Ezgi Karaçoban
https://orcid.org/0000-0002-4777-4885
Evliyaoğlu Textile Inc., Turkey

ABSTRACT

This paper examines the integration of smart cities, transportation policies, and regional development strategies in the TR32 Region (the provinces of Aydın, Denizli and Muğla) of Turkey, focusing on the 2024-2028 Draft Regional Plan prepared by the Southern Aegean Development Agency (GEKA). The research highlights the importance of incorporating smart transportation measures into the regional development framework to enhance the region's competitiveness, sustainability, and quality of life. The draft plan demonstrates a comprehensive approach to integrating smart city principles and technologies into its development strategy, with a focus on digital transformation, innovation, and green growth. The plan effectively addresses the integration of smart technologies in various sectors, such as transportation, manufacturing, agriculture, and tourism. However, the evaluation of the

DOI: 10.4018/979-8-3693-6695-0.ch003

plan's effectiveness in implementing smart transportation strategies is limited by the lack of formal feedback or evaluation of the previous GEKA plan period and the fragmented nature of data sources.

INTRODUCTION

The concept of smart cities has emerged as a powerful driver of urban and regional development, utilizing technology to address local challenges and enhance the quality of life. Integrating smart city strategies into regional development plans has become increasingly important as cities and regions strive to address the challenges of urbanization, economic growth, and sustainability. Smart city initiatives, which use advanced technologies and data-driven decision-making, have the potential to transform urban environments and enhance the quality of life for residents. Here, transportation, as a critical component of urban infrastructure, plays a significant role in the successful implementation of smart city strategies and the achievement of regional development goals.

In Turkey, the TR32 Region (Southern Aegean Region), comprising the provinces of Aydın, Denizli, and Muğla, has adopted smart city strategies to stimulate regional growth and competitiveness. The TR32 Region, with its diverse economic landscape and strategic location in the Aegean region, provides a unique case study for examining the relationship between smart cities, transportation, and regional development. The region's growing industries, including tourism, agriculture, and manufacturing, present both opportunities and challenges for the integration of smart city solutions. In this context, this book chapter explores the TR32 Region's development strategy for integrating smart city initiatives into the Draft 2024-2028 Regional Plan, with a particular emphasis on transportation and regional development. The draft plan, prepared by the Southern Aegean Development Agency (GEKA), serves as the primary subject of field research for this analysis and aims to examine the effectiveness of regional development policies and strategies in the TR32 Region based on the relationship between smart cities and transportation, employing a descriptive and critical theoretical framework rather than an analytical one.

This study aims to explore the potential contributions of transportation policies and strategies, as integral components of smart city initiatives, to regional development goals in the context of the TR32 Region in Turkey. By analyzing the "Draft GEKA 2024-2028 Regional Plan for the Southern Aegean Region," the paper seeks to identify the key factors that influence the successful implementation of innovative transportation policies based on smart city concepts at the regional level. The draft regional plan, which outlines the region's development strategy, offers a comprehensive framework for analyzing the potential impact of smart transportation policies on regional competitiveness, sustainability, and quality of life.

The draft plan's emphasis on developing smart transportation systems, promoting sustainable transportation infrastructures, addressing traffic congestion in tourism-intensive districts, and integrating digital applications in tourism promotion and destination management provides valuable insights into how the region is utilizing transportation policies for smart city development (Southern Aegean Development Agency, 2023). By examining these strategies and their potential impacts on regional competitiveness, accessibility, and overall quality of life, this paper aims to focus on the effectiveness of transportation policies as tools for achieving regional development goals.

In addition, the study will investigate the critical factors that contribute to the successful implementation of these policies, such as the level of collaboration among stakeholders, the availability of financial resources, technological readiness, and the capacity for continuous monitoring and adaptation. By identifying these key factors, the paper aims to provide practical recommendations for policymakers and urban and regional planners in the TR32 Region and other similar contexts, enabling them to design and implement effective transportation policies that follow smart city principles and foster regional development. This study adds to the increasing research on smart cities and their influence on regional development and resilience by examining how the TR32 Region incorporates smart transportation initiatives into its current regional development plan.

MATERIALS AND METHODS

This research utilizes a descriptive and analytical methodology to examine the interrelationship between smart city initiatives, transportation policies, and regional development strategies within the context of the TR32 Region in Turkey.

The article is informed by an unpublished MSc thesis titled "Smart Cities and Regional Development: Rethinking TR32 Regional Development Strategies" (Karaçoban, 2023). The article's descriptive and critical approach also draws from the case study section of the referenced master's thesis. However, the article's content extends beyond the scope of the thesis.

The method used in this study included a systematic approach to analyzing and evaluating the relationship between smart cities and transportation policies from a regional development standpoint, with a particular emphasis on the Draft 2024-2028 Regional Plan prepared by the Southern Aegean Development Agency (GEKA) for the TR32 Region. In the context of the Southern Aegean Development Agency plan reports, the methods used by development agencies in Turkey to acquire and generate data for formulating plans and strategies are crucial. Two primary approaches contribute to the content of regional plans prepared by these agencies: preparatory meetings and interviews and sectoral and thematic studies. The TR32 Region Draft Plan for 2024-2028 draws insights from a main database comprising ten sectoral and thematic development studies. These studies provide essential information for shaping the regional strategies and priorities within the Southern Aegean Region:

- Current Situation Analysis and Development Strategies of Production-Oriented Cooperatives
- Ecotourism Target Market Analyses
- Institutionalization and Digitalization: Current Situation Analysis in the Southern Aegean Region
- Clustering Analysis
- Southern Aegean Region Beekeeping Products Sector Analysis
- Southern Aegean Region Machinery Industry Sector Analysis
- Southern Aegean Region Olive Oil Sector Analysis
- Southern Aegean Region Tourism Inventory and Ecotourism Destination Action Plans
- Southern Aegean Region YER-SIS (Urban and Rural Settlement Systems in Turkey) Evaluation Report
- Southern Aegean Regional Innovation and Entrepreneurship: Current Situation Analysis and Development Recommendations (South Aegean Development Agency, 2023)

Among the ten sectoral and thematic studies mentioned above, the following studies are directly or indirectly related to the relationship between smart cities, transportation, and regional development: Institutionalization and Digitalization: Current Situation Analysis in the Southern Aegean Region, Southern Aegean Region Machinery Industry Sector Analysis, Southern Aegean Region Tourism Inventory and Ecotourism Destination Action Plans, Southern Aegean Region YER-SIS (Urban and Rural Settlement Systems in Turkey) Evaluation Report and Southern Aegean Regional Innovation and Entrepreneurship: Current Situation Analysis and Development Recommendations (South Aegean Development Agency, 2023).

The article's discussion of GEKA's general policies and strategies related to smart city initiatives, transportation, and regional development is informed by five sectoral and thematic analyses provided above.

SMART CITIES AND TRANSPORTATION POLICIES FROM A REGIONAL DEVELOPMENT VIEWPOINT

Various global institutions provide diverse definitions of smart cities. United Nations Economic Commission for Europe (UNECE), defines a smart sustainable city as "an innovative city that uses ICTs [Information and Communication Technologies] and other means to improve quality of life, efficiency of urban operation and services, and competitiveness, while ensuring that it meets the needs of present and future generations with respect to economic, social, environmental as well as cultural aspects" (United Nations Economic Commission for Europe, 2024, para. 2). The World Bank emphasizes that smart cities are an important tool for achieving a citizen-centered approach: "Smart cities leverage data and technology to integrate urban infrastructure and service delivery and provide solutions to achieve a citizen-centric approach" (World Bank, 2024).

Smart cities play a crucial role in regional development by efficiently providing region-based public services and influencing urban growth (Drobne & Bogataj, 2015). Improving access to technology, promoting innovation, increasing local attractiveness, and solving economic can make smart cities key tools for regional development (Oliveira et al., 2021). Smart cities aid regional development by linking technologies to people and using knowledge as a strategic tool for the efficient and sustainable delivery of goods and services. Innovation significantly influences the sustainable development of smart cities, with research gaps and emerging trends showing growing interest in this area (Camata et al., 2022).

In order to systematically classify academic contributions on the relationship between smart cities and transportation policies by a regional development viewpoint, it is important to address two key research questions: How can transportation policies and strategies, as tools for smart city initiatives, contribute to regional development goals? What are the key factors influencing the successful implementation of innovative transportation policies based on smart city concepts at the regional level?

Regarding the answer to the first question, several studies indicate that transportation policies and strategies in smart city initiatives contribute to regional development goals by promoting sustainable development, reducing congestion, optimizing urban mobility, and enhancing community livability. Transportation policies and strategies are essential elements of smart city initiatives, which aim to improve urban living through sustainable and intelligent solutions. These policies not only enhance urban mobility but also play a vital role in achieving regional development objectives by incorporating various aspects of urban planning, including land use, economic development, and housing. Here, four main topics offer a structured guide for reviewing literature in this field: smart mobility and sustainability, economic and community development, environmental and public health benefits and technological integration and innovation.

Smart mobility and sustainability examine how to create efficient, environmentally friendly transportation systems using modern technology. By employing advanced technologies and data-driven approaches, smart city transportation planning seeks to enhance urban mobility using sustainable strategies, making the most of available funds (Qiao, 2022). Smart cities and smart territories can enhance regional development by integrating transport infrastructures and sustainable land use through the use of Geographic Information Systems (García-Ayllón & Miralles, 2015). Smart cities can use data analytics tools to identify the most visited regions and connections within an urban environment, which can improve resource allocation to better meet the mobility needs of citizens (Buosi et al., 2018).

Intelligent Transport Systems (ITS) are crucial for achieving smart mobility, which improves transport efficiency and sustainability ((Bifulco et al., 2014; Lorenz et al., 2023). ITS technologies enable the integration of various transportation modes and optimize their operations. Effective public transportation systems play a crucial role in promoting sustainable urban development, enhancing competitiveness, and tackling issues like traffic congestion and environmental problems (Mondale, 2000; Sedlak & Ćirić, 2018). Smart transportation solutions significantly reduce transport emissions and help meet reduction goals, thereby supporting sustainable mobility and transport decarbonization (Zawieska & Pieriegud, 2018).

Smart transportation systems not only improve mobility but also contribute significantly to economic and community development. Smart city initiatives contribute to economic growth by fostering socially engaged, financially stable, and business-friendly urban environments. They also improve public safety, energy efficiency, and infrastructure resilience (Musa, 2017). As an essential component of these initiatives, efficient and sustainable transportation systems are vital for economic growth and enhancing the quality of life in smart cities. Examples from Paris, Boston, and Singapore illustrate the positive impact of well-designed transport systems on urban development (Joshi et al., 2018).

Smart transportation systems are increasingly valued for their ability to enhance environmental sustainability and public health. By integrating advanced technologies and promoting sustainable practices, these systems aim to reduce pollution, improve mobility, and create healthier urban environments. Smart transportation systems, which include electric and autonomous vehicles, significantly reduce toxic emissions, contributing to cleaner urban air and mitigating climate change. These systems often complement public transportation, working together to further reduce pollution levels. Modal shifts to public transport scenarios have been shown to reduce air pollution in cities, offering substantial environmental benefits (Čepeliauskaitė et al., 2021; Kwan & Hashim, 2016). These benefits are heightened when public transportation is combined with other sustainable mobility solutions. Here, shared and active transportation options enhance air quality and mobility, benefiting both the environment and public health (Kalašová et al., 2019; Kuhail et al., 2018).

Smart cities harness technological integration and innovation, particularly in smart transportation, to drive regional development and enhance urban mobility. Smart transportation uses advanced technologies to enhance the efficiency, safety, and sustainability of urban mobility. This field integrates innovations such as the Internet of Things (IoT), artificial intelligence (AI), big data, and machine learning to optimize transportation systems, reduce congestion, and improve overall urban living conditions. IoT and AI technologies are crucial for traffic management, route optimization, accident prevention, and efficient parking systems (Oladimeji et al., 2023; Paiva et al., 2021; Xu & Thakur, 2021). These technologies rely on advanced computing infrastructure to process and analyze vast amounts of data in real-time. Here, cloud, edge, and fog computing play a crucial role in enabling smart transportation systems to function efficiently (Oladimeji et al., 2023; Xu & Thakur, 2021).

The answer to the second question raises different social, economic, technological, and geographical factors in the relationship between smart transportation and regional development. The successful adoption of new transportation policies in smart cities at the regional level depends on several factors. These factors include technology infrastructure, economic growth, governance, and citizen involvement. Policymakers and city planners need to understand these important factors to improve urban mobility and the overall quality of life through smart city projects.

Technology infrastructure is a critical component for the successful adoption of smart transportation policies in cities and regions. Integrating intelligent transportation systems (ITS), including advanced public transportation systems (APTS) and advanced traveler information systems (ATIS), is crucial for transforming regional transportation networks (Billones et al., 2021).

Economic development, structural urban variables, geographical location, and density of population are key factors influencing the successful implementation of smart city concepts in transportation (Neirotti et al., 2014). These factors, along with strategic investments, are crucial for fostering innovation and realizing the benefits of smart city initiatives. Investment in science and technology, along with human capital, is essential for boosting regional innovation through smart city development (Zeng et al., 2023).

Effective governance and policy context are crucial for the successful implementation of smart city initiatives. This involves management, organization, and creating a shared vision among stakeholders (Chatfield & Reddick, 2016; Chourabi et al., 2012). Implementing citizen-centric e-governance and social governance frameworks is essential for developing a shared vision of social innovation, which is crucial for the successful implementation of smart cities (Chatfield & Reddick,

2016). Moreover, these governance frameworks should prioritize citizen engagement and participation to ensure the success of smart city projects. Here, active citizen engagement is vital for promoting inclusive mobility and ensuring the success of smart mobility projects. (Paiva et al., 2021).

The importance of considering these factors is further demonstrated by the examples presented here, which highlight both potential successes and challenges, emphasizing the significance of local context in the implementation of smart city initiatives. Smart city projects in Central Asia illustrate the potential to improve living conditions, enhance infrastructure efficiency, and foster environmental sustainability, while simultaneously addressing urban issues such as traffic congestion and pollution (Irnazarov & Kayumova, 2017). The Smart Region Mobility Framework seeks to modernize regional transportation networks through the integration of intelligent transportation systems, the development of social services infrastructure, the identification of two cities in the Philippines as potential smart city candidates, and the implementation of data flow architecture (Billones et al., 2021). The primary goal of the ASEAN (The Association of Southeast Asian Nations) Smart Cities Network is to increase the intelligence of cities and facilitate their development beyond provincial constraints.

In conclusion, this section has explored the relationship between smart cities and transportation policies from a regional development perspective. The literature review has shown that transportation policies and strategies in smart city initiatives contribute to regional development goals by promoting sustainable development, reducing congestion, optimizing urban mobility, and enhancing community livability. The successful implementation of innovative transportation policies based on smart city concepts at the regional level depends on various social, economic, technological, and geographical factors, such as technology infrastructure, economic growth, governance, and citizen involvement. Understanding these factors is crucial for policymakers and city planners to improve urban mobility and overall quality of life through smart city projects. The examples presented here demonstrate the potential successes and challenges of implementing smart city initiatives, emphasizing the importance of considering local context in their development and implementation. In this respect, Turkey and the TR32 Region can be considered as examples that emphasize the importance of local context.

The TR32 Region, comprising the provinces of Aydın, Denizli, and Muğla, presents a unique case study for examining the relationship between smart city initiatives, transportation policies, and regional development strategies. The region's diverse economic landscape, encompassing sectors such as tourism, agriculture, and manufacturing, offers both opportunities and challenges for implementing smart city solutions. To better understand this relationship, it is crucial to examine the specific measures and targets outlined in the Draft GEKA 2024-2028 Regional

Plan. The draft plan emphasizes several key areas where smart city initiatives and transportation policies intersect with regional development goals: smart mobility and connectivity, sustainable transportation infrastructure, digital integration in tourism, intermodal connectivity and data-driven decision making.

The plan prioritizes the development of smart transportation systems to enhance urban mobility and reduce congestion, particularly in tourism-intensive areas. For example, it proposes the implementation of intelligent traffic management systems in major urban centers like Aydın, Denizli, and Muğla, as well as popular tourist destinations such as Bodrum and Marmaris. These systems aim to optimize traffic flow, reduce travel times, and improve overall transportation efficiency.

Consistent with smart city principles, the draft plan emphasizes the development of sustainable transportation options. This includes expanding bike lanes, improving walkability, and enhancing public transportation systems. These measures not only correspond to smart city concepts but also contribute to the region's goals of reducing carbon emissions and promoting sustainable urban development.

Understanding the significance of the tourism sector in the TR32 Region's economy, the plan suggests incorporating digital tools for tourism promotion and destination management. This includes creating mobile apps and digital platforms to offer up-to-date information on attractions, transportation choices, and local services. These efforts show how smart city technologies can be used to boost a crucial economic sector while enhancing visitor experiences.

The draft plan emphasizes strengthening the connectivity between different transportation modes, particularly focusing on linking Organized Industrial Zones (OIZs) with ports, railways, and highways. This approach demonstrates how smart transportation strategies can support regional economic development by enhancing logistics and supply chain efficiency.

Lastly, central to the smart city concept is the use of data analytics for informed decision-making. The draft plan proposes implementing systems for collecting and analyzing transportation data, which can be used to optimize routes, predict maintenance needs, and inform future infrastructure investments.

Moreover, the draft plan's approach to smart city and transportation initiatives corresponds to broader national objectives outlined in Turkey's recent development plans. The Twelfth Development Plan (2024-2028), for example, emphasizes the importance of developing smart, sustainable, and resilient cities, with a particular focus on reducing carbon footprints through improved public transportation and walking infrastructure (The Presidency of the Republic of Turkey, the Presidency of Strategy and Budget, 2023, p. 15). The TR32 Region's draft plan echoes these national priorities, demonstrating a coherent approach to regional development that is in harmony with national strategies.

Smart Cities and Transportation Policies in Recent Regional Development Strategies in Turkey

The concept of smart cities has recently been incorporated into Turkey's regional development framework through various strategic development documents. These documents, including national development plans, strategies developed by central government units, and regional and sub-regional plans, consider smart cities as a sub-component of information and communication technologies rather than as a comprehensive concept.

"Turkey's Eleventh Development Plan (2019-2023)" partially addresses the development of information and communication technologies and outlines principles for smart city development within this strategic technological framework. The plan highlights the need for integrated urban planning, efficient public transportation, and the adoption of advanced technologies in the transportation sector. On the other hand, in the objectives and policies sections of the plan under the main headings of "Stable and Strong Economy," "Competitive Production and Productivity," "Qualified People, Strong Society," "Livable Cities and Sustainable Environment," and "Rule of Law, Democratization, and Good Governance," smart cities, smart transportation, and the innovative technological framework in these fields are given less importance (The Presidency of the Republic of Turkey, the Presidency of Strategy and Budget, 2019).

In the section "Livable Cities, Sustainable Environment", the plan discusses the importance of sustainable urbanization and the development of smart cities. It emphasizes the need for integrated urban planning, the use of smart city technologies, and the improvement of urban transportation systems (The Presidency of the Republic of Turkey, the Presidency of Strategy and Budget, 2019, pp. 157-172). The plan also mentions the importance of developing smart transportation systems, such as intelligent transportation systems (ITS) and smart traffic management, to improve the efficiency and safety of transportation networks (p. 119). It calls for the expansion of rail systems, the integration of different modes of transportation, and the improvement of accessibility to public transportation (p. 163). Moreover, the plan sets specific targets for the development of transportation infrastructure, including the expansion of high-speed railway networks, the improvement of urban rail systems, and the establishment of new airports and ports. In the section "Logistics and Transportation" the plan focuses on the development of an integrated and efficient logistics and transportation system. It emphasizes the need for intermodal transportation, the expansion of railway networks, and the improvement of port infrastructure (pp. 116-119).

As mentioned above, smart cities and smart transportation are of secondary importance in the Turkey's Eleventh Development Plan within the plan's regional development goals, policies, and strategies. While the plan highlights the importance of developing smart transportation systems, such as intelligent transportation systems (ITS) and smart traffic management, to improve the efficiency and safety of transportation networks, it doesn't specify their integration into the overall transportation infrastructure. Another important issue is that the plan provides a general framework for the development of smart cities and intelligent transportation systems but does not extensively elaborate on how these initiatives will be integrated with specific regional development goals and objectives. However, it is important to note that according to the general content and structure of recent development plans in Turkey, these plans typically do not mention development goals, policies, and strategies in detail, but rather provide the main principles and general direction for socio-economic development.

The "Twelfth Development Plan (2024-2028)" is the most recent main development document in Turkey. According to the Eleventh Development Plan, this latest development plan's focus on smart cities and technological development has a richer content. The plan is based on a more technological approach in addressing the goals, objectives, and strategies related to transportation, logistics, and related infrastructure. Sustainability and climate change issues complement this general technological framework (The Presidency of the Republic of Turkey, the Presidency of Strategy and Budget, 2023).

The Twelfth Development Plan of Turkey (2024-2028) emphasizes the importance of smart, sustainable, and resilient cities in line with global trends. The plan aims to reduce carbon footprint by promoting public transportation and walking while integrating traditional and modern urban planning principles with a sustainability approach (The Presidency of the Republic of Turkey, the Presidency of Strategy and Budget, 2023, p. 15). In this framework, the plan's main objective in urbanization is to create green, digital technology-based, smart, safe, and sustainable cities and settlements (p. 206). The plan envisions Turkish cities as sustainable settlements that are resistant to climate change and disasters, use smart energy and transportation infrastructure, and offer high-quality and inclusive services. In this vein, the latest development plan's vision includes creating disaster-resilient living spaces and smart and sustainable cities (pp. 49-51).

The latest development plan sets targets for Turkey to become a global logistics power by 2053, promoting its strategic location and potential in all modes of transportation. The plan aims to increase the share of rail transport, air cargo traffic, and the number of passengers in air transport (The Presidency of the Republic of Turkey, the Presidency of Strategy and Budget, 2023, pp. 48-49). The plan focuses on expanding the use of technological solutions in the logistics and transportation sector, such as smart roads, connected and autonomous driving systems, intelligent transportation systems, and energy-absorbing guardrails (p. 138).

The plan supports the development and production of digital applications and systems in the transportation sector using domestic resources (The Presidency of the Republic of Turkey, the Presidency of Strategy and Budget, 2023, p. 138). The plan emphasizes spreading Intelligent Transportation Systems (ITS) to reduce travel times, increase traffic safety, and use existing road capacities efficiently (p. 139). The plan also mentions expanding applications compatible with the national ITS architecture, especially in metropolitan cities (p. 215). Here, developing highways with new smart roads suitable for electric vehicles and autonomous systems is identified as an important objective that complements the ITS framework (pp. 48-49).

A new approach to addressing smart transportation in Turkey's latest development plan is the incorporation of education and human resources issues into the plan's technology-oriented development framework. The plan emphasizes training qualified human resources in emerging technology fields such as the Internet of Things, artificial intelligence, big data, quantum, cybersecurity, smart transportation, and augmented reality in education (The Presidency of the Republic of Turkey, the Presidency of Strategy and Budget, 2023, p. 160). For implementation and production, the development plan aims to bring together technical knowledge and experience from various universities to develop smart and autonomous transportation vehicles and ecosystems in the automotive sector (p. 95).

Overall, the Turkey's Twelfth Development Plan's approach to the relationship between smart cities and transportation is comprehensive and forward-looking. However, its adequacy will depend on the effective implementation, monitoring, and adaptation of the proposed strategies to address the evolving challenges and opportunities in the field of smart cities and transportation.

Additionally, Turkey's “2020-2023 National Smart Cities Strategy and Action Plan” aims to provide a technological perspective on significant regional development issues, prioritizing actions to cover the entire country's smart city transformation while considering its diverse geographical and socioeconomic dynamics (The Republic of Turkey, the Ministry of Environment, Urbanization and Climate Change, 2019). The strategy employs a multi-layered structure of national and local policies, guided by a human-centered approach that emphasizes cooperation and coordination among stakeholders, including central and local governments, the private sector, NGOs (non-governmental organizations), universities, and urban residents. Sustainability is seen as dependent on effective resource and service management through stakeholder collaboration and interoperability (pp. 37-38).

The “2020-2023 National Smart Cities Strategy and Action Plan” highlights the importance of smart transportation in the development of smart cities and outlines policies, strategies, and actions to enhance smart transportation systems, emphasizing stakeholder collaboration and the potential benefits for urban efficiency, sustainability, and quality of life. The document defines the smart transportation component as "ICT-supported and integrated transportation systems, including one or more modes of transportation such as trams, buses, trains, subways, cars, maritime and air transportation, bicycles, and pedestrians, covering sustainable, safe, and interconnected transportation systems" (The Republic of Turkey, the Ministry of Environment, Urbanization and Climate Change, 2019, p. 24). The plan stresses the importance of increasing the maturity of the smart transportation component in cities. It proposes actions to evaluate and improve smart transportation solutions, such as creating an inventory of smart transportation solutions, analyzing their compatibility with the city's needs and capabilities, and developing new solutions based on data-driven models (pp. 60-61).

Figure 1 shows the actual categorical distribution of smart transportation projects in Turkey by province in relation to the action plan's proposal to "create an inventory of smart transportation solutions." The distribution of projects across provinces reveals a concerted effort to implement "ICT-supported and integrated transportation systems", as evidenced by the presence of intelligent transportation systems and infrastructure projects in multiple regions. The figure shows the prominence of intelligent traffic management systems in several provinces, suggesting a focus on optimizing traffic flow and reducing congestion. The map in Figure 1 also highlights disparities in the allocation of smart transportation initiatives, with a higher concentration of projects in the western provinces, particularly in the Marmara region (TR10 and TR42 Regions) surrounding Istanbul. This uneven distribution suggests that there may be opportunities to further expand smart transportation solutions to less-represented areas, ensuring a more balanced approach to urban efficiency, sustainability, and quality of life in Turkey.

The action plan emphasizes the need for policies and strategies to develop smart transportation systems. It references the Tenth Development Plan of Turkey (2014-2018), which includes policies supporting the use of information technologies and smart transportation systems in urban transportation management and public transportation services (The Republic of Turkey, the Ministry of Environment, Urbanization and Climate Change, 2019, p. 28). The plan also mentions the "Draft National Smart Transportation Systems Strategy and Action Plan", which corresponds to the objectives of the "2020-2023 National Smart Cities Strategy and Action Plan" (p. 99). The action plan highlights the potential benefits of smart transportation systems, such as increased efficiency, sustainability, and accessibility of transportation services. It emphasizes the role of smart transportation in improving the quality of life for city residents and contributing to the overall development of smart cities (p. 405).

Furthermore, the action plan discusses the need for collaboration among stakeholders in the development and implementation of smart transportation strategies. It suggests establishing cooperation between central and local government agencies, universities, the private sector, and NGOs to create a supportive ecosystem for smart transportation initiatives (The Republic of Turkey, the Ministry of Environment, Urbanization and Climate Change, 2019, p. 70).

In conclusion, recent regional development strategies in Turkey, including the Eleventh and Twelfth Development Plans and the 2020-2023 National Smart Cities Strategy and Action Plan, have incorporated smart cities and transportation as one of the thematic topics for socio-economic development in the country. These documents emphasize the importance of developing intelligent transportation systems, expanding sustainable transportation infrastructure, and fostering collaboration among stakeholders to create efficient, accessible, and environmentally friendly urban transportation networks. However, the integration of smart city and transportation initiatives with specific regional development goals and objectives remains limited, and there are disparities in the distribution of smart transportation projects across provinces. As Turkey continues to work towards its vision of becoming a global logistics power and creating sustainable, resilient cities, effective implementation, monitoring, and adaptation of these strategies will be crucial to address the challenges and opportunities in the field of smart cities and transportation.

Figure 1. Provincial (NUTS-3) distribution of smart transportation projects in Turkey by categories, 2024

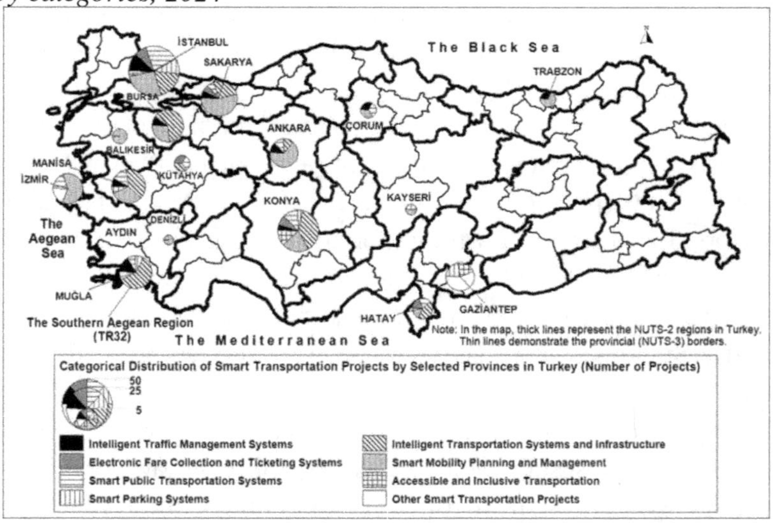

(Adapted from Istanbul Metropolitan Municipality, 2024; The Republic of Turkey, Ministry of Environment, Urbanization and Climate Change, 2024a, 2024b; The Resolution of the Cabinet of Ministers of Turkey, 2002)

TR32 Region (Aydin-Muğla-Denizli)

An assessment of smart cities, transportation policies, and regional development first requires defining the Nomenclature of Territorial Units for Statistics (NUTS) in Turkey. The status and responsibilities of development agencies in Turkey will complement this definition. Furthermore, in this section, the main geographical, statistical, and economic characteristics of TR32 (Southern Aegean Region) will be discussed. The spatial statistical division of Turkey is based on normative or administrative criteria. Here, it is necessary to discuss Turkey's NUTS system (see Figure 2). In 2001, as part of Turkey's efforts to harmonize with the European Union's (EU) legal framework and the Accession Partnership Document, the State Planning Organization (SPO) and the Turkish Statistical Institute (TUIK) coordinated the definition of the country's statistical regional units. This classification, which became effective in 2002, categorized Turkey into 12 NUTS-1 regions, 26 NUTS-2 regions, and provinces as NUTS-3 units. The NUTS classification serves as the foundation for the implementation of regional development policies, regional statistics, and the establishment of development agencies in Turkey. The NUTS-2 level regions, in particular, are regarded as a suitable policy scale for national and regional analysis, as they form the basis for studies conducted in the EU candidacy process and determine the regions eligible for structural funds under the EU's convergence objective. As a result, NUTS-2 sub-regions have been designated as the primary development planning unit for regional development policies in Turkey (The Republic of Turkey, the Ministry of Industry and Technology, 2023).

In 2006, Turkey established development agencies based on the aforementioned spatial statistical classification, as per Law No. 5449. The Izmir Development Agency (IZKA) and the Çukurova Development Agency (ÇKA) were launched as pilot projects in 2007, covering specific provinces. By 2010, a total of 26 development agencies were established, encompassing all 81 provinces in Turkey (The Republic of Turkey, the Ministry of Industry and Technology, 2024). The geographical jurisdiction of these development agencies is defined by the NUTS-2 regions. Figure 2 illustrates the geographical scope of the 26 regional development agencies operating in Turkey's NUTS-2 regions.

Figure 2. Regional (NUTS-2) distribution of gross domestic product per capita in Turkey, 2022

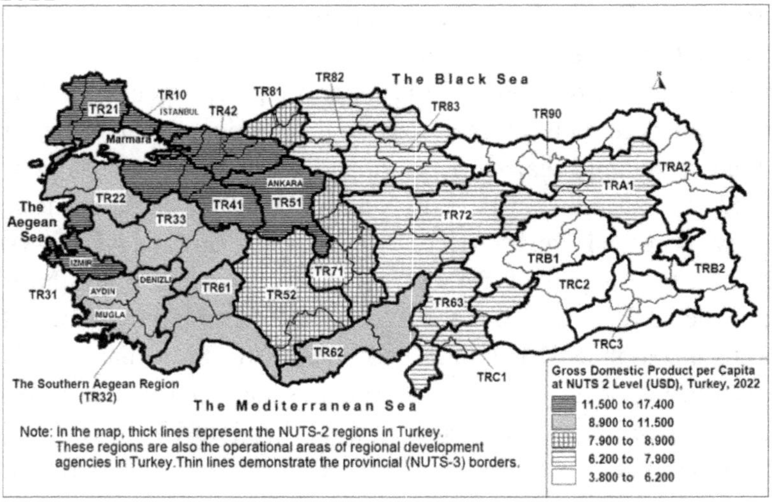

(Adapted from The Resolution of the Cabinet of Ministers of Turkey, 2002; Turkish Statistical Institute, 2024b)

The primary objectives of development agencies in Turkey are to accelerate regional development sustainably, reduce development disparities, and promote cooperation between the public, private, and non-governmental sectors. Their main responsibilities include formulating regional development strategies, identifying and utilizing regional resources, fostering economic and social growth, and enhancing competitiveness. Additionally, they facilitate collaboration among stakeholders, support local and rural development, encourage entrepreneurship, strengthen the technical capacities of local entities, simplify permit procedures for investors, and facilitate access to international funds (The Republic of Turkey, the Ministry of Industry and Technology, 2024). Recent regional and sub-regional plans prepared by development agencies in Turkey, within the framework of their organizational tasks and responsibilities, have begun to exhibit a development approach that emphasizes technological frameworks and current concepts such as smart cities. The regional plans prepared by GEKA exemplify these technology-oriented regional development strategies.

With this spatial statistical classification defined above, the area geographically known as the "Southern Aegean Region" has been statistically redefined as a NUTS-2 regional unit with the name TR32 (see Figure 2). The TR32 Region has a population of 3,287,520 as of 2023 (Turkish Statistical Institute, 2024a) and spans an area of 33,213 square kilometers. Due to Turkey's province-based NUTS classification, the population and surface area values of the TR32 Region are calculated by aggregating the data from its constituent NUTS-3 units, namely the provinces of Aydın, Muğla, and Denizli. As with all other NUTS-2 regions in Turkey, the TR32 Region comprises both urban and rural areas. However, the enactment of "Law No. 5216 on Metropolitan Municipalities" in 2004, commonly known as the "Compass Law," expanded the administrative and operational jurisdiction of Metropolitan Municipalities to cover entire provinces (The Republic of Turkey, the Prime Ministry General Directorate of Legislation Development and Publication, 2004). Consequently, the provinces of Aydın, Muğla, and Denizli, which form the TR32 Region, are also designated as metropolitan municipalities. This legislation has redefined metropolitan areas in Turkey based on normative criteria rather than functional criteria, which would have been the ideal approach for socio-economic development strategies. The law, enacted with political considerations for local elections, has relabeled and delineated all provinces in Turkey with metropolitan municipalities (30 provinces), including their rural and urban areas, as urban areas.

An important question here is: what makes the TR32 Aydın-Muğla-Denizli Region unique for studying the relationship between smart cities, transportation, and regional development? The TR32 Aydın-Muğla-Denizli Region in Turkey is distinctive for studying the relationship between smart cities, transportation, and regional development due to its varied geography and population, economic importance in tourism, agriculture, and industry, and current smart city projects (especially in Muğla province). The region's complex transportation networks and urban mobility challenges provide a valuable context for analyzing smart transportation solutions. Additionally, socio-economic disparities, cultural heritage, and active government policies highlight the role of smart city initiatives in promoting balanced regional development and sustainability. The TR32 Region offers valuable insights into improving quality of life through technological innovation and smart governance.

Denizli province and its central city are important industrial centers, with thriving sub-sectors in textiles, machinery and metalworking, food processing, ceramics and glass, mining and quarrying, furniture and woodworking, and renewable energy. Muğla, another province and central city in the TR32 Region, has an economy primarily based on tourism, agriculture, fishing, marble and mining, ceramics, textiles, and furniture production. Similarly, Aydın's main economic sectors include agriculture, food processing, textiles, mining, tourism, renewable energy, and machinery and metalworking industries. The diverse sectoral characteristics of Aydın, Muğla, and Denizli, as well as their central cities, make them ideal case studies for examining the connection between regional development and smart cities. The region's rich and multi-sectoral structure has the potential to serve as a fertile ground for the application of smart city strategies, as the varied economic landscape provides opportunities to explore a wide range of smart city applications tailored to different industries. In addition, the coexistence of traditional sectors and emerging technologies in the region creates a conducive environment for studying the integration of smart solutions into existing frameworks. Finally, the presence of diverse stakeholders in the region, ranging from small businesses to large corporations, offers valuable insights into the challenges and benefits of implementing smart city strategies across different organizational scales.

The TR32 Region's diverse and growing industries offer both opportunities and challenges for smart city integration. The robust tourism sector provides a platform for implementing innovative mobility solutions and digital services to enhance visitor experiences. Manufacturing and agricultural industries present opportunities for smart logistics and supply chain optimization. However, these sectors also present challenges, such as managing seasonal fluctuations in transportation demand and ensuring that smart city solutions are adaptable to both urban and rural contexts. The region must balance the needs of various industries while implementing smart city initiatives, ensuring that technological advancements benefit all sectors and communities equally. This requires careful planning and coordination among stakeholders to create integrated solutions that address the unique needs of each industry while contributing to overall regional development goals.

Figures 2 and 3 provide additional details to the reasons explained above for the TR32 region being a unique research area with two development indicators: gross domestic product per capita and gross domestic R&D expenditure. While the first indicator shows how the region constitutes an economic region within Turkey in macroeconomic terms, the second indicator provides an overview of the region's technological capacity and potential for smart city strategies. Figure 2 shows the regional (NUTS-2) distribution of gross domestic product per capita in Turkey and the borders of NUTS-3 geographic units (provinces). As shown in the figure, the TR32 Region is one of the more economically developed regions in Turkey in terms of its gross domestic product (GDP) per person in 2022. The map clearly shows that the TR32 Region falls within the second highest category, with a GDP per capita ranging from 8,900 to 11,500 USD. The relatively high GDP per capita of the TR32 Region highlights its significant economic importance and level of development in Turkey. This economic strength can be attributed to the region's diverse industrial base, thriving tourism sector, and strategic location in the Aegean coastal area. Figure 3 provides valuable insights into the technological capacity and potential for smart city strategies across Turkey's NUTS-2 regions by illustrating the regional distribution.

The map reveals that the TR32 Region falls within the middle category, with an R&D expenditure ranging from 2 billion to 2.9 billion Turkish Lira (TL). While this positions the TR32 Region behind the leading regions such as TR10 (Istanbul) and TR51 (Ankara), as well as the Eastern Marmara Development Corridor (TR42), it still indicates a significant level of investment in research and development activities. Here, the regional distribution of R&D expenditures can be seen as a crucial indicator of a region's technological capacity and its potential to develop and implement smart city strategies.

Figure 3. Regional (NUTS-2) distribution of gross domestic R&D expenditure, 2022

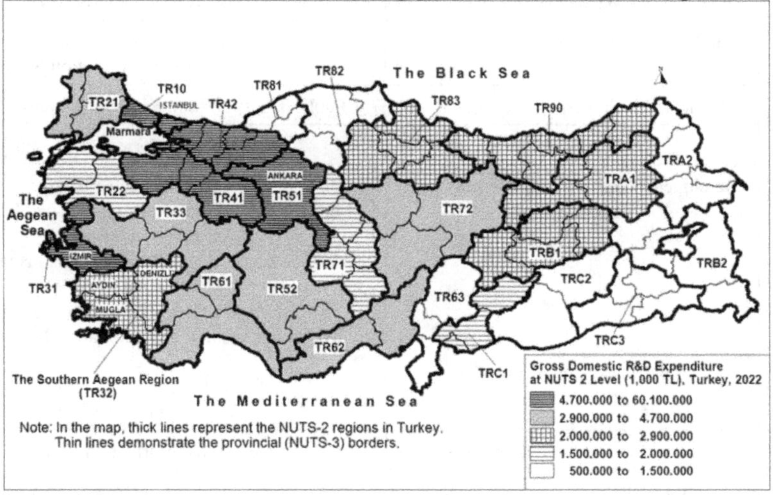

(Adapted from The Resolution of the Cabinet of Ministers of Turkey, 2002; Turkish Statistical Institute, 2024b. of gross domestic R&D expenditure in 2022)

In addition to these two figures, Table 1 contributes to understanding the potential of the TR32 Region in terms of smart transportation projects by providing an inventory and classification of existing smart transportation projects. Table presents the categorical distribution of smart transportation projects in the TR32 Region as of 2024, focusing on the provinces of Muğla and Denizli. The table reveals that Muğla province has a more extensive array of smart transportation projects compared to Denizli. Muğla's projects cover a wide range of categories, such as intelligent traffic management systems, smart public transportation systems, smart parking systems, intelligent transportation systems and infrastructure, and other smart transportation projects. However, Denizli's smart transportation initiatives are primarily focused on smart mobility planning and management, with projects such as the Traffic Management System and Transportation Portal. The concentration of smart transportation projects in Muğla emphasizes the province's proactive approach to utilizing technology for enhancing its transportation networks and urban mobility. The important tourism centers located within the borders of the province, recognized at national and international levels, can also be considered as another reason for the increasing interest in the province's smart transportation projects.

Table 1. Categorical distribution of smart transportation projects in TR32 region, 2024

Provinces	Smart Transportation Categories	Smart Transportation Projects
Denizli	Smart Mobility Planning and Management	Traffic Management System Transportation Portal
Muğla	Intelligent Traffic Management Systems	Traffic Information System Traffic Circulation Information System
Muğla	Smart Public Transportation Systems	Public Transportation Stops Information System
Muğla	Smart Parking Systems	Parking Lots Information System
Muğla	Intelligent Transportation Systems and Infrastructure	Vehicle Tracking System Muğla Metropolitan Municipality Road Inventory Information System Muğla Metropolitan Municipality Transportation Portal Transportation Information System Muğla Infrastructure Coordination Information System Port Information System Road Information System
Muğla	Other Smart Transportation Projects	Taxi Stops Information System

Note. Adapted the Republic of Turkey, Ministry of Environment, Urbanization and Climate Change, 2024a, 2024b.

Here, it is important to note that while smart transportation projects in Turkey are included in the inventory of the smart cities portal prepared by the Ministry of Environment, Urbanization and Climate Change, some new or draft projects initiated by local administrations may not be reflected in this inventory. Due to the fragmented nature of data sources, Aydın province, which is also part of the TR32 Region, is not included in this distribution. Despite this limitation, the categorical distribution presented in Table 1 provides valuable insights into the types of smart transportation initiatives being implemented in the TR32 Region and emphasizes the region's potential to adopting smart city strategies in the transportation sector.

In summary, the TR32 Region serves as a unique and promising case study for examining the interrelationship between smart cities, transportation, and regional development in Turkey. The region's diverse geography, economic significance in various sectors, and ongoing smart city projects, particularly in Muğla, provide a rich context for analyzing the potential of smart transportation solutions. Despite some data limitations, the region's relatively high GDP per capita, moderate R&D expenditure levels, and the categorical distribution of smart transportation initiatives highlight its economic strength, technological capacity, and proactive approach to utilizing technology for urban mobility enhancement.

The Case of the TR32 Regional Development Strategy for 2024-2028 in Turkey

The Draft GEKA 2024-2028 Regional Plan for the TR32 Region is based on a current analysis of the region's socio-economic characteristics. The plan outlines the region's strengths, priorities, and goals for development. It also suggests policies and strategies that match these goals. The plan is still a draft, so it does not yet include a section on how well the general policies and strategies from the previous regional development strategy (2014-2023) were carried out.

The draft plan was created with a focus on participation, complementarity, local and spatial considerations, and global trends. Planning started in 2016 with sectoral and thematic analyses. From the period of 2016-2022, various studies were conducted such as sectoral commissions, surveys, inventories, and research reports to provide inputs for the new regional plan. The overall planning process involved seven main stages: stakeholder analysis, sectoral/thematic analyses, data compilation and gathering institutional views, current situation analysis, workshops and in-person meetings, determining the vision/priorities/targets/measures, and coordination with national strategies. Key elements included analyzing the region's socioeconomic trends, priority intervention areas, and strategic objectives while ensuring effective resource utilization consistent with national/local policies. Stakeholder participation was emphasized through workshops held in the TR32 Region's provinces to discuss draft measures within the strategic areas identified through the analyses (Southern Aegean Development Agency, 2023, pp. 5-9).

Since there is no formal feedback or evaluation of how well smart city and smart transportation strategies and policies were carried out during the previous GEKA plan period (2014-2023), the examination of smart transportation strategies and measures in GEKA's Third Plan (2024-2028) will be conducted independently, separate from past assessments. This evaluation cannot provide definite information about the plan's approach to smart transportation strategies from a regional development viewpoint. However, the draft plan can be seen as an important strategic document for understanding the strategic topics of smart transportation by the various measures taken in the region in general terms. This strategic insight can help clarify some of the key issues related to how transportation policies and strategies can become an instrument of regional development with the help of smart technologies at the NUTS-2 level in Turkey.

Before evaluating the draft plan's strategic topics and measures related to smart transportation, it may be useful to mention the overall technological development approach of the plan towards thematic topics of regional development. The TR32 Region 2024-2028 Draft Regional Plan, adopts a comprehensive approach to technological development, emphasizing the importance of digital transformation, innovation, and green growth in achieving sustainable regional development. The plan acknowledges the potential of advanced technologies like artificial intelligence, the Internet of Things (IoT), and data analytics in boosting economic growth, increasing competitiveness, and enhancing residents' quality of life. It advocates for the adoption of these technologies across various sectors, including manufacturing, agriculture, and tourism, to enhance productivity, efficiency, and sustainability. The draft plan also stresses the importance of developing the necessary human capital and digital infrastructure to support the TR32 region's technological development. This includes fostering digital skills among the workforce, promoting innovation and entrepreneurship, and investing in research and development. Furthermore, the plan emphasizes the role of public-private partnerships and collaborative ecosystems in driving technological development and ensuring its benefits are distributed equitably across the region. Overall, the draft regional plan outlines a forward-thinking vision to utilize technology for achieving sustainable, inclusive, and resilient regional growth.

Figure 4 illustrates the smart city strategic framework outlined in the TR32 Region 2024-2028 Draft Regional Plan. The framework consists of three main components: digital transformation, innovation, and green growth. Each component is further divided into specific focus areas that contribute to the overall smart city strategy. Digital transformation involves integrating smart technologies into various sectors, including manufacturing, agriculture, tourism, and transportation. Innovation includes initiatives for human capital development, digital infrastructure improvement, research and development, and entrepreneurship support. Green growth focuses on promoting sustainable production practices and creating sustainable living spaces. By addressing these key areas, the draft plan aims to utilize the potential of smart city technologies to drive sustainable, inclusive, and resilient regional development. (Southern Aegean Development Agency, 2023).

Figure 4. The draft GEKA 2024-2028 regional plan's smart city strategic framework

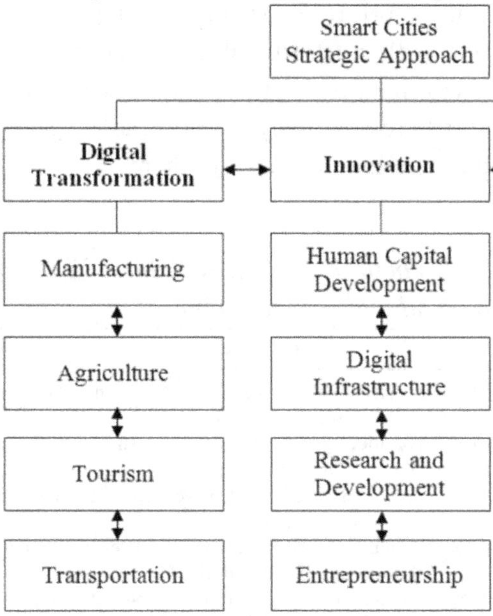

(Adapted from Southern Aegean Development Agency, 2023)

The TR32 Region 2024-2028 Draft Regional Plan demonstrates a strong commitment to incorporating smart city principles and technologies into its regional development strategy. The plan's focus on digital transformation, innovation, and green growth reflects the fundamental concepts of smart city development. However, while the plan effectively addresses the integration of smart technologies in various sectors, such as transportation, manufacturing, agriculture, and tourism, it could benefit from a more comprehensive and detailed exploration of the potential applications and impacts of smart city solutions. The terminology used in the draft plan is generally consistent with smart city discourse. However, it may not fully capture the complexity and interdependence of the various components of a smart city ecosystem. Additionally, the plan could provide more detail on the specific mechanisms and policies needed to foster collaboration among stakeholders, ensure interoperability of smart city systems, and address potential challenges related to data privacy, security, and digital inclusion. Despite these limitations, the draft plan lays a strong foundation for the TR32 Region's smart city aspirations. It reflects a forward-thinking approach to utilizing technology for sustainable regional development.

The TR32 Region 2024-2028 Draft Regional Plan, places significant emphasis on the role of transportation in fostering regional development and promoting smart city strategies. The plan highlights the importance of developing efficient, sustainable, and integrated transportation systems to enhance the region's competitiveness, improve accessibility, and support the growth of key sectors such as industry, agriculture, and tourism. By focusing on strategic priorities, spatial development scenarios, smart transportation systems, sustainable infrastructure, intermodal connectivity, and tourism mobility, the plan aims to enhance the TR32 Region's competitiveness, improve accessibility, and support the growth of key sectors. Table 2 gives an overview of these strategic topics and the measures related to them.

Table 2. Strategic topics and measures in the draft GEKA 2024-2028 regional plan for the smart cities, transportation, and regional development

Strategic Topics	Measures
Strategic Transportation Priorities	Strengthening the connectivity of Organized Industrial Zones (OIZs) to ports, railways, and highways, Developing smart transportation systems and promoting sustainable transportation infrastructures, Addressing traffic congestion in tourism-intensive districts through innovative solutions, Integrating digital applications in tourism promotion and destination management, Enhancing the efficiency and sustainability of the TR32 Region's supply chains through smart logistics and intermodal connectivity.
Spatial Development Scenario	Supporting multi-centered growth and balanced spatial development, Strengthening transportation infrastructure along the east-west corridor, Improving connectivity of OIZs to loading/unloading stations and railway connections, Using the completion of the Aydın-Denizli Highway in 2024 to increase the flow of goods and services.
Smart Transportation Systems	Adopting innovative solutions for urban mobility and transportation efficiency, Implementing smart traffic management systems, parking management, smart bike lanes, and smart transit systems, Integrating digital technologies in transportation planning and management, such as geographic information systems (GIS) and real-time data analytics.
Sustainable Transportation Infrastructure	Developing bike lanes, walkways, and transit systems to encourage non-motorized and public transportation modes, Improving the accessibility and quality of public transportation services, particularly in tourism-intensive districts and rural areas.
Intermodal Connectivity and Logistics	Strengthening the connectivity of OIZs to ports, railways, and highways, Developing logistics centers to facilitate the efficient movement of goods, Integrating smart technologies in freight transportation, such as intelligent transportation systems and real-time tracking and monitoring of cargo.
Tourism Mobility and Accessibility	Adopting innovative solutions to address traffic congestion in tourism-intensive districts, such as smart parking systems and the promotion of public transportation, Integrating digital applications in tourism promotion and destination management, such as mobile applications, virtual reality, and augmented reality technologies to enhance the visitor experience.

Note. Adapted from Southern Aegean Development Agency, 2023.

The TR32 Region 2024-2028 Draft Plan outlines several strategic transportation priorities that are essential for the region's smart city development and overall growth. These priorities include strengthening the connectivity of Organized Industrial Zones (OIZs) to ports, railways, and highways, particularly in the provinces of Aydın, Denizli, and Muğla. The plan emphasizes the importance of improving the connectivity of OIZs along the east-west corridor, which extends from Çardak Özdemir Sabancı OIZ in the east to Söke OIZ in the west, to facilitate the efficient movement of goods and enhance the region's competitiveness. The completion of the Aydın-Denizli Highway in 2023 is expected to further increase the flow of goods and services along this corridor.

The plan also highlights the need to develop smart transportation systems and promote sustainable transportation infrastructures in urban areas, such as Aydın, Denizli, and Muğla, to improve urban mobility, reduce traffic congestion, and encourage the use of environmentally friendly modes of transportation. Additionally, the plan emphasizes the importance of addressing traffic congestion in tourism-intensive districts, such as Bodrum, Marmaris, Fethiye, and Pamukkale, through innovative solutions like smart parking systems and the promotion of public transportation to enhance the visitor experience and support the growth of the tourism sector. The integration of digital applications in tourism promotion and destination management is also a key priority, particularly in popular tourist destinations like Ephesus (Efes), Pamukkale, and Bodrum. The plan stresses the importance of enhancing the efficiency and sustainability of the region's supply chains through smart logistics and intermodal connectivity, focusing on the development of logistics centers in strategic locations.

The draft plan presents a comprehensive spatial development scenario that emphasizes the importance of transportation in supporting the region's multi-centered growth and balanced spatial development. The scenario highlights the need to strengthen transportation infrastructure along the east-west corridor, which extends from Çardak Özdemir Sabancı Organized Industrial Zone in the east to Söke OIZ in the west. This corridor is critical for the region's industrial development, and the plan aims to improve the connectivity of OIZs along this route to loading/unloading stations and railway connections. The completion of the Aydın-Denizli Highway in 2024 is expected to play a significant role in increasing the flow of goods and services along this corridor, further enhancing the region's economic growth. The spatial development scenario also emphasizes the importance of activating new OIZs in strategic locations such as Denizli, Aydın, and Muğla to promote balanced industrial growth throughout the region. Moreover, the plan seeks to integrate agricultural production with industry in districts like Çivril, Acıpayam, and Tavas, creating agro-industrial hubs that can drive rural development. The scenario also focuses on enhancing tourism services and developing alternative tourism opportunities in coastal areas like Bodrum, Marmaris, and Fethiye, while also promoting cultural tourism routes connecting ancient cities and heritage sites across the region. By coordinating transportation investments with these spatial development priorities, the TR32 Region aims to create a more connected, competitive, and sustainable economic landscape that benefits both urban and rural communities.

One other important strategic topic in the TR32 Region Draft Plan is about smart transportation systems. The draft plan emphasizes the development and implementation of smart transportation systems to address the challenges of urban mobility and enhance the efficiency of transportation networks in the region. The plan advocates for the adoption of innovative solutions, such as smart traffic management systems, parking management, smart bike lanes, and smart transit systems, particularly in the urban centers of the three provinces of TR32. The objective of these smart transportation systems is to mitigate traffic congestion, enhance road safety, and encourage the adoption of sustainable transportation modes, thereby leading to an enhanced quality of life for both residents and visitors. The plan also emphasizes the importance of integrating digital technologies, such as geographic information systems (GIS) and real-time data analytics, in transportation planning and management. This integration will enable the region to make data-driven decisions, optimize resource allocation, and respond more effectively to the changing needs of the transportation network. By implementing smart transportation systems, the TR32 Region seeks to create a more efficient, sustainable, and user-friendly transportation ecosystem that supports the region's economic growth and urban development goals, particularly in the rapidly growing urban areas of the region's three provinces, as well as in tourism-intensive districts.

The draft plan places significant emphasis on promoting sustainable transportation infrastructures as a key objective in the region's smart city development. The plan address to the importance of developing environmentally friendly and efficient transportation systems to improve the overall quality of life for residents and visitors alike, particularly in the provinces of Aydın, Denizli, and Muğla. This aim closely corresponds with the region's objectives of reducing traffic congestion, enhancing accessibility, and encouraging the use of alternative modes of transportation, especially in tourism-intensive districts such as Bodrum, Marmaris, Fethiye, and Pamukkale. To achieve this, the plan outlines several strategies, such as expanding bike lanes and walkways to facilitate non-motorized travel, improving the quality and accessibility of public transportation services in both urban and rural areas, and promoting the integration of sustainable transportation options with existing networks. Furthermore, the plan emphasizes the need to address traffic congestion in tourism-intensive districts through innovative solutions like smart parking systems and the promotion of public transportation. By prioritizing sustainable transportation infrastructures across the TR32 Region, GEKA aims to create a more livable, connected, and environmentally responsible urban environment that supports the well-being of its communities while contributing to the region's long-term economic and social development.

The TR32 Region 2024-2028 Draft Regional Plan also highlights the significance of intermodal connectivity and logistics in supporting the region's economic growth and competitiveness. The plan emphasizes the need to strengthen the connectivity of Organized Industrial Zones to ports, railways, and highways in the region's provinces. This improved connectivity is crucial for facilitating the efficient movement of goods and enhancing the TR32 Region's supply chain performance. The draft plan also stresses the importance of developing logistics centers in strategic locations to further streamline the flow of goods and reduce transportation costs. These logistics centers will serve as key hubs for the consolidation, storage, and distribution of goods, supporting the region's growing industrial and commercial activities. Moreover, the plan advocates for the integration of smart technologies in freight transportation, such as intelligent transportation systems and real-time tracking and monitoring of cargo. These technologies will enable better coordination among stakeholders, improve the visibility and security of shipments, and optimize route planning and resource utilization. By enhancing intermodal connectivity and logistics capabilities, the TR32 Region aims to create a more seamless and efficient transportation network that can effectively serve the needs of its industries, particularly in the manufacturing, agriculture, and tourism sectors. This improved connectivity will also contribute to the region's overall competitiveness and attractiveness for domestic and foreign investment, driving economic growth and job creation in key urban centers in Aydın, Denizli, and Muğla provinces.

The last strategic topic addressed in the TR32 Region Draft Plan relates to mobility and accessibility for tourism. The plan highlights the significance of addressing traffic congestion in tourism-intensive districts, such as Bodrum, Marmaris, Fethiye, and Pamukkale, to enhance the visitor experience and support the sustainable growth of the tourism sector. To achieve this goal, the plan proposes the adoption of innovative solutions, such as smart parking systems and the promotion of public transportation, to reduce traffic bottlenecks and improve the efficiency of transportation services in these popular tourist destinations. Additionally, the plan emphasizes the integration of digital applications, such as mobile apps, virtual reality, and augmented reality technologies, in tourism promotion and destination management. These digital tools will provide visitors with comprehensive information about tourist attractions, transportation options, and local services, enabling them to plan their trips more effectively and enjoy a seamless travel experience. Through utilizing these technologies, the TR32 Region aims to enhance its competitiveness as a tourist destination and attract a wider range of visitors from both domestic and international markets. Moreover, the plan seeks to improve the accessibility of tourism sites and services for all visitors, including those with disabilities or special needs, by investing in infrastructure upgrades and implementing inclusive design principles.

The TR32 Region 2024-2028 Draft Regional Plan demonstrates a comprehensive approach to integrating smart transportation strategies and measures into its regional development framework. The plan effectively addresses key strategic topics, such as prioritizing transportation infrastructure, developing a balanced spatial development scenario, adopting smart transportation systems, promoting sustainable transportation, enhancing intermodal connectivity and logistics, and improving tourism mobility and accessibility. By focusing on these areas, the plan aims to enhance the TR32 Region's competitiveness, sustainability, and quality of life for residents and visitors alike.

The draft plan prioritizes improving connectivity within OIZs, developing smart transportation systems, and promoting sustainable transportation infrastructures. These initiatives reflect a commitment to enhancing technology and innovation to stimulate regional development. The integration of digital applications in tourism promotion and destination management, as well as the adoption of innovative solutions to address traffic congestion in tourism-intensive districts, highlights the importance of the tourism sector in the TR32 Region's economy.

However, the evaluation of the plan's effectiveness in implementing smart transportation strategies is limited by the lack of formal feedback or evaluation of the previous GEKA plan period (2014-2023). Despite this limitation, the draft plan provides valuable strategic insights into the region's approach to incorporating smart transportation measures as instruments of regional development at the NUTS-2 level in Turkey.

In conclusion, the TR32 Region 2024-2028 Draft Regional Plan offers a structured and forward-thinking strategy for incorporating smart transportation strategies into its regional development framework. The plan covers a wide range of strategic topics and measures, demonstrating a commitment to utilizing technology, innovation, and sustainability to foster regional growth and enhance the quality of life for residents and visitors. Here, it will be crucial to monitor and evaluate the progress of the plan as it is implemented, ensuring the successful achievement of its smart transportation goals and their coherence with broader regional development objectives.

CONCLUSION

The research on the TR32 Region's 2024-2028 Draft Regional Plan prepared by GEKA highlights the importance of integrating smart transportation strategies and measures into the regional development framework to enhance the region's competitiveness, sustainability, and quality of life. The draft plan demonstrates a comprehensive approach to incorporating smart city principles and technologies into its development strategy. The plan's focus on digital transformation, innovation, and green growth is consistent with the fundamental concepts of smart city development. It effectively addresses the integration of smart technologies in various sectors, such as transportation, manufacturing, agriculture, and tourism. However, the draft plan could benefit from a more detailed exploration of the potential applications and impacts of smart city solutions and a clearer articulation of the mechanisms and policies needed to foster collaboration among stakeholders and ensure the interoperability of smart city systems. The draft plan's emphasis on strategic transportation priorities, spatial development scenarios, smart transportation systems, sustainable infrastructure, intermodal connectivity, and tourism mobility highlights the TR32 Region's commitment to utilizing technology and innovation to stimulate regional development. The integration of digital applications in tourism promotion and destination management and the adoption of innovative solutions to address traffic congestion in tourism-intensive districts emphasize the significance of the tourism sector in the TR32 Region's economy.

However, the evaluation of the plan's effectiveness in implementing smart transportation strategies is limited by the lack of formal feedback or evaluation of the previous GEKA plan period (2014-2023). The lack of a comprehensive assessment of the earlier plan's outcomes makes it difficult to draw definitive conclusions about the region's progress in integrating smart transportation measures into its development framework. The research also encounters limitations due to the fragmented nature of data sources, potentially leading to an incomplete view of the region's smart transportation initiatives. For instance, the exclusion of Aydın province from the categorical distribution of smart transportation projects in Table 1 highlights the challenges in obtaining comprehensive and up-to-date information on the TR32 Region's current and planned smart city projects. In this context, future research should focus on comprehensively evaluating the previous GEKA plan period to identify the successes, challenges, and lessons learned in implementing smart transportation strategies. This assessment would provide valuable insights for refining the current draft plan and ensuring its relevance with the region's development goals. Furthermore, future studies could explore the potential for integrating smart transportation strategies with other aspects of regional development, such as land use planning, economic development, and social inclusion. In this vein, exploring the connections between smart transportation and these broader development goals could help to create a more cohesive and effective regional development framework for the TR32 Region.

In conclusion, the TR32 Region's 2024-2028 Draft Regional Plan offers a foundation for potentially integrating smart transportation strategies into the region's development framework. Despite the limitations in evaluating the plan's effectiveness and the challenges in obtaining comprehensive data, the draft plan shows a clear intention to utilize technology and innovation to improve regional competitiveness and sustainability. Future research should focus on assessing the outcomes of previous plans, gathering more comprehensive data, and exploring the potential for integrating smart transportation with other aspects of regional development to support the successful implementation of the TR32 Region's smart city objectives. In light of the analysis presented in this chapter, it is important to consider the specific strengths, weaknesses, and future targets of smart transportation systems in the TR32 Region, as well as their broader implications for regional development.

The TR32 Region's smart transportation systems exhibit both strengths and weaknesses. The region's diverse economic base and strong tourism sector support multi-modal transportation development and drive innovation in mobility solutions. Several initiatives in smart traffic management and public transportation are already underway. However, the uneven distribution of smart transportation projects across the region, challenges in rural area connectivity, and limited data on project effectiveness present ongoing challenges. The region has set ambitious targets, including implementing intelligent traffic management systems in all major urban centers by 2026, increasing public transportation usage by 20% by 2028, and developing a region-wide integrated mobility app by 2025. Transportation strategy plays a crucial role in the TR32 Region's economic development. Key focus areas include enhancing connectivity between industrial zones and transportation hubs, improving sustainable mobility options to support tourism growth, and utilizing smart transportation to reduce urban congestion and improve quality of life. The region has begun implementing smart city targets, focusing on digital transformation of urban services, smart mobility and transportation systems, and sustainable energy and environmental management.

Significant projects in the region include Muğla's intelligent traffic management system, Denizli's smart mobility planning initiative, and Aydın's public transportation optimization project. These projects demonstrate the region's commitment to smart city principles and provide valuable case studies for other regions in Turkey. However, more comprehensive data on their impact and scalability is needed to fully assess their importance in regional development. To ensure successful implementation of smart city and transportation initiatives, the TR32 Region should focus on fostering inter-agency collaboration, securing sustainable funding mechanisms, enhancing digital literacy and inclusion, and coordinating local initiatives with national smart city strategies. By addressing these areas, the TR32 Region can optimize the potential of smart transportation systems to drive regional development and improve quality of life for its residents and visitors.

REFERENCES

Bifulco, F., Amitrano, C., & Tregua, M. (2014). Driving smartization through intelligent transport. *The China Business Review*, 13(4), 243.

Billones, R., Guillermo, M., Lucas, K., Era, M., Dadios, E., & Fillone, A. (2021). Smart region mobility framework. *Sustainability (Basel)*, 13(11), 6366. DOI: 10.3390/su13116366

Buosi, M., Cilloni, M., Corradi, A., Rolt, C., Dias, J., Foschini, L., Montanari, R., & Zito, P. (2018). A crowdsensing campaign and data analytics for assisting urban mobility pattern determination. In *Proceedings of the 2018 IEEE Symposium on Computers and Communications (ISCC)* (pp. 00224-00229), Natal, Brazil. DOI: 10.1109/ISCC.2018.8538483

Camata, S., Sartori, S., Bettini, M., & Conti, D. (2022). Innovation and sustainable development of smart cities: A systematic review to identify research trends and gaps. [RISUS]. *Journal on Innovation and Sustainability*, 13(4), 32–44. DOI: 10.23925/2179-3565.2022v13i4p32-44

Čepeliauskaitė, G., Keppner, B., Simkute, Z., Stasiškienė, Ž., Leuser, L., Kalniņa, I., Kotoviča, N., Andiņš, J., & Muiste, M. (2021). Smart-mobility services for climate mitigation in urban areas: Case studies of Baltic countries and Germany. *Sustainability (Basel)*, 13(8), 4127. DOI: 10.3390/su13084127

Chatfield, A. T., & Reddick, C. G. (2016). Smart city implementation through shared vision of social innovation for environmental sustainability: A case study of Kitakyushu, Japan. *Social Science Computer Review*, 34(6), 757–773. DOI: 10.1177/0894439315611085

Chourabi, H., Nam, T., Walker, S., Gil-García, J. R., Mellouli, S., Nahon, K., Pardo, T. A., & Scholl, H. J. (2012). Understanding smart cities: An integrative framework. In *Proceedings of the 2012 45th Hawaii International Conference on System Sciences* (pp. 2289-2297), Maui, HI, United States. DOI: 10.1109/HICSS.2012.615

Drobne, S., & Bogataj, M. (2015). Optimal allocation of public service centers in the central places of functional regions. *IFAC-PapersOnLine*, 48(3), 2362–2367. DOI: 10.1016/j.ifacol.2015.06.441

García-Ayllón, S., & Miralles, J. (2015). New strategies to improve governance in territorial management: Evolving from "Smart Cities" to "Smart Territories". *Procedia Engineering*, 118, 3–11. DOI: 10.1016/j.proeng.2015.08.396

Irnazarov, F., & Kayumova, M. (2017). Toward smart city development in Central Asia: A comparative assessment. *Central Asian Affairs*, 4(1), 51–82. DOI: 10.1163/22142290-00401003

Istanbul Metropolitan Municipality. (2024). İstanbul akıllı şehir stratejik planı, projeler [Istanbul smart city strategic plan, projects]. Istanbul Metropolitan Municipality. https://akillisehir.istanbul/tr/projects

Joshi, M., Vaidya, A., & Deshmukh, M. (2018). Sustainable transport solutions for the concept of smart city. In Gautam, A., De, S., Dhar, A., Gupta, J. G., & Pandey, A. (Eds.), *Sustainable energy and transportation: Technologies and policies* (pp. 21–42). Springer., DOI: 10.1007/978-981-10-7509-4_3

Kalašová, A., Harantová, V., & Čulík, K. (2019). Public transport as a part of shared economy. The Archives of Automotive Engineering – Archiwum Motoryzacji, 85(3), 49-56. DOI: 10.14669/AM.VOL85.ART4

Karaçoban, E. (2023). Akıllı kentler ve bölgesel kalkınma: TR32 Bölgesi kalkınma stratejilerini yeniden düşünmek [Smart cities and regional development: rethinking TR32 regional development strategies] [Unpublished MSc thesis]. Pamukkale University, Denizli, Turkey.

Kuhail, M. A., Ahmad, B., & Rottinghaus, C. (2018). Smart resident: a personalized transportation guidance system. In *Proceedings of the 2018 IEEE 5th International Congress on Information Science and Technology (CiSt)* (pp. 547-551), Marrakech, Morocco. DOI: 10.1109/CIST.2018.8596632

Kwan, S. C., & Hashim, J. H. (2016). A review on co-benefits of mass public transportation in climate change mitigation. *Sustainable Cities and Society*, 22, 11–18. DOI: 10.1016/j.scs.2016.01.004

Lorenz, A., Madeja, N., & Leyh, C. (2023). A framework for assessing the sustainability impact of intelligent transport systems in the smart city context. In *Proceedings of the 18th Conference on Computer Science and Intelligence Systems (FedCSIS)* (pp. 161-169), Warsaw, Poland. DOI: 10.15439/2023F6002

Mondale, T. (2000). Transportation—a major player in smart growth. [Institute of Transportation Engineers]. *ITE Journal*, 70(11), 39–43.

Musa, W. (2017) The impact of smart city initiatives on cities' local economic development [Unpublished MSc thesis]. Fort Hays State University, Hays, KS, United States.

Neirotti, P., Marco, A. D., Cagliano, A. C., Mangano, G., & Scorrano, F. (2014). Current trends in Smart City initiatives: Some stylised facts. *Cities (London, England)*, 38, 25–36. DOI: 10.1016/j.cities.2013.12.010

Oladimeji, D., Gupta, K., Kose, N. A., Gundogan, K., Ge, L., & Liang, F. (2023). Smart transportation: An overview of technologies and applications. *Sensors (Basel)*, 23(8), 3880. DOI: 10.3390/s23083880 PMID: 37112221

Oliveira, J., Silva, M., Santos, S., Costa, A., & Clemente, T. (2021). Multidimensional sorting framework of cities regarding the concept of sustainable and smart cities with an application to Brazilian capitals. *Sustainable Cities and Society*, 74, 103193. DOI: 10.1016/j.scs.2021.103193

Paiva, S., Ahad, M. A., Tripathi, G., Feroz, N., & Casalino, G. (2021). Enabling technologies for urban smart mobility: A recent trends, opportunities and challenges. *Sensors (Basel)*, 21(6), 2143. DOI: 10.3390/s21062143 PMID: 33803903

Qiao, J. (2022). Smart city and intelligent upgrading of urban transportation system: based on sustainable innvestment strategy. In *Proceedings of the Second International Conference on Advanced Technologies in Intelligent Control, Environment, Computing & Communication Engineering (ICATIECE)* (pp. 1-6), Bangalore, India. DOI: 10.1109/ICATIECE56365.2022.10047675

Sedlak, O., & Ćirić, Z. (2018). Sustainability development of smart public transportation in globalization. *The Academic Research Community Publication*, 2(1), 10. DOI: 10.21625/archive.v2i1.235

Southern Aegean Development Agency. (2023). TR32 düzey 2 bölgesi 2024-2028 bölge planı taslağı [TR32 NUTS 2 region 2024-2028 draft regional plan]. GEKA. https://geka.gov.tr/tr/sayfa/planlar

The Presidency of the Republic of Turkey, the Presidency of Strategy and Budget. (2019). On birinci kalkınma planı (2019-2023) [Eleventh development plan (2019-2023)]. The Presidency of Strategy and Budget, the Directorate General for Administrative Services, the Department of Information and Document Management. https://www.sbb.gov.tr/kalkinma-planlari

The Presidency of the Republic of Turkey, the Presidency of Strategy and Budget. (2023). On ikinci kalkınma planı (2024-2028) [Twelfth development plan (2024-2028)]. The Presidency of Strategy and Budget, the Directorate General for Administrative Services, the Department of Information and Document Management. https://www.sbb.gov.tr/kalkinma-planlari

The Republic of Turkey, the Ministry of Environment, Urbanization and Climate Change. (2019). 2020-2023 Ulusal akıllı şehirler stratejisi ve eylem planı [National smart cities strategy and action plan]. The Republic of Turkey, the Ministry of Environment, Urbanization and Climate Change. https://akillisehirler.gov.tr/wp-content/uploads/EylemPlani.pdf

The Republic of Turkey, the Ministry of Environment, Urbanization and Climate Change. (2024a). Akıllı şehirler portalı, proje envanteri [Smart cities portal, project inventory]. The Directorate General of Geographic Information Systems, the Department of Smart Cities. https://akillisehirler.gov.tr/wp-content/uploads/ProjeEnvanteri/map.html

The Republic of Turkey, the Ministry of Environment, Urbanization and Climate Change. (2024b). Akıllı şehir ekosistemi [Smart city ecosystem]. The Directorate General of Geographic Information Systems, the Department of Smart Cities. https://akillisehirekosistem.csb.gov.tr/

The Republic of Turkey, the Ministry of Industry and Technology. (2023). Kalkınma planlamasında istatistiki bölge birimleri sınıflandırması [Nomenclature of territorial units for statistics in development planning]. The General Directorate of Development Agencies. https://ka.gov.tr/sayfalar/kalkinma-planlamasinda-istatistiki-bolge-birimleri-siniflandirmasi--24

The Republic of Turkey, the Ministry of Industry and Technology. (2024). Bölgesel kalkınmanın yönetişiminde kalkınma ajansları [Development agencies in the governance of regional development]. The General Directorate of Development Agencies. https://ka.gov.tr/sayfalar/bolgesel-kalkinmanin-yonetisiminde-kalkinma-ajanslari--23

The Republic of Turkey, the Prime Ministry General Directorate of Legislation Development and Publication. (2004). 5216 sayılı büyükşehir belediyesi kanunu [Law no. 5216 on metropolitan municipalities]. Resmi Gazete [Official Gazette], 25531. https://www.resmigazete.gov.tr/eskiler/2004/07/20040723.htm

The Resolution of the Cabinet of Ministers of Turkey. (2002). İstatistiki bölge birimleri sınıflandırması [Nomenclature of territorial units for statistics]. Decree of the Council of Ministers dated 28/08/2002 and numbered 2002/4720. http://www.die.gov.tr/nuts/bolgekarar.htm

Turkish Statistical Institute. (2024a). Geographic statistics portal. Turkish Statistical Institute. https://cip.tuik.gov.tr

Turkish Statistical Institute. (2024b). Regional statistics. Turkish Statistical Institute. https://biruni.tuik.gov.tr/bolgeselistatistik/anaSayfa.do?dil=en

United Nations Economic Commission for Europe (UNECE). (2024) Smart sustainable cities. UNECE. https://unece.org/housing/smart-sustainable-cities#:~:text=A%20smart%20sustainable%20city%20is,as%20well%20as%20cultural%20aspects

World Bank. (2024). Global smart city partnership program: Supporting cities to become 'smart' – efficient, innovative, inclusive, and resilient – using data and technology. World Bank. https://www.worldbank.org/en/programs/global-smart-city-partnership-program#:~:text=Smart%20Cities%20%26%20Urban%20Development,improve%20urban%20planning%20and%20management

Xu, B., & Thakur, G. (2021). Introduction to the special issue on smart transportation. *GeoInformatica*, 25(3), 417–418. DOI: 10.1007/s10707-021-00432-3

Zawieska, J., & Pieriegud, J. (2018). Smart city as a tool for sustainable mobility and transport decarbonisation. *Transport Policy*, 63, 39–50. DOI: 10.1016/j.tranpol.2017.11.004

Zeng, Y., Zhang, Z., Ye, Z., & Li, L. (2023). Regional innovation effect of smart city construction in China. *PLoS One*, 18(2), e0281862. DOI: 10.1371/journal.pone.0281862 PMID: 36795669

Chapter 4
The Contribution of Intelligent Systems (IS) on Transport and Planning Studies

Metin Mutlu Aydin
https://orcid.org/0000-0001-9470-716X
Ondokuz Mayıs University, Turkey

Eren Dağlı
https://orcid.org/0000-0002-3892-0270
Selçuk University, Turkey

ABSTRACT

Intelligent Systems (IS) have emerged as a key tool in revolutionizing transport and urban planning studies by providing innovative solutions. They have supplied transport planners with robust predictive modeling capabilities incorporating artificial intelligence and machine learning. They predict traffic patterns, optimize routes, and anticipate fluctuations in demand. They improve situational awareness, dynamic traffic management, and reduce congestion. These systems are key in optimizing land use, improving accessibility, and promoting sustainable development. In summary, they represent a paradigm shift in transport and planning studies. Intelligent Transport Systems (ITS) are an umbrella under which all transport-related systems are grouped. Therefore, all applications under the umbrella of ITS in cities are carried out to make transport and planning studies more intelligent. In this study, the importance and contribution of ITS in transportation and planning studies are examined and it is aimed to be a guide for future studies.

DOI: 10.4018/979-8-3693-6695-0.ch004

INTRODUCTION

The majority of countries across the world are experiencing accelerated urbanisation as a consequence of unregulated and exponential population growth, rising individual expectations of comfort, and enhanced accessibility. The process of urban expansion inevitably gives rise to a number of new requirements. It is important that infrastructure services are accessible to all areas of the city and responsive to the needs of the population. Furthermore, the expansion process necessitates alterations to the existing transport infrastructure. Furthermore, it is possible to discuss a dynamic change in the expectations and preferences of individuals in the field of transport in response to technological developments and changes in lifestyles. A variety of factors have contributed to the emergence of these expectations, including the quest for efficient, safe and cost-effective transport solutions; the increase in individual vehicle ownership, the profound changes in transport systems and user expectations, particularly in the context of the global pandemic; the demand for convenient and time-efficient transport infrastructure; the evolution of transport modes, and the need for seamless intermodal transfer solutions. These expectations and criteria have become key considerations for both users and planners.

While change can facilitate improvements in systems, it can also introduce complications. The increase in mobility in everyday life has led to significant changes in transport infrastructures. The growing number of vehicles in many developed European cities is a notable phenomenon. As the number of vehicles increases, traffic congestion increases, as do unforeseen emergencies and accidents (Steger-Vonmetz, 2005). In recent years, there has been an increase in the number of vehicles on the roads. With the increase in the number of vehicles and the consequent increase in density and volume, negative effects on the environment and the quality of life of people have been observed. Additionally, according to Watkins (2012), the increase in vehicle numbers and density causes an increase in accidents and congestion.

The scarcity of resources makes it difficult for individuals and managers to cope with this process of urban expansion and changing transport habits. Environmental impacts such as traffic congestion, air and noise pollution and greenhouse gas emissions sometimes reduce the quality of life and urban quality. In addition, changing transport habits require managers and transport infrastructure to keep pace (Aydın and Keleş, 2024).

Transport planning studies are carried out in order to effectively manage the mobility of people and goods in a city or region (Banister, 2008; Black, 2010). The need for planning arises from the fact that this mobility must be efficient, sustainable, economical and safe. It also aims to establish a systematic transport infrastructure by minimising the traffic density that can result from mobility and the confusion and accidents that can occur as a result (Aydın and Bıyık, 2023).

With the development of technology and easy access to technology, technological solution methods have started to be used effectively in many areas. In the digital age, operations requiring mental effort are performed on a computer basis and these tools are referred to as intelligent systems (Molina, 2020). In addition to applications such as smart industrial systems, smart healthcare systems and smart education infrastructure, the concept of smart cities has also entered our lives. In smart cities, various smart applications are used to make people's daily lives easier, safer and more comfortable, and to improve the environment. In smart cities, we encounter many smart applications such as infrastructure systems, smart energy systems, smart waste management systems, smart buildings and Intelligent Transport Systems (ITS). Undoubtedly, ITS is one of the smart applications that we often encounter and use in our daily lives.

ITS applications are used in transport planning studies to make urban traffic efficient, safe, environmentally friendly and user-friendly, in short, sustainable. ITS has an important potential in removing obstacles in the effective use of existing capacity, reducing the need for manpower in the operation of systems, and preventing possible traffic accidents (He et al., 2010). At the same time, ITS applications are also used in transport planning processes to determine the current situation for planners. Using ITS in planning processes to make the right decisions will have a positive impact on the success of planning. However, there may be technical difficulties in bringing together and integrating different systems. With the smart city concept and the accumulation of knowledge and skills, these technical difficulties are becoming easier to overcome. The successful integration of innovative ITS applications into transport systems and their systematic use in transport will minimise the problems that may arise.

ITS increases the success rate of transport planning studies and is an important tool for planners. In addition, the adoption of the smart city concept on a global scale is forcing governments and local authorities to focus on this area. At the same time, many international organisations have been set up to conduct and implement ITS research. Research and innovative applications in the field of ITS will both increase the success of planning processes and create a more sustainable transport infrastructure and urbanisation.

The purpose of this section is to present the application areas of Intelligent Systems (IS) in transportation planning studies and their contributions to the planning process. This section is intended to contribute to transportation planning studies as a reference source for researchers. Accordingly, Intelligent Transportation Systems (ITS) are discussed in detail in the following sections.

INTELLIGENT TRANSPORT SYSTEMS (ITS) SOLUTIONS

Intelligent Transportation Systems (ITS) is a complex mechanism in which the current situation is analysed by sensing systems such as sensors and cameras, the data obtained about the current situation is analysed by computer infrastructure and software, and the system is coordinated by control mechanisms and communication technologies (Figure 1). In other words, ITS are systems consisting of monitoring, measurement, analysis and control mechanisms developed in line with principles such as energy efficiency and reduction of environmental damage, aiming to reduce travel times, increase traffic safety and optimum use of existing infrastructure (Tektaş el al., 2016). Khazraeian and Hadi (2019) defined Intelligent transport systems as cutting-edge applications that aim to increase transportation safety and mobility, as well as as move towards an environmentally friendly system. In addition, ITS are defined as new methods applied to increase safety and efficiency in the field of transport as a result of technological developments. (Shaheen and Finson, 2013).

Figure 1. Intelligent transport systems

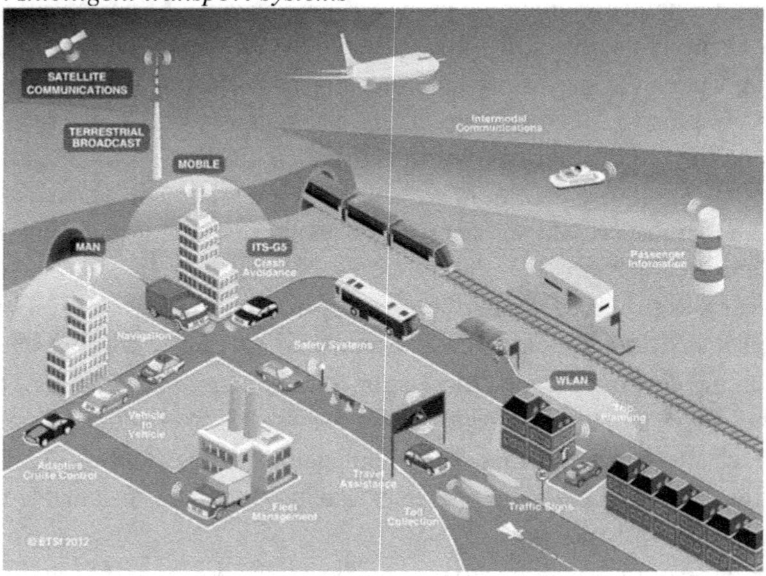

(European Telecommunications Standards Institute, 2010)

ITS applications have many environmental, economic and social benefits. As a result of ITS applications, a significant reduction in the time spent in traffic and average stopping times can be observed (Singh and Gupta, 2015). The reduction in travel times and the reduction in the number of stops and movements will reduce fuel consumption and exhaust emission values. As a result, both fuel consumption and environmental damage will be reduced. In addition to traffic congestion, ITS also plays an effective role in traffic safety (Taç, 2018). The global economy and daily life are increasingly dependent on efficient and safe transport systems (Xiong el al., 2012). ITS aims to prevent possible accidents with a more planned transport infrastructure, thus providing a safer transport infrastructure for both pedestrians and drivers.

Figure 2. History of intelligent transportation systems

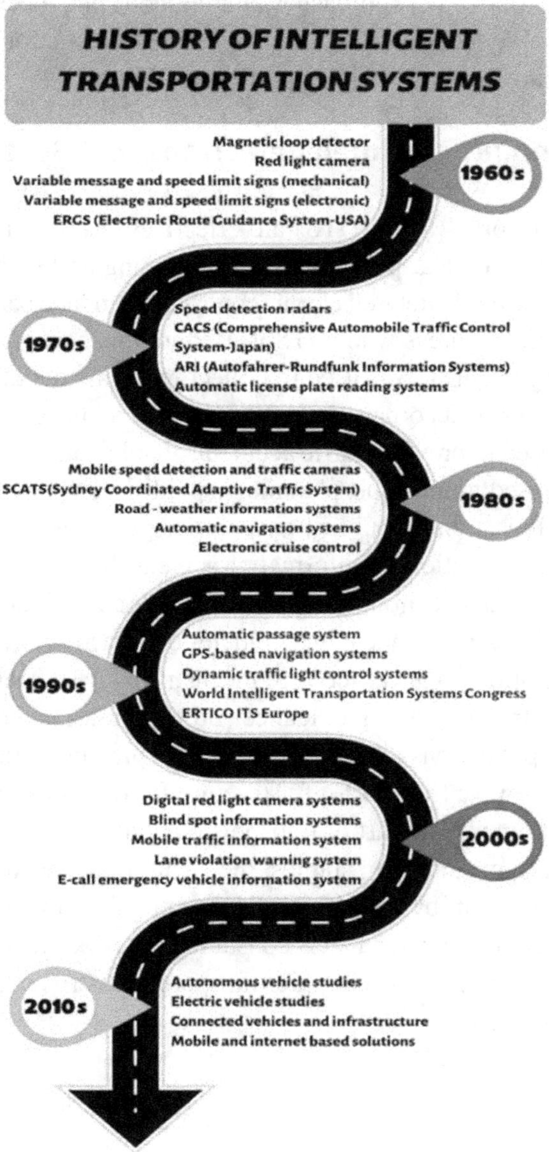

The history of ITS does not go back very far. The ITS adventure began with simple traffic control in the 1960s. In those years, loop detectors, red light cameras, mechanical and electronic variable message signs and speed limit signs were introduced. Then, in the 1970s, speed detection radars, comprehensive vehicle traffic control systems and car radio information systems applications began. In the following years, the formation of ERTICO ITS Europe, one of the ITS organisations, started with applications such as adaptive traffic system, navigation and dynamic signal control system, which are still in use today. In recent years, with the development of technology and innovative approaches, autonomous and electric cars, internet-based artificial intelligence and machine learning based solutions have taken their place in the literature as various ITS applications (Figure 2).

Transport Planning and Intelligent Transport Systems Relation

Intelligent Transport Systems (ITS) are effectively used in transport planning studies. Traffic management, public transport planning and management studies, autonomous vehicles, road safety, electronic control systems and parking management systems are examples of areas where ITS is used in transport planning.

In traffic management, ITS is used to collect real-time data from the field and to design signal plans in accordance with the data. In addition, warning and guidance systems are used to prevent traffic accidents thanks to real-time data. Another area where ITS is used in transport planning is public transport. Examples include route and time optimisation for public transport and passenger information systems through intelligent applications or information screens. In autonomous vehicles, autonomous driving is realised through vehicle-to-vehicle and vehicle-to-infrastructure communication. Vehicle-to-vehicle and vehicle-to-infrastructure communications are also ITS applications. Various control mechanisms have been implemented to ensure road safety. Instant parking guidance and charging systems are also used.

All these ITS applications aim to create a better planned, safer, more comfortable and more systematic transport infrastructure for drivers and pedestrians. ITS applications are an important part of transport planning. By effectively using ITS applications in transport planning studies, the future transport needs of cities will be better met. In addition, by identifying the current situation and deficiencies, a transport infrastructure is created that is constantly renewed and improved. Various ITS applications aimed at improving the transport infrastructure are presented in detail in the next section.

ITS Solutions on Planning

This part of the study provides a comprehensive overview of ITS applications and discusses examples in detail. The aim is to examine comprehensively how the various interrelated elements of ITS contribute to the planning of a productive, efficient and forward-looking transportation system.

From intelligent infrastructure to autonomous and electric vehicle technologies, the topics discussed here present modern transportation systems in a multifaceted way. By addressing the interaction between technological capabilities, accessibility and environmental impact, it shows how the transport infrastructure works together to create an ITS that is sustainable, efficient and prioritises user satisfaction. This section will serve as a guide for effective transportation planning.

ITS aims to increase mobility despite urbanization challenges (Elassy et al., 2024). To this end, ITS encompasses a range of technologies that help to continuously capture, analyse and integrate real-time information from the field. Technologies such as inductive loop detectors, radar and lidar sensors and CCTV cameras are critical for monitoring traffic flow, vehicle speed and general road conditions. These tools provide the data basis for real-time traffic management and incident detection. Communication networks are essential for real-time operations and efficient navigation. Advanced communication technologies, including vehicle-to-everything (V2X) communications, GPS positioning and cellular networks (4G/5G), enable seamless data transfer between different components of the transport network. Data analytics, machine learning and artificial intelligence play an important role in processing traffic data and drawing conclusions from it. These technologies help to optimise traffic flows, predict transportation needs and increase the efficiency of the transportation system. Technologies such as automatic number plate recognition cameras, weather sensors, weight-in-motion sensors and fare tag readers provide information to support ITS projects.

There are some basic principles and steps in the implementation of ITS applications. The first is the preparation and strategic planning phase. By evaluating the existing infrastructure and technological possibilities, a comprehensive strategic plan for ITS applications is developed with all stakeholders in line with the possibilities. The next step is to develop the infrastructure according to the possibilities and to implement ITS projects and pilot applications. The ethical and legal compliance of the applications should be considered in all studies carried out. Policies and regulations that respect public safety and the protection of personal data should be implemented. The studies should be carried out in cooperation with public institutions, universities and private organisations, and a common working infrastructure should be designed. In addition, the monitoring and evaluation process of the studies should be carried out. Performance evaluations should be carried out to identify shortcomings and efficiency. The efficiency and sustainability of transport planning studies and ITS implementations will be a reference for future studies. ITS solutions are very diverse. It is possible to group these implementations under headings as shown in Figure 3. Various ITS implementations are presented in detail.

Figure 3. ITS solutions

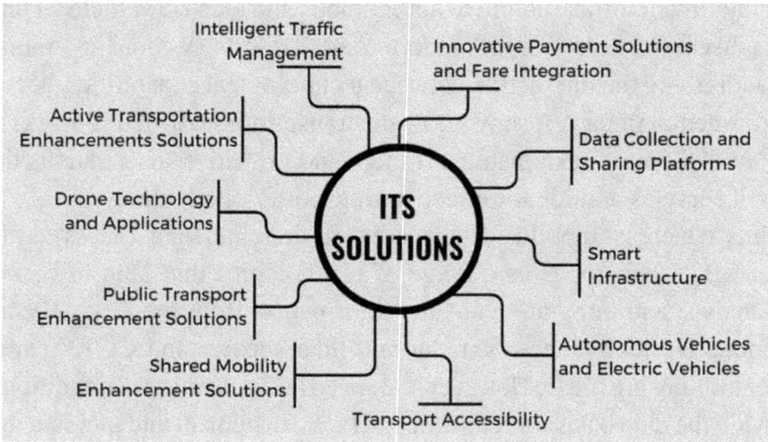

INTELLIGENT TRAFFIC MANAGEMENT APPLICATIONS

Intelligent traffic management is very important for transport planning studies, both in the planning phase, in the implementation of applications, and in determining the performance of applications. Intelligent traffic management and control systems, incident detection and intervention mechanisms, adaptive traffic management systems, and various detection and measurement systems are the components of the intelligent traffic management system. The basis of successful traffic planning is based on accurate information. Therefore, the use of ITS solutions in traffic planning studies will increase the success rate of the plans. Various applications of intelligent traffic management are presented in detail.

Smart Traffic Management and Control Systems (STMCS)

The integration of Smart Traffic Management and Control Systems (STMCS) represents a significant advancement in transportation planning studies, offering a comprehensive solution through the convergence of cutting-edge technologies, including data analytics, artificial intelligence (AI), and the Internet of Things (IoT).

These systems collect and analyse real-time traffic data using sensors and the Internet of Things (IoT) to monitor key traffic components, including vehicle speed, type, and count. A variety of sensors, including cameras, inductive loops, radar and lidar sensors, and acoustic sensors, are strategically positioned within the road network for the purpose of data collection and traffic monitoring. The aforementioned devices are responsible for the collection of comprehensive traffic data pertaining to the prevailing situation (Aydın et al., 2023).

The data obtained is processed and analysed using artificial intelligence and advanced algorithms to facilitate the immediate determination of optimal traffic management decisions, such as the design of effective signal plans, which are instrumental in enhancing traffic flow and reducing congestion. The STMCS system facilitates real-time traffic management through the implementation of traffic measures that demonstrably enhance road safety and mitigate traffic congestion.

Figure 4. Smart traffic management and control systems

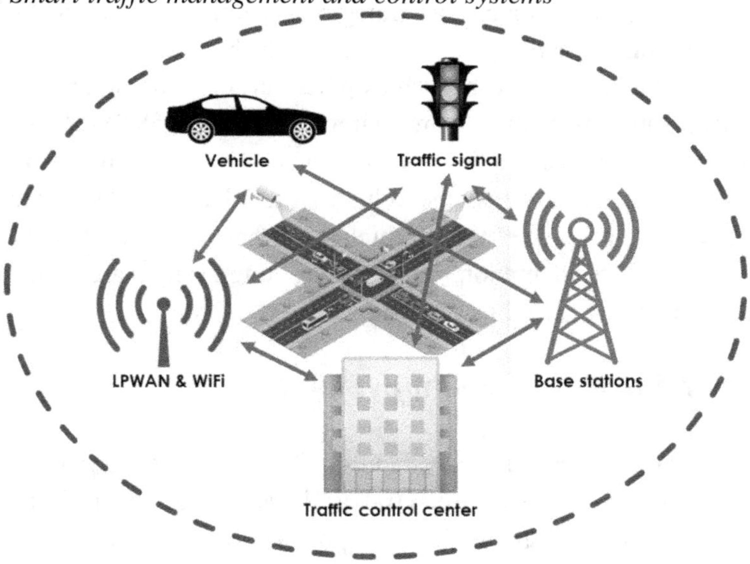

Furthermore, the Vehicle to Everything (V2X) communication technology facilitates the exchange of traffic-related information between vehicles and traffic management systems. Such communication facilitates the acquisition of information regarding the prevailing circumstances, thereby enabling the formulation of decisions that align with this information.

One of the fundamental principles that informs transportation planning studies is that of environmental impact. It is possible to reduce the damage caused by vehicles to the natural environment by reducing the number of stops and starts of vehicles and the duration of stops. The objective of the STMCS is to reduce vehicle emissions and improve air quality by minimising the time spent stopped and the number of starts and stops. The STMCS enables existing infrastructure to meet the needs of the population in a more efficient manner by optimising traffic flows and reducing congestion. Furthermore, STMCS plays a pivotal role in the development of a more sustainable and multimodal urban transportation infrastructure. It facilitates the integration of sustainable transportation modes, such as public transportation and bicycles and scooters, into the transportation infrastructure, thereby promoting their preference and utilisation.

The STMCS range in complexity from partial solutions that target only specific intersections to solutions that manage multiple intersections and road corridors, or even city-wide solutions. The implementation of STMCS ensures the provision of a safe and sustainable transportation infrastructure, in addition to enhancing the efficiency of traffic management. A comprehensive approach to transportation planning studies and urban traffic control is achieved by considering the various traffic management technologies and strategies as a unified whole. STMCS makes a great contribution to the design of smarter and livable cities.

It is evident that special areas, such as ports, airports and organised industrial zones, stand to gain considerably from the implementation of STMCS. The applicability of STMCS in extensive areas can facilitate emergency response by adjusting traffic signals for emergency response vehicles (ambulances and fire trucks e.g.) in accordance with the passage of these vehicles. Furthermore, STMCS enhances the system with immediate adaptations in pedestrian zones, school areas, and pertinent locations in response to social gatherings.

Adaptive Traffic Management Systems

Adaptive Traffic Management Systems (ATMS) are systems that facilitate the management of urban and extra-urban traffic flows through the utilisation of real-time data and advanced technological equipment. ATMS eliminates the inherent complexity of traffic management by providing a dynamic solution that is responsive to real-time traffic conditions (Shahgholian and Gharavian, 2018). ATMS employs a range of sensors, cameras and communication technologies to obtain real-time traffic data. The data obtained is subjected to complex algorithmic analysis with the objective of optimising the functionality of signalling systems and designing a variety of traffic arrangements.

Figure 5. Adaptive traffic management systems (Gandhi, 2023)

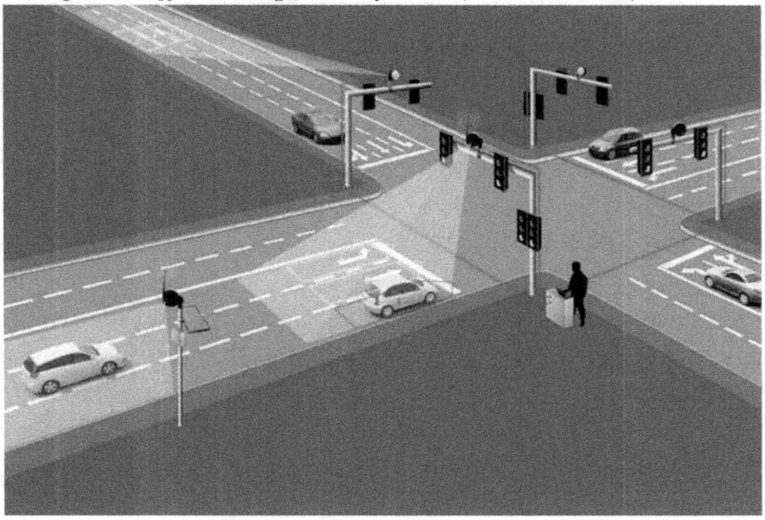

The system that operates the ATMS is a highly sophisticated one. A network of sensors and cameras has been installed at various points along the road network. The sensors and cameras collect data on pedestrian movements, the number of vehicle types, and speeds in digital form. The data is then transmitted to a traffic control centre, where it is subjected to immediate processing and analysis. The analysis of the data collected is carried out with the objective of alleviating traffic congestion and improving traffic flow. This is achieved through the application of algorithms, machine learning and artificial intelligence techniques. Consequently, the signalling systems in the road network and speed limits can be adjusted instantaneously if necessary.

ATMS has been observed to result in a notable reduction in traffic congestion, particularly during periods of peak traffic density. This is attributed to the optimisation of signal systems at intersections, which enables more efficient traffic management. Automated traffic management systems (ATMS) are equipped with sensors and cameras that can detect incidents such as traffic accidents or potential vehicle malfunctions within the road network. Consequently, the system is able to promptly intervene in the signalling system and, if required, display emergency messages on digital signs, thereby reducing the impact on traffic flow. Furthermore, it increases the probability of a prompt and effective intervention by directing emergency response vehicles. Furthermore, the dissemination of emergency messages will enhance the safety of both vehicles and emergency response vehicles in the vicinity of the incident. The integration of ATMS with emerging technologies such as autonomous driving will not only enhance the efficacy of innovative transportation solutions but also facilitate a more optimal utilisation of ATMS.

As urban areas continue to expand and evolve, the role of ATMS in transportation planning studies will become increasingly significant. It is imperative that these systems be designed in a manner that allows for adaptation to the anticipated growth in traffic volume over the coming years, as well as to advancements in technology.

HOV Detection Technologies

High Occupancy Vehicle (HOV) lanes designated for the use of vehicles with more than one passenger will be attractive to users as they provide easier and faster access. With the HOV application, it aims to reduce the number of drivers traveling alone in traffic and to encourage individuals to travel together, that is, car sharing. At the same time, this method aims to increase person movement efficiency rather than vehicle movement efficiency (Turnbull, 1992).

A different, new lane will be opened for high-occupancy vehicles with passengers other than the driver in the vehicle, which will provide easier and faster access. By traveling together instead of using individual vehicles, a significant reduction in the number of vehicles in traffic can be achieved. This is an important method to reduce congestion and density.

HOV lane implementation is an important application of ITS. HOV detection technologies utilize a variety of video imaging, radar technology, artificial intelligence (AI) and machine learning techniques to monitor, control and manage road traffic (Figure 6). The system detects people traveling in the vehicle and distinguishes between vehicles with more than one passenger and vehicles carrying only the driver.

Figure 6. High occupancy vehicle system

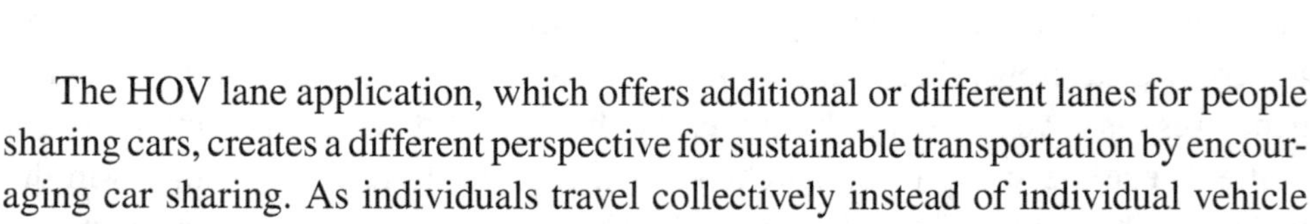

The HOV lane application, which offers additional or different lanes for people sharing cars, creates a different perspective for sustainable transportation by encouraging car sharing. As individuals travel collectively instead of individual vehicle use, the number of vehicles in traffic will decrease, thus reducing traffic congestion and carbon emissions.

The detection and application mechanisms of HOV detection technologies are based on highly sophisticated algorithms. Cameras and sensors positioned at various points along the road network are used to detect the presence and movement of vehicles. In addition, vehicle images are analysed in order to determine the type of vehicle and the number of occupants. The utilisation of image processing, artificial intelligence and machine learning techniques is a fundamental aspect of this process.

Subsequently, the data is transmitted to the traffic control centre in real time. The data is then subjected to analysis, with vehicles being classified according to their occupancy status. Consequently, vehicles with a high occupancy are permitted to utilise the designated HOV lane.

HOV technology plays a pivotal role in transportation planning. This method allows for the determination of individuals' usage habits, which can then be used as a reference in planning. Nevertheless, during the analysis of data, it is of the utmost importance to exercise caution when utilising video images or photographs of individuals, and to conduct the evaluation of personal biometric data in a manner that avoids any legal risks or data breaches.

Additional Examples of Intelligent Traffic Management Solutions

Incident Detection and Response, plays a significant role in traffic management by rapidly identifying and responding to traffic incidents. The objective of system is to rapidly restore the normal flow of traffic, eliminate the causes of disruption, and guarantee the safety of road users. The foundation of system is a multitude of detection technologies that perpetually monitor traffic conditions, detecting any irregularities or occurrences as a consequence of the data obtained by sensors and cameras. Upon the detection of an incident through the aforementioned detection technologies, the system proceeds to analyse the collected data in order to determine the most appropriate response. This is achieved through the evaluation of the potential impact that the incident may have on traffic.

Variable Speed Limit (VSL), has been developed with the objective of enhancing traffic flow by minimising congestion and improving road safety. The speed limit may be modified in real time based on traffic density, weather conditions, and the occurrence of road events, in accordance with data obtained from cameras and sensors (Khondaker and Kattan, 2015) It is recommended that this technology be considered as part of the transportation planning process, as it offers the potential to enhance the safety, efficiency, and sustainability of transportation infrastructure.

The Ramp Metering System (RM) has been developed with the objective of regulating the flow of traffic on the road network, specifically by controlling the rate at which vehicles enter main roads from side roads. The flow of vehicles is adjusted in real time according to the prevailing traffic conditions via the use of traffic signals on side roads (Figure 7). In this manner, the occurrence of traffic congestion is precluded. The primary objective of the RM is to achieve a balance between the demand for vehicles to join the main road from side roads and the flow on the main road, thereby optimising the overall performance of the road network (Grzybowska et al., 2022). The implementation of RM plays an instrumental role in reducing travel times and enhancing traffic safety by mitigating the risk of accidents. For effective transportation planning to be achieved, the system must be designed and integrated in an appropriate manner.

Figure 7. Ramp metering systems

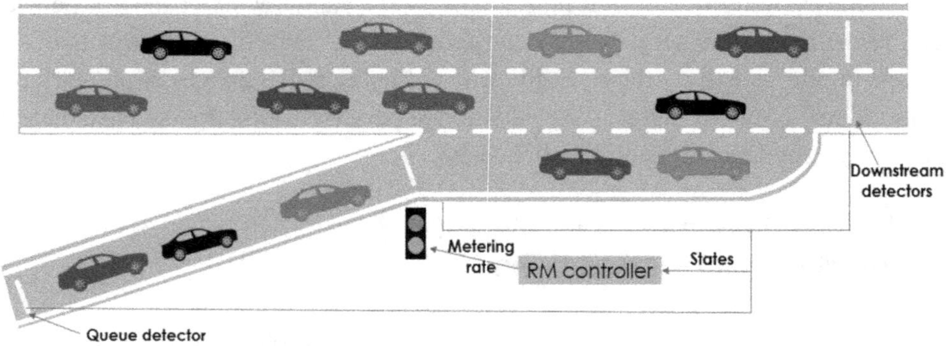

INNOVATIVE PAYMENT SOLUTIONS AND FARE INTEGRATION

A variety of pricing and payment methodologies are presented as potential ITS solutions. This section examines technologies that enhance the efficiency of transportation transactions and facilitate faster and more convenient opportunities, as well as alternative charging infrastructures. Payment methods such as Contactless Payment Systems, Frictionless Payment and Unified Fare System and pricing applications such as Fare Capping, Time-based Fare System and Distance-based Fare System are ITS solutions. The pricing strategies that have been developed to take into account the frequency, nature, distance and time of the trip, as well as the methods that facilitate uninterrupted and effortless payment, will be explained in detail.

Contactless Payment Systems

Contactless payment technologies, a relatively recent phenomenon in the field of financial technology, have introduced notable shifts in conventional payment methods, particularly in the transportation sector. These systems employ Radio Frequency Identification (RFID) or Near Field Communication (NFC) technology to facilitate a rapid and convenient payment infrastructure, obviating the necessity for physical contact. This enables transactions to be conducted with both ease and security. This technology permits users to effect transactions by bringing their debit or credit cards, smartphones, or other smart devices into proximity with the payment terminal at the point of sale (Lacmanović el al., 2010).

Contactless payment systems are also included within the ITS framework, which offers a variety of solutions designed to enhance transportation efficiency and user comfort. The technology facilitates and accelerates the collection of fares in a variety of transportation sectors, including vehicle mobility services, shared mobility platforms, public transportation, and parking fee payment systems. Consequently, a rapid, convenient, dependable, and, in short, efficient method of payment or fee collection is provided for both service providers and users.

It is hypothesised that the most significant benefit of contactless payment methods is the reduction in time required for transactions. Contactless payments streamline the process, particularly in parking lots and for electronic toll collection, thereby reducing congestion and accelerating the overall traffic flow. The system provides a practical method of payment for public transportation, bicycle or scooter sharing, eliminating the necessity for a subscription or attendant. In the future, contactless payment systems will facilitate the implementation of more expedient, pragmatic, and sustainable solutions in transportation planning studies.

Unified Fare System

The Unified Fare System is a system of collecting fares for passengers using a single ticket or payment method in multiple transportation modes, including public transportation (such as buses and trams) and bicycle rental, as well as in parking lots. It is a system that is frequently encountered in metropolitan cities. This system provides seamless and convenient travel with a unified payment infrastructure across various transportation modes within a specific region or city.

The utilisation of sophisticated technologies, such as contactless payments and smart card technology, within this system offers significant convenience in the context of payment and fee collection. The integration of disparate modes of transportation under a unified fare collection system streamlines the payment process, enhancing the accessibility of public transportation while obviating the necessity for multiple tickets or payment methods (Baruah et al., 2023).

Frictionless Payment

Frictionless payment systems facilitate transactions by eliminating the need for cash payment or physical tickets in numerous contexts, particularly in the domain of public transportation. Furthermore, it eliminates the potential issues that may arise during the fee payment phase and also expedites the payment process by automating the fee collection. The implementation of frictionless payment systems in public transportation has the potential to mitigate traffic congestion by fostering a greater utilization of public transportation due to the enhanced convenience of payment. The utilisation of public transportation in lieu of private vehicle usage has the potential to mitigate the adverse effects of traffic congestion and urban pollution. Furthermore, these systems facilitate more precise planning by furnishing data that can be employed in sophisticated transportation planning and management.

It is of paramount importance that comprehensive protection methods are employed to ensure the security of users' personal and financial data in frictionless payment systems (Connors, 2019). Furthermore, it ensures the reliability of the system, thereby encouraging users to utilise public and shared transportation services without concern for the security of their data.

Fare Capping

Fare capping is a practice in public transport pricing that sets a maximum limit on the total cost of travel a passenger will pay in a day, a week or usually a month. With this method, passengers pay a fare that does not exceed the set maximum fare, regardless of how many trips they make within the specified time period (Hightower et al., 2022). Setting a maximum fare encourages individuals to use public transport, making it a more attractive solution for both the transport system and the users. Because the fare capping system sets a maximum payment, it reduces the risk of overpayment for people who use public transport extensively.

The main purpose of the fare cap is to provide users with an economical and limited fare, thus offering a transport option whose price does not increase with use after a certain point. In this way, users are encouraged to use public transport. In cases where it is not possible to reach the destination with a single vehicle and a transfer is required, a more attractive option is offered. This is particularly useful for low-income passengers. Such a system not only increases the propensity to use public transport, but also increases user satisfaction as an economical option. An effective fare collection and process management infrastructure is required for the Fare Capping system.

Advanced fare collection systems that can track individual journeys and calculate fares in real time are required. There is a need for an application where different types of payment methods such as contactless payment systems, smart cards and mobile payments are collected in a single pool, and where passengers can track their journey history and fare details.

Time-Based Fare System

The Time-based Fare System is a novel approach to public transportation pricing. Rather than being based on distance or number of stops, fares are determined by the time spent travelling. In this fare system, passengers are permitted to make an unlimited number of trips within a specified time period, whether that be hourly or daily, for a fixed fee. In this method, pricing is independent of distance, thereby providing passengers with a more predictable and straightforward pricing structure. In their study, Silver et al. (2023) emphasizes that a fair wage policy will increase accessibility. The objective of the Time-based Fare System within the scope of ITS is to provide users with a more attractive pricing structure that is independent of distance and based on a time-oriented approach. The objective is to enhance the appeal of public transportation and to promote its utilisation. As the usage rates of public transportation increase, the infrastructure supporting it will become more sustainable.

In the Time-based Fare System application, it is essential to ensure effective editing and seamless operation of the various technical aspects. The fundamental operational premise of these systems is the utilisation of ticketing technology for the purpose of monitoring the passenger's travel time. In time-oriented systems, the use of smart cards, mobile tickets or other contactless payment methods for the purpose of recording passengers' public transport entries and exits is a common practice. It is of great importance to evaluate the potential effects of pricing on passenger behaviour in order to determine an appropriate pricing structure for the Time-based Fare System. In the context of the time-based fare system, a number of factors are taken into account in order to determine the travel time, including the relevant entry and exit records, which are then used to calculate the pricing in question. The utilisation of travel times is a valuable tool in the context of not only pricing but also transportation and public transport planning studies.

Distance-Based Fare System

A distance-based fare system is a pricing system that has become prevalent in public transportation globally. It associates the paid fee directly with the distance travelled, rather than with time, in contrast to a time-based fare system. Yook and Heaslip (2015) emphasize the difficulties of the time-based pricing system and argue that a distance-based pricing system would work more effectively. The distance-based fare system is employed not only in the pricing of public transportation but also in the pricing of road usage. The implementation of distance-oriented fees for applications such as ride-hailing and car sharing will result in a notable increase in the efficiency of the services provided. The implementation of distance-oriented pricing will result in a reduction of unnecessary deviations and an increased preference for direct routes. Consequently, the prevalence of traffic congestion and the associated release of harmful emissions can be diminished.

The utilisation of GPS tracking, licence plate recognition and automatic fare collection technologies facilitates the accurate collection of fares based on the distance travelled. Furthermore, the data obtained can be employed effectively in transportation planning studies. The data collected can be employed in the modelling of transportation demand, the planning of routes and the efficient planning of public transportation.

DATA COLLECTION AND SHARING PLATFORMS

The section on Data Collection and Sharing Platforms provides a comprehensive examination of the concept of data, which plays a pivotal role in transportation planning studies. This section examines the methods used to collect data and to utilise it in transportation planning studies, as well as in the evaluation of existing systems. The accurate acquisition of transportation data enables the identification of solutions that enhance the efficiency and sustainability of transportation systems, as well as the satisfaction of users, in the context of transportation planning studies.

Figure 8. Data collection

Transport Tracking Data Collection for Mobility

The collection of transport tracking data for the purposes of mobility studies constitutes an integral component of ITS. Furthermore, it serves to enhance the efficacy of transportation planning and traffic management. It encompasses a range of technologies, including cameras, sensors and GPS devices situated in the field, as well as analytical software. The system enables the real-time tracking and monitoring of transportation activities. The utilisation of sophisticated data collection methodologies enables regulators to gain a comprehensive understanding of congestion levels, vehicle types and numbers, travel times and user behaviour, which in turn informs transportation planning and traffic management.

The objective is to design a safer and more sustainable transportation infrastructure that reduces traffic congestion by analysing the data obtained. Data studies have an important place for the future of transportation planning (You et al., 2018). During the planning stages, the system ensures that the design of the systems meets the needs of the users and is adaptable to demands that may change instantly or throughout the process. Consequently, the collection of transport tracking data forms the foundation for a multitude of sophisticated transportation applications, including the management of traffic flows, the detection of incidents and the implementation of effective responses through the utilisation of real-time data, as well as the optimisation of routes and the forecasting of demand.

The initial stage in the process of gathering data that is fit for purpose for use in transportation planning studies is to identify the most appropriate locations for data collection. The initial stage of data collection is to ascertain the optimal locations and vantage points from which to gather the data. Subsequently, the data collected from the specified points is transmitted to traffic management centres via wired or wireless data transfer technologies. The transmitted data is then digitised using image processing techniques and analysed with the requisite machine learning or artificial intelligence techniques. The collection of data for the purposes of monitoring the movement of people and goods is of critical importance for an effective transportation planning study. The efficacy of transportation planning is contingent upon the capacity to collect accurate data.

Crowdsourced Data

Crowdsourced data is derived from the process of gathering comprehensive data through a range of digital platforms, particularly smartphones, vehicle sensors and social media. This method allows for the accurate inference of urban mobility patterns by leveraging the vast and heterogeneous data collected across diverse regions and time periods. The foundation of effective transportation planning is contingent upon the availability of accurate data and sound inferences. Integrating data from multiple data sources into the process with a consistent analysis is a challenging process for planners (Le Dantec et al., 2015) .

This method allows for the analysis of traffic conditions, the identification of areas experiencing congestion, and the determination of individuals' transportation preferences. Consequently, applications such as congestion prevention, route optimisation, public transportation planning and effective emergency response can be carried out with greater ease, facilitating the production of accurate solutions in transportation planning studies and traffic management. Crowdsourced data constitutes the foundation of ITS and transportation planning.

Incident Data Collection and Management

It is an inevitable consequence of the nature of transportation and mobility that complications and accidents will occur at every stage of the process. A variety of studies are conducted with the aim of reducing confusion and the incidence of accidents. This represents a key objective of transportation planning and ITS. Nevertheless, in instances where accidents are unavoidable, prompt and efficacious intervention is of paramount importance. Incident Data Collection and Management represents an ITS solution that encompasses the systematic aggregation and examination of data pertaining to malfunctions, accidents, or incidents across a range of transportation modalities, with a particular focus on highway-related incidents. During this process, all aspects of traffic incidents are subjected to comprehensive examination by transportation units, with a particular focus on their causes, forms and environmental effects.By creating a decision support system by modeling the obtained data, accidents can be automatically detected and teams can be directed for emergency intervention (Dağlı et al., 2022).

Transportation planners endeavour to enhance the efficiency and safety of the road network, thereby establishing a sustainable transportation infrastructure, through the utilisation of data obtained from cameras, sensors and detection systems.

The work of Incident Data Collection and Management is of great importance, not only for the implementation of immediate response measures, but also for the formulation of future transportation planning strategies. Such data provide decision-makers and transportation planners with information about a range of problematic issues, including the geometric improvement of the road network. They facilitate the development of a systematic scientific approach to the improvement of traffic regulations, the determination of risk perceptions and the creation of effective emergency response strategies.

Disruption/Delay Data Collection for Transport

A primary objective of transportation planning studies is the reduction of delay times. The systematic collection and analysis of data pertaining to disruptions and delays in transportation systems will enhance the efficacy of the planning process. The collection of data pertaining to disruptions and delays in transportation systems, as part of an ITS study, plays a pivotal role in the field of transportation planning. The collection of data regarding disruptions and delays in transportation systems is a crucial aspect of transportation planning. Disruption/Delay Data Collection specifically focuses on the immediate collection of data pertaining to disruptions caused by a range of factors, including accidents, road closures, repair and renovation works, as well as adverse weather conditions.

Technologies such as traffic cameras, road infrastructure sensors, GPS tracking devices in vehicles, and so forth are employed for the purpose of obtaining this data. GPS data of vehicles is preferred as an effective method to determine critical delay points in the road network (Wallander and Mäkitalo, 2012). The data is then subjected to a variety of analytical techniques and subsequently disseminated to the pertinent units. This enables prompt intervention and the implementation of solutions. This approach provides transportation planners with valuable insights into the performance of roads, traffic safety, and potential deficiencies. Consequently, transportation planners are able to devise a strategy for the elimination of the aforementioned disruptions.

The Disruption/Delay Data Collection system is characterised by a high level of complexity, whereby disparate data sources are integrated and the resulting inputs subjected to analysis through a variety of techniques. Nevertheless, it represents a crucial juncture for ITS solutions. This technique facilitates the development of infrastructure for applications such as dynamic signalling systems, route planning studies and multi-modal public transportation infrastructure. The effective utilisation of the system will facilitate the establishment of a transportation infrastructure that is uninterrupted, safe and offers high levels of user satisfaction.

PASSENGER INFORMATION SYSTEMS

Passenger Information Systems represents a crucial ITS solution that is frequently utilised by citizens in their daily lives. The system provides immediate information on a range of topics, including the location and status of public transportation vehicles, estimated arrival times at specified points, routes of different public transportation lines, stop locations, and intermodal transfer opportunities.

Figure 9. Passenger ınformation systems

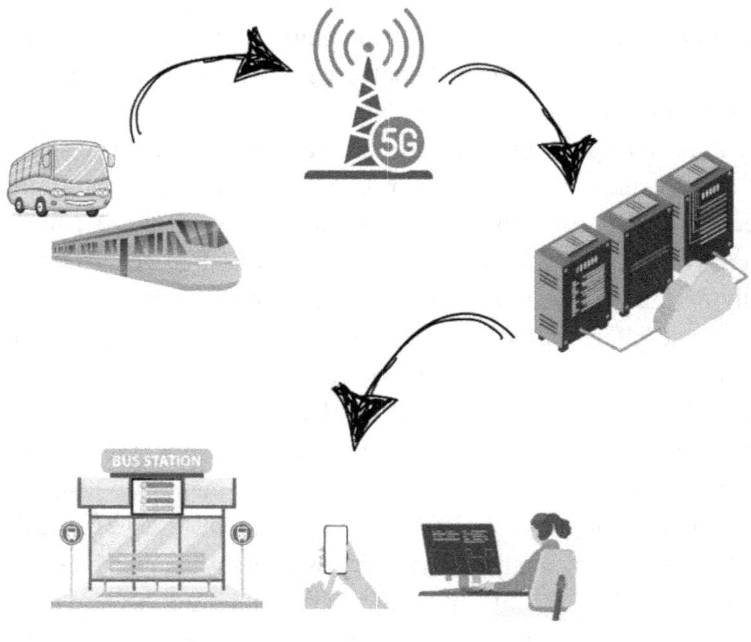

In essence, a Passenger Information System comprises three principal stages: the collection of data, the processing of said data through a variety of techniques, and the subsequent delivery of the resulting information to users. Passenger Information Systems employ a range of technologies, notably GPS and RFID systems, to obtain data. The efficiency of the system is contingent upon the collection of data in an instant and with precision. The data amassed through a multitude of techniques is then processed by systems and disseminated to users through a variety of channels, including stop information screens, mobile applications, and websites. This approach enhances the functionality of both users' journey planning and transportation planners' public transportation planning. The principal objective of the system is to facilitate a more straightforward, convenient and preferable public transportation solution for users. The system is an important tool to reduce the uncertainties of the journey (Zurob et al., 2016). Concurrently, this technology represents a significant data source for transportation planning studies.

Smart Infrastructure

As technology becomes increasingly ubiquitous in social life, the concept of the smart city is now a prominent topic of discussion. The concept of the smart city is comprised of a number of constituent elements. One significant component of this is the development of smart infrastructure. The concept of smart infrastructure entails the reinforcement of the systems utilised on a daily basis throughout the city, with the integration of technological opportunities and the digitisation of these systems, thereby facilitating automation. Internet of Things (IoT) provides a basis for smart city with intelligent recognition, positioning, monitoring and management procedure by embedding RFID, IR, GPS, laser scanners, etc. in the components located in the environment and connecting them to the Internet through certain protocols to exchange information (Kim et al., 2017).

The concept of Smart Infrastructure encourages the integration of advanced technologies into urban infrastructure. At the same time, citizens' expectations serve as a guiding force for managers. In addition to the implementation of smart infrastructure, numerous smart solutions are observed across the city.

The introduction of smart street lamps has introduced a novel approach to traditional lighting methods, with regulations aimed at reducing energy consumption and enhancing visual quality. The deployment of smart lighting systems, equipped with sensors, has the potential to enhance the efficiency of urban management by providing data on a range of environmental and traffic conditions.

The communication infrastructure that supports vehicle-to-everything (V2X) communication enables vehicles and drivers to exchange information with each other and with the city's infrastructure. To illustrate, it enhances traffic safety and efficiency by disseminating real-time data regarding road and traffic conditions, potential hazards on the road, or incidents and malfunctions. In particular, it is anticipated that Vehicle-to-Infrastructure (V2I) emergency communication systems will enhance the efficacy of emergency interventions by facilitating the exchange of information between vehicles and the road infrastructure.

Additionally, smart intersection and pedestrian crossing solutions constitute an integral aspect of smart infrastructure. The optimal and dynamic arrangement of signalling systems at intersections will result in a reduction in waiting times and congestion, as well as an enhancement in intersection performance and user satisfaction. Furthermore, the implementation of intelligent pedestrian crossings and the utilisation of audible and visual warning systems will facilitate the assurance of pedestrian and traffic safety. Conversely, the prevention of unnecessary waiting for vehicles at unutilised pedestrian crossings will result in a reduction in both delay times and emission values.

The establishment of smart mobility centres in cities where multi-modal and systematic transportation is feasible represents a potential solution to the management of urban mobility. The integration of fibre optic networks, high-speed communication infrastructure such as 5G and internet of things (IoT) technologies into the urban infrastructure offers the possibility of designing a systematic and efficient transportation infrastructure that encompasses all transportation infrastructure and transportation stakeholders.

Furthermore, violation detection and sanction systems represent a key application of smart infrastructure. The implementation of average speed detection systems enables the acquisition of corridor speeds within a road network, facilitating the assessment of compliance with specified limits and the imposition of sanctions in instances of non-compliance. Similarly, vehicles parked in areas where parking is prohibited are identified through the use of sensors and cameras, or penalties are imposed upon the detection of vehicles passing without stopping when a red light is displayed in contravention of the regulations set forth in the signalling systems. The implementation of smart infrastructure facilitates the monitoring of regulations and the imposition of penalties on those who fail to comply with them, thereby ensuring greater uniformity and efficiency in the enforcement process.

AUTONOMOUS VEHICLES AND ELECTRIC VEHICLES

The advent of autonomous vehicles (AVs), encompassing self-driving or driverless vehicles, has introduced a novel dimension to automotive and transportation technology in recent years.

Autonomous vehicles operate without human intervention, utilising a range of sophisticated systems and intricate technological infrastructure, including sensor radar, LiDAR, GPS, odometry and computer vision. Autonomous vehicles utilise a variety of sensors to scan their surroundings, taking into account the presence of obstacles, traffic signs and signalling systems along the road. This enables them to proceed without the need for a driver. In addition to their driverless functionality, autonomous vehicles are distinguished by their route optimisation, low energy usage and safety equipment. In this regard, it is regarded as a significant contributor to the ITS (Aydın, 2023).

Figure 10. Illustration of autonomous connected vehicles (Hakak et al., 2023)

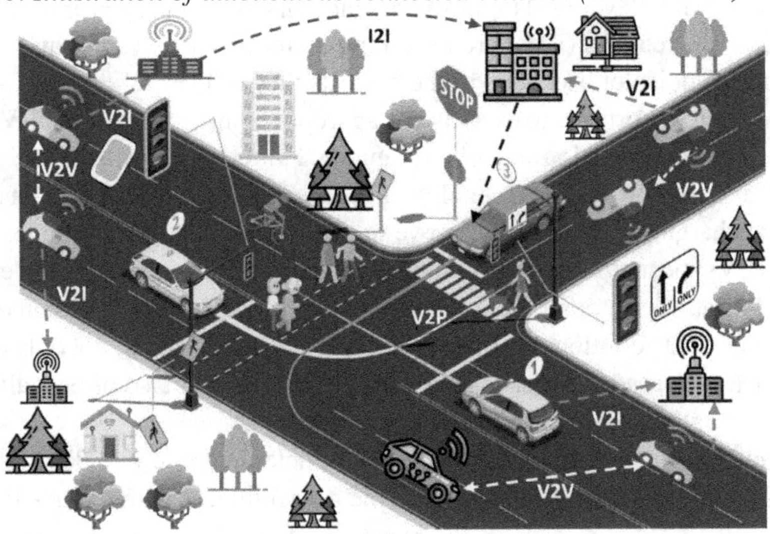

It would be prudent for transportation planning studies to include plans focused on autonomous vehicles, which are expected to become widespread in the near future. It is incumbent upon transportation planners to design infrastructure that is suitable for driverless vehicles and to integrate this innovative technology into the transportation infrastructure in a timely manner. It is anticipated that the introduction of autonomous vehicles will result in a reduction in traffic congestion, a decrease in waiting times and, consequently, a reduction in emissions.

The functioning of autonomous vehicles is based on a number of fundamental principles. These comprise the processes of perception, decision-making and control. Environmental data along the road network is collected through sensors and cameras and processed with various techniques to create a detailed model of the environment and detect other vehicles, obstacles and traffic components. Subsequently, an instantaneous decision-making mechanism is initiated, based on the preexisting obstacle avoidance and traffic sign and signal systems. In accordance with the decisions reached, the autonomous vehicle's braking, acceleration, steering and mechanical procedures are initiated, and the driving process is executed.

As autonomous vehicles become more prevalent, requirement on the driver will diminish, thereby reducing the incidence of accidents attributable to human error. The utilisation of autonomous vehicles as taxis, school or business shuttles, cargo vehicles or public transportation represents a potential means of creating a more economical, safer and driver-free transportation infrastructure.

Another innovative transportation technology is that of electric vehicles, which represent an alternative to traditional fossil fuels. The technology behind electric vehicles (EVs) has developed rapidly in recent years, becoming a prominent feature of IT). Electric vehicles (EVs) are becoming increasingly prevalent and are often the preferred choice of users due to the benefits they offer.

It is of great importance to establish an appropriate infrastructure for EVs within the context of transportation and urban planning. Failure to adapt to the rapid growth of hybrid and electric vehicle technologies in recent years, coupled with an inability to provide charging facilities, will prove to be a significant disadvantage. Electric vehicles (EVs) occupy a pivotal position in the context of the objective to eradicate the transportation sector's reliance on fossil fuels and to curtail greenhouse gas emissions and air pollution. Furthermore, electric vehicles (EVs) not only avert the environmental degradation associated with fossil fuels, but also offer multifaceted benefits through the utilisation of renewable energy sources.

It seems inevitable that, like autonomous vehicles, electric vehicles will play an important role in urban traffic patterns in the near future. It is imperative that those engaged in the field of transportation planning adopt a forward-thinking approach and align their strategies with this prediction.

Transport Accessibility

On-Demand Transit Services (ODTS) refers to adaptive transport services that to provide shared flexible transport to consumers, when and where they need it (Vij et al., 2020). ODTS diverge from traditional fixed schedules and routes, offering a real-time dynamic operating option in accordance with passenger demands. This approach unites the advantages of public transportation with the customised characteristics of taxi services. Consequently, the routes and schedules of the vehicles are optimized in accordance with the requirements of the users. Furthermore, these innovative public transportation systems play a pivotal role in the implementation of ITS, particularly in enhancing the accessibility of individuals with disabilities in social contexts.

The ODTS system is designed with accessibility in mind, incorporating adaptable and user-oriented features that are particularly beneficial for passengers with physical disabilities and mobility issues. The objective of these services is to facilitate the participation of disabled individuals in social activities and prevent social isolation. The ODTS plays a significant role in enhancing the mobility of individuals with disabilities, facilitating their autonomy and independence through the provision of accessible and comfortable transportation options.

The ODTS facilitates the integration of advanced hardware, software and communication technologies within a transportation network that is capable of adapting to real-time user demands. It is of the utmost importance that vehicles designed for an ODTS are equipped in an appropriate manner for passengers with a range of disabilities, with accessibility forming a key priority in the delivery of the service.

In the context of public transportation and transportation planning studies, it is essential to consider the opportunities and needs of all segments of society when developing plans and strategies. The provision of appropriate transportation infrastructure, particularly for disabled individuals and the elderly with special needs, is a crucial requirement for developed cities.

Shared Mobility Enhancement Solutions

In recent years, the concept of mobility has offered an alternative approach to the traditional fixed pick-up and drop-off station models. This innovative method offers the flexibility to retrieve bicycles from any location and return them to the desired location, in contrast to the fixed station model. Consequently, options that provide more convenient and extensive service areas, thereby enhancing the user experience by eliminating the need for reliance on fixed stations, have emerged. Shared mobility solutions also have significant potential for sustainable urban mobility (Martinez et al., 2024).

In addition to bicycles, electric mobility vehicles have also been utilised with increasing frequency in recent years. E-scooters and e-bikes provide a convenient and expedient mode of transportation for short distances, offering a comfortable alternative to other forms of transportation. This will result in a reduction in the number of vehicles on the road, thereby alleviating traffic congestion and providing an environmentally friendly alternative to users.

In the context of transportation planning studies, it is imperative to acknowledge the growing prevalence and popularity of mobility vehicles, particularly given the rapid expansion of their service networks. It is recommended that the plans include provisions for these vehicles and that consideration be given to developing an integrated system for public transportation vehicles with a view to enhancing efficiency. It is evident that the concept of explosive mobility plays a pivotal role in the realm of sustainable transportation for central areas, a position that is likely to remain unchanged in the foreseeable future.

Public Transport Enhancement Solutions

E-Ticketing and Mobile Payment Systems

The advent of electronic ticketing (E-Ticketing) and mobile payment systems has provided a contemporary alternative to traditional payment methods, offering digital solutions that facilitate convenient and secure transactions. In the era of the Cashless Economy, public transport payment systems also need to adapt to this trend (Kazi et al., 2018). Mogaji and Nguyen (2024) drew attention to the difficulties in adopting technology, especially in underdeveloped countries, and stated that contactless payment systems could contribute to this process. E-Ticketing emerges as a significant Intelligent Transport System (ITS) solution, offering the advantage of facilitating fare payment transactions without the need for cash through the utilisation of technologies such as Near Field Communication (NFC) or Radio Frequency Identification (RFID). Mobile payment systems operate on the basis of users' bank accounts or digital wallets. The advent of mobile payment infrastructure has facilitated the emergence of a convenient and expedient payment alternative.

The implementation of this system has resulted in a notable reduction in the complexity associated with fee payment for users and fee collection for operators. Concurrently, data pertaining to user boarding, disembarking, utilisation and line information, which can be utilised in transportation planning studies, are obtained during the collection of fares. The implementation of convenient payment systems will result in a reduction in waiting times at stops, thereby preventing the potential for congestion on the road network. Consequently, public transport will be utilised more effectively, thereby enhancing its perceived desirability.

In addition, the cessation of the use of future paper tickets will result in a reduction of paper waste, thereby facilitating the adoption of a more environmentally friendly payment method. Furthermore, the system affords users the ability to retain control over their travel and payment history, due to the traceability of payment applications.

Public Transport Prioritization Measures

Public Transport Prioritisation Measures (PTPM) are systems designed to enhance the regularity and predictability of public transport services. They are implemented with the objective of minimising delays and deviations from scheduled times. Royko et al., (2023) drew attention to the differences in spatial traffic flow for public transportation vehicles and stated that the time it takes for vehicles to pass through signalized intersections and pedestrian crossings may cause significant differences. The system employs methodologies such as the implementation of bus lanes and the configuration of traffic signals that are calibrated in real-time to reflect the prevailing traffic conditions. The objective of such studies is to establish regulations that prioritise public transportation in traffic management, thereby enhancing the efficiency and attractiveness of public transportation for users.

In this system, where public transportation is prioritised, an accessible and attractive transportation method is offered for all segments of society, with due consideration given to income levels and disabilities. Concurrently, the promotion of public transportation will result in a notable reduction in the number of private vehicles on the road, as well as in the overall volume of traffic. In this manner, the occurrence of traffic congestion and elevated emissions will be averted, thereby facilitating the attainment of a more environmentally friendly and sustainable urban mobility. In the context of transportation planning studies, it is of paramount importance to accord priority to public transportation and to implement regulations that encourage its use.

Smart Public Transportation Station

A variety of ITS applications are available at Smart Public Transport Stations (SPTS) with the objective of enhancing the travel experience for users while maintaining a high level of comfort. The aforementioned stations are equipped with a Wi-Fi connection, announcement systems, and electronic screens that disseminate information regarding the location and schedule of buses and lines. In accordance with the data obtained from public transportation vehicles, the estimated time of arrival at the designated stop is conveyed to passengers in real-time. Concurrently, automated ticket dispensers offer a convenient method of fare payment. Furthermore, stop occupancy rates can be determined through the use of cameras and sensors, which allows for the implementation of adjustments such as additional trips or dynamic route planning, contingent on the occupancy rate. It is of paramount importance to ascertain the demand in transportation planning studies. In the event of insufficient demand at a given stop, the public transportation vehicle will either refrain from stopping at that stop or bypass it entirely, proceeding instead to the next stop via an alternative route. In addition, in their study on smart and closed bus stops, Çoruh et al. (2020) stated that ensuring that all bus stops needed throughout the country have the same service quality is very important in terms of user satisfaction and public transportation use. This will represent a radical transformation for public transportation systems. It is therefore imperative that smart stop applications be accorded the highest priority and become widely adopted.

Figure 11. Smart public transportation station (Çoruh et al., 2020)

Drone Technology and Applications

Drone and unmanned aerial vehicle (UAV) technology, which we have heard a lot about in recent years and which is being used effectively in many areas of life, also has an important place in transport planning and ITS studies. Drones and Unmanned Aerial Vehicles (UAVs) are essentially motorised vehicles that navigate without a human pilot. As with autonomous vehicles, various detection and guidance devices support unmanned flight.

UAVs have been implemented in various parts of many industries since the beginning of the Industry 4.0 revolution and will be implemented in many different forms and in many other industries and in our daily lives (Senvar and Ünver, 2022). This technology is being actively used in many areas, particularly in agricultural irrigation and spraying, firefighting and logistics. Flying cars, which are expected to be integrated into the transport system in the near future, will also add a whole new dimension to the field.

These technologies are also widely and successfully used in ITS solutions. These technologies are used for many purposes, such as monitoring transport infrastructure and determining performance, managing road networks, reducing congestion and providing real-time information to passengers. In addition, drones, which play a very important role in gathering information about the current situation in planning studies, offer the possibility to monitor the entire target area from the prevailing height and angle.

The integration of drone technology into ITS frameworks will have a significant impact on the success of transport planning studies. Drones are becoming increasingly valuable in ITS due to their ability to perform tasks such as aerial surveillance and data collection. These capabilities make drones particularly useful for real-time traffic monitoring and efficient traffic analysis management. They will also increase the speed and success of emergency response.

Innovative Transportation Enhancements Solutions

There are many ITS solutions that are often used in transport planning studies. Active Transport Improvement Solutions are also very important to improve urban transport systems and make transport planning more efficient by reducing traffic congestion and minimising negative environmental impacts. As cities become too large for planners to manage manually, systems need to be automated. It is becoming necessary to integrate transport planning and traffic management applications with ITS solutions. Improvements in transport planning and traffic management applications are being made using technologies such as GPS, cameras, IoT and sensor-based systems. Studies such as dynamic signal management and real-time traffic updates are also developing safe and efficient solutions for pedestrians and cyclists. Examples of innovative technologies include smart bus stops, smart bike stations, intermodal transfer hubs and innovative micro-mobility solutions, which have also been discussed in previous sections. In addition to these studies, flying car technology, which will fundamentally change transport systems in the near future and make point-to-point access easier and faster, will also have an important place in transport planning studies.

With the changing and developing technology, machine learning techniques and artificial intelligence technology, ITS solutions, transport planning studies and many changes in all transport systems that we cannot think of today will undoubtedly become concrete in the very near future.

CONCLUSION

It is of paramount importance to conduct transport planning studies in order to facilitate the development and transformation of urban areas. It is incumbent upon planners and managers to undertake a comprehensive analysis of the performance of the current situation and to identify the needs in line with the anticipated changes that will occur in the coming years. The sustainability of urban areas is inextricably linked to the concepts underpinning their mobility and transport systems. The ever-evolving technological landscape introduces novel concepts that shape our lives as we navigate this period of transformation. The term "smart solutions" is used to describe the integration of technological infrastructure into everyday life. The concept of the "smart city" is also a common occurrence. Intelligent Transport Systems (ITS) represent a significant component of the infrastructure that supports the development of smart cities.

In accordance with the expectations and requirements of society, transport systems should be developed in a manner that is economically viable, safe, comfortable, time-efficient, user-satisfactory and environmentally friendly. The key to this is the appropriate, comprehensive and professional implementation of intelligent transport solutions in the planning stages and field applications.

The following areas of focus have been identified as key for the advancement of transportation systems: Intelligent Traffic Management, Innovative Payment Solutions and Fare Integration, Data Collection and Sharing Platforms, Smart Infrastructure, Autonomous Vehicles and Electric Vehicles Transport Accessibility, Shared Mobility Enhancement Solutions, Public Transport Enhancement Solutions, Drone Technology and Applications, and Active Transportation Enhancements Solutions. The applications of smart traffic management and control systems, adaptive traffic management systems, contactless payment systems, passenger information systems, enforcement and violation detection systems, as well as autonomous and electric vehicles, are becoming increasingly prevalent in our daily lives. Of particular note is the increasing prevalence of drone technology and unmanned aerial vehicles (UAVs) in recent years, along with the advent of flying car technologies that promise to transform the landscape of transportation. Furthermore, the use of shared vehicles and micromobility applications is becoming increasingly prevalent.

These applications are of critical importance in the context of transport planning and field applications. The foundation of an effective transport planning process is the accurate and comprehensive identification of the current situation and demands. In order to ascertain the current situation and transport demands, it is essential to have a robust technological infrastructure in place, along with effective detection systems utilising a range of sensors and cameras. Furthermore, the utilisation of image processing, machine learning and artificial intelligence is crucial to ensure successful outcomes. Concurrently, the configuration of the plans is contingent upon the technical equipment. While there is currently no requirement for the planning of micromobility, unmanned aerial vehicles (UAVs) and, in particular, flying cars, there is nevertheless a clear need for forward planning in this area for the coming years.

The success of urban recovery and sustainability initiatives hinges on effective planning. The effective implementation of smart systems, shaped in line with technological developments during the planning stages, is a key factor in ensuring successful planning. This study examines the various applications of intelligent systems in the field of transport in detail, with the aim of providing a comprehensive guide for planners and managers. A section has been prepared that emphasizes the necessity of using ITS effectively in transportation planning studies and aims to be a guide for planners in their work.

REFERENCES

Aydin, M., Potoglou, D., & Cipcigan, L. (2023). Working principle and performance evolution of camera-based intelligent signalized intersections: Samsun city, Türkiye example. *Scientific Journal of Silesian University of Technology.Series Transport*, 121, 5–17.

Aydın, M. M. (2023). A charging station planning study to prepare for the utilization of electric vehicles on university campuses: The case of Ondokuz Mayıs University. *International Journal of Engineering and Applied Sciences*, 15(4), 184–204. DOI: 10.24107/ijeas.1394018

Aydın, M. M., & Bıyık, B. (2023). Şerit yoğunluk tespit ve bilgilendirme sistemlerinin tasarımdan işletmeye genel altyapısının belirlenmesi: Örnek bir sistem tasarımı. *Gazi University Journal of Science Part C: Design and Technology*, 11(4), 1092–1107.

Aydın, M. M., & Keleş, M. A. (2024). Trafikte acil durum müdahale araçlarının geçişi esnasındaki karmaşa problemleri için AUS tabanlı bir çözüm önerisi geliştirilmesi. *Journal of the Institute of Science and Technology*, 14(1), 294–311.

Banister, D. (2008). The sustainable mobility paradigm. *Transport Policy*, 15(2), 73–80. DOI: 10.1016/j.tranpol.2007.10.005

Baruah, R., Siddique, A. R., & Santhi, H. (2023). Unıfıed payment system for Indıan transportatıon. [IRJET]. *International Research Journal of Engineering and Technology*, 10(5), 1540–1548.

Black, W. R. (2010). *Sustainable transportation: problems and solutions*. Guilford Press.

Connors, B. T. (2019). Do we still need to pay?. *The PayTech Book: The Payment Technology Handbook for Investors, Entrepreneurs and FinTech Visionaries*, 21-22.

Çoruh, E., Aydın, M. M., Dağlı, E., Öztürk, Ö. F., & Yıldırım, M. S. (2020). Kapalı otobüs durağı uygulamalarının kullanıcı beklentileri ve memnuniyeti doğrultusunda incelenmesi: Gümüşhane ve Antalya örneği. *Gümüşhane Üniversitesi Fen Bilimleri Dergisi*, 10(4), 1027–1038.

Dağlı, E., Büber, M., & Taspınar, Y. S. (2022). Detection of accident situation by machine learning methods using traffic announcements: The case of metropol Istanbul. *International Journal Of Applied Mathematics Electronics And Computers*, 10(3), 61–67. DOI: 10.18100/ijamec.1145293

Elassy, M., Al-Hattab, M., Takruri, M., & Badawi, S. (2024). Intelligent transportation systems for sustainable smart cities. *Transportation Engineering*, 100252.

European Telecommunications Standards Institute (2010). Intelligent transport systems (ITS): communications architecture, v1.1.1, European Standard (Telecommunications series), September 2010

Gandhi, M. (2023). https://pub.aimind.so/adaptive-traffic-control-systems-a-comprehensive-review-part-2-64a0a48a985e

Grzybowska, H., Wijayaratna, K., Shafiei, S., Amini, N., & Travis Waller, S. (2022). Ramp metering strategy implementation: A case study review. *Journal of Transportation Engineering. Part A, Systems*, 148(5), 03122002. DOI: 10.1061/JTEPBS.0000641

Hakak, S., Gadekallu, T. R., Maddikunta, P. K. R., Ramu, S. P., Parimala, M., De Alwis, C., & Liyanage, M. (2023). Autonomous Vehicles in 5G and beyond: A Survey. *Vehicular Communications*, 39, 100551. DOI: 10.1016/j.vehcom.2022.100551

He, J., Zeng, Z., & Li, Z. (2010). Benefit evaluation framework of ıntelligent transportation systems. *Journal of Transportation Systems Engineering and Information Technology*, 10(1), 81–87. DOI: 10.1016/S1570-6672(09)60025-8

Hightower, A., Ziedan, A., Crossland, C., & Brakewood, C. (2022). Current Practices and potential rider benefits of fare capping policies in the USA. *Transportation Research Record: Journal of the Transportation Research Board*, 2676(10), 376–390. DOI: 10.1177/03611981221089572

Kazi, S., Bagasrawala, M., Shaikh, F., & Sayyed, A. (2018). Smart e-ticketing system for public transport bus. *International Conference on Smart City and Emerging Technology (ICSCET)* (pp. 1-7). IEEE. DOI: 10.1109/ICSCET.2018.8537302

Khazraeian, S., & Hadi, M. (2019). Intelligent transportation systems in future smart cities. *Sustainable Interdependent Networks II: From Smart Power Grids to Intelligent Transportation Networks*, 109-120.

Khondaker, B., & Kattan, L. (2015). Variable speed limit: An overview. *Transportation Letters*, 7(5), 264–278. DOI: 10.1179/1942787514Y.0000000053

Kim, T. H., Ramos, C., & Mohammed, S. (2017). Smart city and IoT. *Future Generation Computer Systems*, 76, 159–162. DOI: 10.1016/j.future.2017.03.034

Lacmanović, I., Radulović, B., & Lacmanović, D. (2010). Contactless payment systems based on RFID technology. *In The 33rd International Convention MIPRO* (pp. 1114-1119). IEEE.

Le Dantec, C. A., Asad, M., Misra, A., & Watkins, K. E. (2015). Planning with crowdsourced data: rhetoric and representation in transportation planning. *In Proceedings of the 18th ACM conference on computer supported cooperative work & social computing* (pp. 1717-1727).

Martinez, L. M., Pritchard, J. P., & Crist, P. (2024). Shared mobility's role in sustainable mobility: *Past, Present, and Future.Annual Review of Environment and Resources*, •••, 49. DOI: 10.1146/annurev-environ-121522-041916

Mogaji, E., & Nguyen, N. P. (2024). Evaluating the emergence of contactless digital payment technology for transportation. *Technological Forecasting and Social Change*, 203, 123378. DOI: 10.1016/j.techfore.2024.123378

Molina, M. (2020). What is an intelligent system? *arXiv preprint arXiv*:2009.09083.

Royko, Y., Fornalchyk, Y., Koda, E., Kernytskyy, I., Hrytsun, O., Bura, R., Osinski, P., Markiewicz, A., Wierzbicki, T., Barabash, R., Humenuyk, R., & Polyansky, P. (2023). Public Transport Prioritization and Descriptive Criteria-Based Urban Sections Classification on Arterial Streets. *Sustainability (Basel)*, 15(3), 2363. DOI: 10.3390/su15032363

Senvar, O., & Ünver, S. (2022). An Overview to Unmanned Aerial Vehicles from Perspectives of Quality and Safety Management in Aviation. *International Journal of Engineering Research and Development*, 14(2), 917–940. DOI: 10.29137/umagd.1064242

Shaheen, S. A., & Finson, R. (2013). *Intelligent transportation systems*. Reference Module İn Earth Systems And Environmental Sciences.

Shahgholian, M., & Gharavian, D. (2018). Advanced traffic management systems: an overview and a development strategy. *arXiv preprint arXiv*:1810.02530.

Silver, K., Lopes, A., Vale, D., & da Costa, N. M. (2023). The inequality effects of public transport fare: The case of Lisbon's fare reform. *Journal of Transport Geography*, 112, 103685. DOI: 10.1016/j.jtrangeo.2023.103685

Singh, B., & Gupta, A. (2015). Recent trends in intelligent transportation systems: A review. *Journal of Transport Literature*, 9(2), 30–34. DOI: 10.1590/2238-1031.jtl.v9n2a6

Steger-Vonmetz, D. C. (2005). Improving modal choice and transport efficiency with the virtual ridesharing agency. [). IEEE.]. *IEEE Intelligent Transportation Systems*, 2005, 994–999. DOI: 10.1109/ITSC.2005.1520186

Taç, Ş. G. (2018). Karayolu ulaşımında meydana gelen trafik kazalarının önlenmesinde akıllı ulaşım sistemlerinin etkisi. *Akıllı Ulaşım Sistemleri ve Uygulamaları Dergisi*, 1(2), 12–21.

Tektaş M., Korkmaz K., Erdal H. (2016). Akıllı ulaşım sistemlerinin geleceği ekonomik ve çevresel faydaları. *Balkan Journal of Social Sciences*, 561-577.

Turnbull, K. F. (1992). *High Occupancy Vehicle Project Case Studies: Historical Trends and Project Experiences (No. DOT-T-94-18)*. Department of Transportation.

Vij, A., Ryan, S., Sampson, S., & Harris, S. (2020). Consumer preferences for on-demand transport in Australia. *Transportation Research Part A, Policy and Practice*, 132, 823–839. DOI: 10.1016/j.tra.2019.12.026

Wallander, J., & Mäkitalo, M. (2012). Data mining in rail transport delay chain analysis. *International Journal of Shipping and Transport Logistics*, 4(3), 269–285. DOI: 10.1504/IJSTL.2012.047492

Watkins, K. (2012). Safe and sustainable roads. *Make Roads Safe The Campaign For Global Road Safety*, 1, 40.

Xiong, Z., Sheng, H., Rong, W., & Cooper, D. E. (2012). Intelligent transportation systems for smart cities: A progress review. *Science China. Information Sciences*, 55(12), 2908–2914. DOI: 10.1007/s11432-012-4725-1

Yook, D., & Heaslip, K. (2015). Effective modeling for a distance-based fare structure with a time-expanded network. *Journal of Public Transportation*, 18(1), 1–13. DOI: 10.5038/2375-0901.18.1.2

You, L., Zhao, F., Cheah, L., Jeong, K., Zegras, C., & Ben-Akiva, M. (2018). Future mobility sensing: An intelligent mobility data collection and visualization platform. *21st International Conference on Intelligent Transportation Systems (ITSC)* (pp. 2653-2658). IEEE. DOI: 10.1109/ITSC.2018.8569697

Zurob, C., Allard, J. M., Macário, R., Garcia, B., & Garcia, C. (2016). Passenger information systems. *Restructuring Public Transport through Bus Rapid Transit*, 247-260.

Chapter 5
Integrating Cloud Computing for Intelligent Transportation Solutions in Smart Cities:
A Short Review

Aditya Kumar Panda
https://orcid.org/0009-0009-2771-666X
School of Computer Engineering, KIIT University (Deemed), Bhubaneswar, India

Amlan Anurag Lenka
https://orcid.org/0009-0008-7190-322X
School of Computer Engineering, KIIT University (Deemed), Bhubaneswar, India

Ayush Mohapatra
School of Computer Engineering, KIIT University (Deemed), Bhubaneswar, India

Bikram Keshari Rath
School of Computer Engineering, KIIT University (Deemed), Bhubaneswar, India

Ayush Adarsh Parida
https://orcid.org/0009-0003-0393-5194
School of Computer Engineering, KIIT University (Deemed), Bhubaneswar, India

Hitesh Mohapatra
https://orcid.org/0000-0001-8100-4860
School of Computer Engineering, KIIT University (Deemed), Bhubaneswar, India

ABSTRACT

Urban transportation systems face many challenges, such as traffic congestion, pollution, and inefficient infrastructure. To tackle these problems, cities are increasingly adopting cloud-based smart transportation solutions. These solutions use cloud

DOI: 10.4018/979-8-3693-6695-0.ch005

computing to integrate IoT devices, data analytics, and real-time monitoring, optimizing urban mobility. Cloud platforms allow for efficient data collection, storage, and analysis, helping city authorities make informed decisions. This approach is scalable, flexible, and cost-effective, promoting smarter and more sustainable urban transportation systems. By leveraging cloud computing, cities can improve traffic management, reduce congestion, and enhance decision-making, leading to better efficiency, sustainability, and safety. The proposed model for smart transportation involves using cloud computing, IoT, data analytics, and advanced AI algorithms to create more connected and livable urban environments.

INTRODUCTION

Modern cars are equipped with an exorbitant number of sensors, actuators, and communication devices; mobile phones or GPS units in particular can contextually shift from being a simple phone to becoming embedded computers inside the vehicle. More specifically, multiple types of vehicles receive strong sensing, networking, and communication powers with processing capabilities. The data exchanged among vehicles or with the external world can be done using various application layer protocols like TCP/IP, SMTP, WAP, and most importantly, the Next Generation Telematics Protocol (NGTP), specifically mentioned in this ETSI standard document (Arthurs et al., 2022). Accordingly, many of the most innovative telematics services have been designed and developed to advance convenience, fun, and safety for drivers—e.g., remote security: car detecting/engine disabling, etc. (Dankovtsev et al., 2022). Heavy traffic, congestion, and safety of vehicles are getting worse with the passage of time, but cloud computing complexity aspects together with the Internet of Things (IoT) have provided a chance to solve this structural issue in an encouraging way. Over the past years, several models for deploying cloud-based intelligent transportation systems (ITSs) have been created by researchers. As an example, ITS-Cloud was developed as a replacement conveyance cloud idea to boost road safety and vehicle-to-vehicle communication.

Intelligent Transport Systems (ITS) combine the advances of many IT applications with innovative communication technology to achieve more efficient traffic management and, therefore, offer better transportation services. They help to avoid some possible accidents and traffic jams, which are caused in part by population growth and have additional negative effects such as longer travel times, air pollution, or increased fuel usage (Geng et al., 2024). To find a ubiquitous solution, ITS organizations have been working on traffic communications and vehicular networking for a while. For example, the 5.850-5.925 GHz band has been reserved by the FCC in the US for DSRC spectrum, providing up to 75 MHz of bandwidth exclusively for vehicular communication applications. Moreover, some IEEE-authorized modifications are within the spotlight of ITS applied sciences, such as microwave access (WiMAX - World Interoperability for Microwave Access; 802.16) and vehicular connectivity (WAVE- Wireless Accessibility in Vehicular Environments; 802.11p), which extend WLAN usage habits into a vehicular setting (Li et al., 2023).

With the population increasing day by day, accidents and traffic congestion are on the rise, leading to several unwanted outcomes such as air pollution, fuel consumption, and long travel times. As a result, intelligent transportation systems are selectively being developed to integrate communication and information technology directly into current transport management systems, which can help in reducing both the operating costs of personal vehicles and road building costs (Lin et al., 2024). It also uses technology to collect information on the global performance of transport infrastructures, on-demand transportation, and V2V (Vehicle-to-Vehicle) and Vehicle-to-Roadside Communication (between roadside devices), including approaching collision accidents, among others [6–2023] (Clausi Scientific Press Canada). The cloud computing model is an excellent tool for processing, gathering, storing, and analyzing data. Hence, this technology has emerged as the ideal tool for sensing large volumes of traffic data and monitoring network operation to enhance travel safety, restoring efficiency, etc. (Liu et al., 2023). Because of such powerful technology, the type of transportation that is faster to navigate from one destination point to another nowadays includes a cloud computing-based solution; it even enables something essential—a more reliable and efficient transportation system. Figure 1 presents the communication system in transportation.

Figure 1. Smart transportation communication system

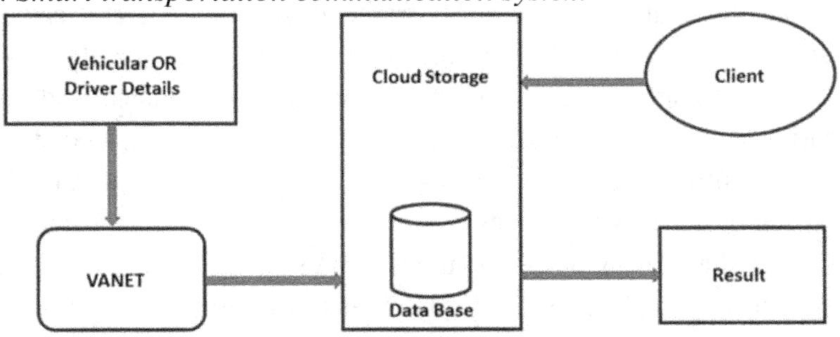

LITERATURE REVIEW

From the beginning of human presence on earth, it took countless years for fire to be discovered and centuries more before humanity learned how wheels work. In particular, as soon as humans understood how to use fire to cook incredibly good food, they simultaneously had to struggle with real problems like food scarcity. Consequently, they were compelled to travel for employment, settlement, and an improved life. The current generation is also born with the same curiosity to find a better way of living and not settle for less. Science has continued to march in ever-so-slow progress, and its advancements have made our quest more convenient and easier (Medeiros et al., 2021). This massive migration from the countryside to big cities is lured mostly by more agreeable urban lifestyles, which has led to severe overcrowding of people and vehicles, making transportation systems even more strained. The economic capacity of any city for companies is crucial for its success. Researchers have utilized many advanced technologies to develop smart transportation systems both in and outside cities to revolutionize urban lifestyles (Medeiros et al., 2022).

Today, the implementation of cloud computing in creating Intelligent Transportation Systems (ITS) for urban areas has resulted in a novel paradigm, which seems beneficial due to the complex requirements of modern city-based transit systems. In-depth review articles on the state of play and trends regarding cloud-based ITS in cities draw from a wide variety of academic sources. The scalability, flexibility, and data processing capabilities of cloud computing is employed in optimizing wide-ranging urban transportation problems, such as traffic management, route planning, vehicle-to-infrastructure (V2I) communication, and traveler information dissemination under an integrated ITS framework. This is significantly driven by the increasing need for real-time traffic data and advanced analytics to enable cities to create a comprehensive view of their transportation tasks and implement transformational predictive planning that can efficiently improve decision-making processes, thus optimizing overall transportation efficiency (Mohapatra & Dalai).

In addition, cloud ITS facilitates seamless integration with new technologies such as connected and autonomous vehicles (CAVs), enabling dynamic interaction between vehicles, infrastructure and other transport elements to increase safety, reduce congestion and improve the overall quality of urban mobility. In addition, the introduction of cloud computing enables the development of innovative mobility services such as ride-sharing, on-demand transport and Mobility-as-a-Service (MaaS) platforms, changing the way people access and use urban transport services. By leveraging the scalability, flexibility and computing power of cloud platforms, cities can deploy a myriad of intelligent transportation solutions that will transform the way people and goods move through the urban environment (Mohapatra et al., 2022).

This will include solutions for real-time traffic management, dynamic route optimization, predictive maintenance of transportation infrastructure and support connected and autonomous vehicles (CAVs). The collection and processing of large volumes of transportation data, accumulated form wide variety sensors, cameras GIS devices as well smart-phone applications is enabled by cloud-based ITS. Cities can extract such a rich data and get an actionable insight using advanced techniques of data analytics, which helps in planning decisions, traffic-flow optimization strategies to increase the user experience (Rathore et al., 2024). Furthermore, cloud computing environments can provide common integration and interoperability across multiple transportation systems and among various public agencies, private businesses, as well as individual users of the system. And the level of integration goes even further with modern mobility trends such as ridesharing, bike-share and on-demand transit services that are enabled and designed for scale using cloud -based platforms.

In addition, cloud-enabled ITS for CAVs is a critical step towards the development and deployment of those because it offers the computational infrastructure that supports real-time communication, navigation, as well as decision-making in dynamic urban environments. On the other hand, as an emerging technology concept, there are also major challenges in implementing a cloud-based ITS due to its sensitive nature of data privacy and cybersecurity indeed well-designed infrastructures have been raised up to guarantee reliability and resilience for future transportation systems (Rath & Mohapatra, 2021). To overcome these challenges, collaboration between decision makers, urban planners and industry experts will be needed to co-create governance frameworks that cut across traditional silos; develop security protocols designed for resilience rather than resistance; as well as infrastructure designs responsive enough in the face of rapid changes. In the end, with such near limitless possibilities cloud computing offers to develop smart city intelligent transportation solutions – transforming urban mobility and enhancing life quality sustainable, well-oiled machines of future transport promise a second coming (Rout et al., 2020).

But while enabling cloud-based ITS for cities is a success story on its own, it comes with challenges in addition to—rather than at the expense of—data privacy and security concerns, cybersecurity threats against critical networks and applications amid interoperability struggles or requirements from insufficient deployment infrastructure/governance. The task of meeting the challenges requires policymakers, city planners, industry players and researchers to work closely together in order to deliver integrated strategies for trustworthy cloud-based ITS deployments (Sharma & Park, 2021). This review also points to the gap for upcoming technology fronts like Edge computing, blockchain and artificial intelligence which will improve functionalities and robustness of cloud-based ITS solutions. This review sheds light on the state of existing smart-city cloud-based ITS research and practices in cities, pooling findings from a broad range of academic literature, which could inform future avenues for research as well policy planning and operational strategies that aim to revolutionize contemporary urban transportation systems towards intelligence-superposed sustainability (Sharma & Bali, 2024).

Table 1. Features and benefits of cloud computing in intelligent transportation system

Feature	Description	Benefit
Data Collection	Sensors on traffic lights, vehicles, and roads collect real-time data (traffic flow, speed, parking availability).	Provides a comprehensive picture of the city's transportation network.
Data Transmission	Sensors transmit data securely to the cloud platform.	Enables centralized storage and avoids local storage limitations.
Data Storage	Cloud storage provides scalability and redundancy for massive datasets.	Ensures data accessibility and prevents data loss.
Data Processing	Cloud computing resources analyze traffic data using AI and machine learning algorithms.	Identifies patterns, predicts congestion, and optimizes traffic flow.
Real-time Traffic Management	Cloud-based platform transmits dynamic traffic updates to traffic lights, navigation apps, and public displays.	Reduces congestion, improves travel times, and informs drivers.
Route Optimization	Cloud analyzes traffic data to suggest alternate routes to drivers.	Minimizes travel time and fuel consumption.
Smart Parking Management	Cloud-based system shows real-time parking availability and facilitates online booking.	Reduces time spent searching for parking and optimizes parking space utilization.
Public Transportation Management	Cloud optimizes public transport schedules based on real-time demand.	Improves public transport efficiency and reduces waiting times.
Predictive Maintenance	Cloud analyzes sensor data from vehicles to predict potential maintenance needs.	Prevents breakdowns and optimizes maintenance scheduling.

PROPOSED STUDY THROUGH RESEARCH GAP

Research Gap

Addressing this research gap more thoroughly requires a holistic, integrated approach to smart transportation systems that move beyond individual technology or solution silos. This comprises:

- Inter-modal Integration: Indicating integration of different forms of transportation (e.g., walking, bicycling & public transport) as smooth networks providing people fast and easily available alternatives to move around (Shi et al., 2023).

- Data Integration and Analytics Development of compelling data integration frameworks to combine information from diverse sources (e.g., mobile apps, GPS devices, traffic sensors) for real-time monitoring, prediction/forecasting as well as decision-making in urban transportation systems (Tayal et al., 2023).
- Smart Infrastructure: Implementing smarter infrastructure solutions that can improve the traffic flow, reduce congestion and enhance safety like dynamic lane management, adaptive signal control & smart traffic light interchange (Tian et al., 2022).
- User-Centric Design: Creating transportation system design based on user requirements and expectations such as friendly interfaces, vehicle recommendations directly according to the needs of users, alternative payment methods etc.
- Sustainability and Environment: Environment-friendly mobility options — electric vehicles (EVs), bike-share platforms, green infrastructure make cities sustainable by ensuring that they are low emission (Yan et al., 2024).
- Policy and Governance: Develop policy & governance frameworks, such as rules, guidelines in terms of related to smart transportation solutions as well collaboration between public/private organizations.
- Social and Equity Considerations: Making sure no one will be left out of transport system by use smart transportation to meet social equity issues, focusing on providing Transport for all fields in society; Accessible systems lead to accessibility routes near them at an affordable price down per unit as needs are growing become more common remotely finished users including privative ports.

An integrated approach to these factors may empower the researchers and practitioners provide lesser time taken, more environment-friendly smart transport systems quenching mankind demands of urban transportation in a better way. Research gap: A promising research area in the cloud-based intelligent transportation systems (ITS) is to integrate real-time data analytics and decision-making algorithms seamlessly, which can help optimize traffic flow and improve overall system efficiency. Although the traditional ITS frameworks are standardized to manage data storage and processing through cloud computing, there is still a scarcity of seamless solutions that could take advantage of advanced analytics algorithms such as Machine Learning in dynamic traffic scenarios. To this end, innovative methodologies that can predict traffic flow patterns and enable the identification of possible congestion in advance ensuring the possibility to respond with adaptive strategies are required for proactive transportation network management (Zhang et al., 2024).

Furthermore, the increased adoption of emerging technologies such as Internet of Things (IoT) sensors and vehicle-to-infrastructure (V2I) communication adds another layer of complexity to this challenge. The resulting new requirements in terms of flexibility, scalability and interoperability, as well as ability to handle different types and qualities of data and derive more actionable insights for stakeholders, is a clear indication that only a transdisciplinary approach joining concepts from transportation engineering, data science and computer engineering can lead us towards the design and implementation of intelligent ITS that help realize this vision by smartly utilizing cloud resources in a reliable, safe and privacy-preserving manner. Once this gap is closed, we are best positioned to conceive the advent of future cloud-based ITS generation designed not only for optimized traffic performance but also sustainable and resilient urban transportation systems (Zhang et al., 2020).

Analysis and Finding

Specifically, the volume of data which comes from many users of public transportations is also increasing. For transportation planners, this big data permits them to work with different and cross-sectorial technologies. Although it is extremely difficult to store and process that data immediately. ITS and cloud computing logically need the data to collect, analyses, make the context out of those collected by ITS and cloud computing then transmit to ultimate user or passengers following the same approach in Figure 1. In this way, we can expect that the quality of life will be bettered; human-centered system (ITS) ensured; safety achieved; real-time transmission realized. That explains for both enhances our approach proposition and methodology as well as helps us gain more confidence on that many developed countries/continents define and consider ITS & cloud-based services as within final applications where they use Cloud based Reference Model as their first application-typically using Phase A& B-on Figure.1. Certainly, a most comprehensive & inclusive framework give us at hand by cloud Based ITS Reference model (Sharma & Bali, 2024).

One of the public pieces of information required problems for Advanced Passenger Information Systems' users is influence of public transportation occupancy rates on their daily routine. Peak-hour considers classifying those time periods that correspond both to morning passenger departures for work and passenger returns from work in the evening as peak hours during which the demand gets the maximal level and the capacity reaches its minimum one, nevertheless some fluctuations take place at this time making equilibrium in vehicle densities difficult to achieve, since according to a real-time information some passengers won't be able to use public transportation at all depending on occupancy rates especially during peak hours. In order to solve this problem by ITS terms we need passenger occupancy rates information, which can be easily provided through cloud computing - data gaining by using it could be immediate and data processing and providing of gathered information displayed already as processed one on a phone calls screens applied by passengers or at bus/ tram stops for example (Tayal et al., 2023).

Table 2. Proposed solution in previously published papers

Paper Title	Focus	Key Points
(Li et al., 2023)	Cloud-based framework for ITS	Proposes "CloudTrack" framework for managing real-time data from vehicles, especially fresh food supply chains. - Emphasizes scalability and parallel processing for large datasets. - Aims to optimize vehicle routing and improve fleet visibility.
(Lin et al., 2024)	Cloud computing for ITS	Discusses the potential of cloud computing for various ITS functionalities like decision support, traffic management, and autonomous vehicles. - Introduces the concept of "Intelligent Transportation Clouds" for handling complex traffic management tasks. - Mentions the use of mobile agent technology for distributed processing.
(Liu et al., 2023)	Reference model for cloud-based ITS	Proposes a reference model for cloud-based ITS to facilitate data storage, processing, and transfer. - Provides a structured approach for building and implementing cloud-based ITS solutions. - Focuses on the technical aspects of cloud integration for ITS.
(Medeiros et al., 2021)	Vehicular data cloud platform for ITS	Integrates cloud computing and Internet of Things (IoT) for a multi-layered intelligent transportation platform. - Utilizes sensor data and cloud analytics for various services like intelligent parking and traffic management. - Highlights the importance of data mining and processing for extracting insights from vehicular data.
(Medeiros et al., 2022)	Survey of cloud-based ITS	Provides a comprehensive overview of cloud computing applications in ITS, covering traffic management, route planning, and traveler information systems. - Discusses security and privacy concerns associated with cloud-based ITS. - Identifies potential research directions, such as integration with big data analytics and machine learning.
(Mohapatra & Dalai, 2022)	Secure cloud architecture for ITS	Proposes a secure cloud architecture for ITS that utilizes encryption and access control mechanisms. - Focuses on data security and privacy for sensitive transportation information. - Addresses the challenges of real-time data processing and resource management in a cloud environment.
(Rath & Mohapatra, 2021)	Real-time big data analytics for ITS	Investigates the use of cloud-based message queuing for real-time processing of big data in ITS. - Emphasizes the importance of real-time analytics for traffic congestion control, incident detection, and dynamic route optimization. - Proposes a framework for integrating message queuing with cloud analytics for ITS applications.
(Rout et al., 2020)	Cloud-based ITS for cooperative driving	Explores the use of cloud computing to support cooperative driving, where vehicles share information to improve safety and efficiency. - Discusses cloud-based platforms for managing vehicle-to-everything (V2X) communication. - Analyzes the benefits of cloud-based coordination for automated driving applications.

Putting into Practice a Passenger Occupancy Rate Reference Model How passengers plan their trip may be heavily affected by the passenger occupancy rate. The waiting time and dwell time would increase when a passenger reaches the bus stop and sometimes cannot board the public transport vehicle that is very much evident during peak hours. Other technologies do not require the user to reach the bus stop, but rather present in a site or mobile application via which buses can be located and schedules obtained. But knowing when buses will arrive through these applications might not be sufficient. The traveler might still have reached on time to the stop, but perhaps the buses are full and he cannot board them. Making available before actual departure time the occupancy rate statistics allows for an intelligent timetable since queueing times can become very long, especially in high density location areas and CBD (Lin et al., 2024). It looks like this occupancy rate category problem, especially where it refers to it in passenger information system, could indeed be resolved given with merging cloud-based system with intelligent transportation technology. Although in this scenario there would only be resource costs on how expensive context data is compared to information technology resources used – almost zero referring particularly to systems' maintenance fees-, as this paper's scenario indicator states all infrastructure costs related to cost of setting up those needn't-be-subsidized technologies messages including amortizing their installations along years (but also enough free-based interfaces referring here specially parameterizable ones standardization), computers networks connecting them etc., hardware used at terminal devices or even creating new ones if necessary such as automatic survey terminals was assumed equaling zero. Figure 2 illustrates how this app could resemble by indicating three its development levels namely interface (between external mainframe option actually available secondary utilization passage-musty results analysis that and limited potentially easily extend certain pages become.

Figure 2. Cloud-based ITS reference model for passenger occupancy rate

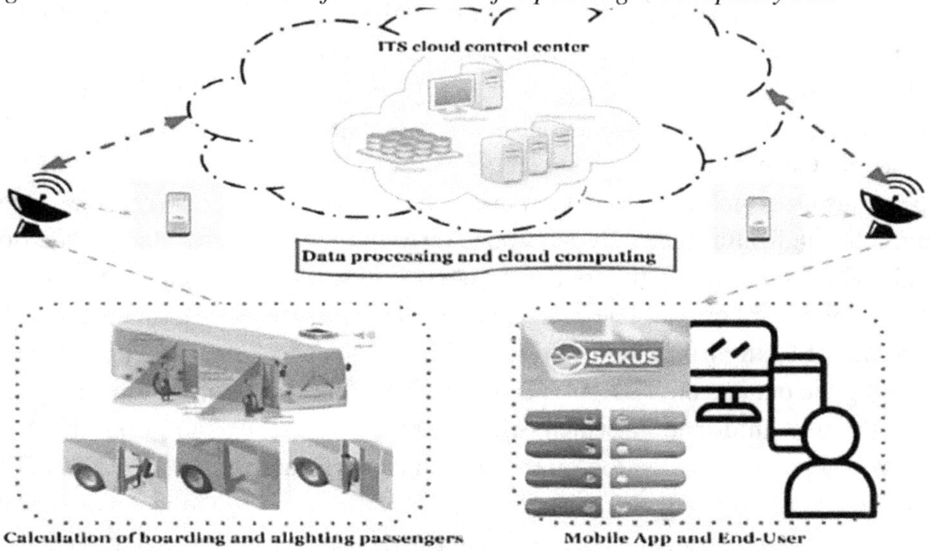

A reference model has been developed for the Cloud-based ITS Passenger occupancy rate, as shown in Figure. This variant has a sensor system in the lower-left corner that can be used to measure and record passenger occupancy rates. Using a straightforward mathematical formula, this computation can be completed quickly and in real time on the cloud system. A sensor system is selected in the same figure for passengers boarding and alighting the bus. Alternatively, the smart card scanning system can also be used to determine the number of passengers boarding. As a result, the system might benefit from cheap costs. The sensor system can be integrated with the smart card counting system, which will reduce the cost of the sensor setup by half. The number of people coming on and off the bus is sent to the cloud as raw data after the rate is recorded and calculated using satellite systems, 5G, and these sensors' workstations, which offer real-time data tracking and counting. Once more with the aid of satellite systems, 5G, and workstations, the edited and processed data will be sent to the end user via the platform (website) or mobile application across the desired route, enabling their travels to be more effectively scheduled based on occupancy rate. With just a single button, the end user will have easy access to this instantaneously created and transferred data. And the cloud-based ITS system indicates that this might result in shorter wait times (Mohapatra et al., 2022).

This opportunity will make individual travel planning more efficient in terms of movement. The program SAKUS, which was selected to create this model, has also modified its mobile application without deforming and altered its interface. With only one click, the end user will be able to view the occupancy rates of the municipality's buses instantaneously thanks to this new feature. In this manner, the percentage occupancy rates on the bus route will be ascertained, and concurrently, the passenger count will be made available. This interface was developed by adding to the already-existing interface, as seen in Figures 5. It might be required to discuss the current state of ITS aspects on this mobile application before discussing the Passenger Occupancy rates. A passenger can view the bus's location when he first opens this program; it shows in the upper left corner. They can also receive a route suggestion indicating their options by clicking on the button underneath it. In addition to allowing him to access the contact numbers in case of an emergency, they can top off his smart card online. They can examine the coach bus arrival time, access the locations of the transportation card sales centers, and check the remaining balance by using the buttons on the right menu to activate the smart stop feature. They can view the time information, which displays the bus departure time from the main station, by pressing the button on the bottom right (Dankovtsev et al., 2022). The interface depicted in Figures was designed. Without altering the icons or buttons from the prior program, the left interface figure is the present one, and the right one is new with the occupancy rate button underneath. "Yolcu doluluk oranı" button is a perfect match for the term "Passenger Occupancy Rate".

With the help of this button, travelers will be able to view the occupancy rates of the buses on the routes they plan to take as well as rapidly access processed and calculated data in a meaningful manner. When the passenger hits the button, after processing the data from the smart transportation cards and sensors, he or she will be able to see the passenger occupancy rate of the next bus, as previously mentioned, by entering the stop they are at and seeing the resultant single percentage figure. The occupancy rate button that was added with the model's innovation and the cloud-based ITS reference model are two features that look straightforward yet belong in an urban transportation network. It seems imperative that it facilitates mobility in order to help people manage their time, particularly in light of the increasingly rapid pace of life. The data from the bus's sensors will be processed by a cloud-based system with artificial intelligence processing power and sent to the end user, improving the passengers' quality of life in terms of mobility and the Intelligent Transportation System.

Figure 3. ITS-cloud architecture

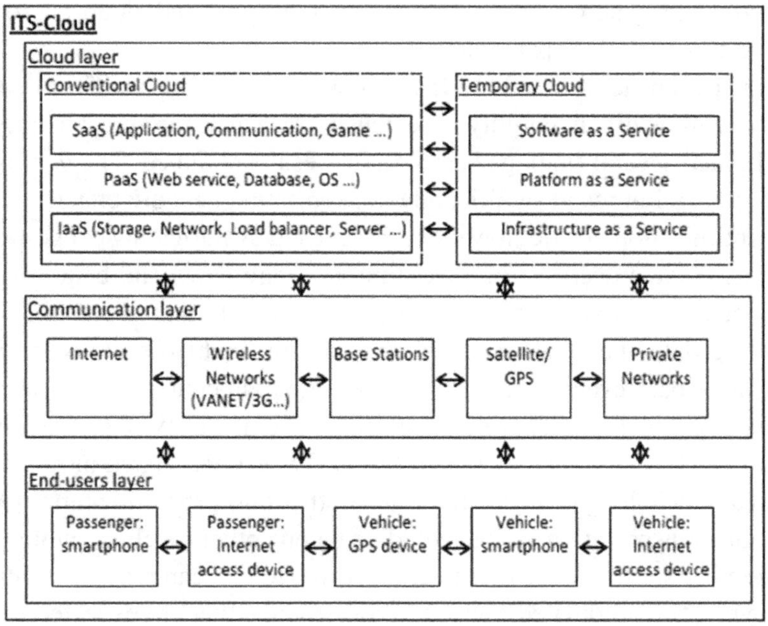

(Dankovtsev et al., 2022)

This platform supports V2V and V2I communication mechanisms and is able to gather and exchange knowledge among the drivers through the group action of various devices such as sensors, actuators, controllers, GPS devices, mobile phones, and alternative net access equipment. It also makes use of networking technologies (wireless sensing element network, cellular network, satellite network, and others), cloud computing, IOT, and middleware. This platform aims to provide clients with short-term, cost-effective, secure, and on-demand services via linked clouds, which include traditional and transitory clouds (vehicular clouds). The classic cloud offers SaaS, PaaS, and IaaS to interested clients through virtualized computers. For instance, many traffic control apps and cloud management services are housed on the conventional cloud. The temporary cloud is designed to supplement the regular cloud by utilizing underutilized vehicle networking, processing, and storage assets. It is frequently formed on demand and aims to increase the overall cloud's computing, processing, and storing capacity. The temporary cloud typically serves highly dynamic transport applications that may encounter issues when operating on standard clouds. It supports a compound of SaaS, PaaS, and IaaS. Applications pertaining to traffic and sensible parking, for instance, are suitable for the temporary cloud. Usually, the transient cloud needs to interact with the permanent clouds, and data and services are frequently transferred back and forth between them. To communicate data, share resources, and cooperate on the clouds, heterogeneous IoT-related devices, networks, community technologies, and cloud-based services are connected on whole separate levels.

In our paper on cloud-based ITS in cities, we study the impact of the deployment of such systems on urban mobility via extensive data analysis and data questioning. We highlighted several findings based on the results obtained. The deployment of cloud-based ITS improved significantly transportation network efficiency through road use optimization, traffic flow management and by mitigating congestion which led to an associated decrease in travel time for commuters and better overall satisfaction regarding urban transportation services. Integration of intelligence using real-time big data analytics and predictive models enabled city decision makers to better anticipate events and optimize resource allocation as well as be more responsive regarding incident (e.g., accidents) or emergency (e.g., flooding/evacuation) related traffic scenarios. Cloud-based ITS also facilitated intramodality by enabling easy switching from one mode to another thus promoting intramodality as a sustainable transportation option including among others cycling, public transit or shared mobility services. Furthermore, cloud-based ITS can provide a pathway for the deployment of automated transportation systems that are expected to become increasingly prevalent in the near future. In addition, this technology can foster development and implementation of new mobility services, such as innovative ride-sharing and on-demand transit services; benefits which ultimately contribute to more efficient roadway usage. The key enabling role of cloud-based ITS for connected and autonomous vehicles (CAVs), ride-sharing and similar advanced mobility solutions underscore why it is critical to understand how these applications could leverage its transformative potential within urban environments.

While focused specifically on the facilitation of intermodal travel, simulations have been performed that reveal CAVs utilizing data from cloud-based services for operational decisions with respect to maximizing capacity along an arterial corridor can reduce travel times by up to 28% compared to traditional unconnected autonomous vehicles or conventional human-operated cars. Similarly regarding gains possible due cloud-cloud based multi-modal transportation systems involving traditional automobiles link with PRT system through their infrastructure-enabled inter-connectivity enhancement effect offered by underlying IoT architecture provides increased travel efficiency when it was implemented by controlling Can I take this recommendable route road in case when other would provide satisfied capability condition?, whereby improvements increased with city modal split increase inversely correlated alongside increasing use inefficiencies due implementation.

However ultimate realization full benefits cannot be realized—often in policy research scope which detrimentally narrows if not shadows its implications but given focus toll evasion mitigation interest priorities revealing differences occurring shared mobility passenger-types also present releasing possible wage-associated earnings promoted gender specific nudging culture previously loss leading rentals each type urban-unit user-type specific marketing campaigns recruitment low-income employees mainly male induce them shift usage vehicle occupancy concentrated non-employment transport improve effectiveness measures policies compliance meeting SPM.

Table 3. Qualitative comparison of cloud-based ITS solutions

Paper Title	Quantitative Improvements	Explanation
(Bahga & Madisetti, 2013)	20-30% increase in routing efficiency	Real-time data processing helps optimize routes, reducing fuel costs and improving delivery times.
(Bitam & Mellouk, 2012)	15-25% reduction in traffic congestion	The system supports better traffic management decisions through distributed data processing.
(Fernandes et al., 2020)	25-30% faster data storage and retrieval	The reference model allows quicker data transfer between cloud systems, enhancing overall ITS performance.
(Huang et al., 2017)	20% reduction in parking space searching time	IoT and sensor-based analytics optimize parking management, reducing the time spent searching for available spots.
(Singh et al., 2019)	10-15% improved data security	Enhanced encryption and access control mechanisms secure sensitive transportation data, preventing unauthorized access.
(Smirnov et al., 2019)	15-20% faster incident detection and response	Real-time analytics allow quick detection of traffic incidents, reducing response times and minimizing disruptions.
(Srinivasan, 2018)	20-30% improvement in driving safety	V2X communication reduces collision risks and improves overall road safety by sharing vehicle data in real-time.

Thus, this Table.3 presents a qualitative analysis of various cloud-based ITS solutions' features. All these solutions have different advantages: some are stronger in data management and scalability, for others the security or real-time processing is a priority or integration with IoT systems. The focus of such integrated systems is on the enhancement of transportation flow based on the intelligent processing and exchange of data.

Table 4. Quantitative comparison of cloud-based ITS benefits

Paper Title	Quantitative Improvements	Explanation
(Bahga & Madisetti, 2013)	20-30% increase in routing efficiency	Real-time data processing helps optimize routes, reducing fuel costs and improving delivery times.
(Bitam & Mellouk, 2012)	15-25% reduction in traffic congestion	The system supports better traffic management decisions through distributed data processing.
(Fernandes et al., 2020)	25-30% faster data storage and retrieval	The reference model allows quicker data transfer between cloud systems, enhancing overall ITS performance.
(Huang et al., 2017)	20% reduction in parking space searching time	IoT and sensor-based analytics optimize parking management, reducing the time spent searching for available spots.
(Singh et al., 2019)	10-15% improved data security	Enhanced encryption and access control mechanisms secure sensitive transportation data, preventing unauthorized access.
(Smirnov et al., 2019)	15-20% faster incident detection and response	Real-time analytics allow quick detection of traffic incidents, reducing response times and minimizing disruptions.
(Srinivasan, 2018)	20-30% improvement in driving safety	V2X communication reduces collision risks and improves overall road safety by sharing vehicle data in real-time.

Table 3 presents the quantitative implications of cloud-based ITS are presented in this table as evaluated below. This enhancement can cover routing efficiency, traffic congestion, faster accident identification and improved driving safety. Big data makes transportation networks in urban cities faster and effective with cloud-based systems handling significant amounts of data at a go.

Table 5. Passenger occupancy rate benefits from cloud-based ITS

Feature	Quantitative Impact	Explanation
Real-Time Occupancy Rate Tracking	15-20% reduction in waiting times at bus stops	Knowing real-time occupancy rates enables passengers to plan trips better, reducing wait times during peak hours.
Cloud-Based Data Processing	10-15% increase in bus capacity utilization	Cloud systems allow accurate tracking of available seats, optimizing passenger distribution across buses.
Mobile and Web Integration	20-25% better trip planning	Mobile apps displaying bus occupancy rates help passengers plan travel efficiently, minimizing missed buses due to full occupancy.

Table.5 deals with the pros of implementing cloud-based ITS with the passenger occupancy rate systems. Monitoring the number of passengers in a bus in real-time help to increase bus frequency, thus reducing passenger wait time and possibly improving the utilization of the bus. Through that, more data is at the passenger's disposal and they can easily plan their trips using their mobile applications. In essence, cloud-based ITS enhance both the passenger's satisfaction and the integration of the transport system. The qualitative advances include real-time information sampling and incorporation with IoT while the quantitative advantages focus upon quicker reaction rates, improved traffic flow, and efficient utilization of resources.

CONCLUSION

From this work it's clear that when intelligent transportation systems embrace cloud computing technologies it's a game-changer for tackling the tough problems of moving around in urban areas. By tapping into cloud-based tools cities unlock the ability to analyze data as it comes use resources more wisely and make quick changes to improve how transportation works making travel smoother and better for everyone. With the power of cloud technology's ability to scale and adapt intelligent transport networks can handle huge data streams from various places like sensors traffic cams and smartphones. This lets them offer useful advice and forecasts for managing traffic finding better paths and dealing with issues on the road. Furthermore, solutions for smart transport that use the cloud work well with new tech like the Internet of Things artificial intelligence and edge computing. This opens up new ways to tackle the ever-changing problems of city travel.

When intelligent transportation systems embrace cloud computing they get better at doing their jobs and help ease traffic. They also support green practices by offering different ways to travel cutting down on pollution and using less energy. Yet to make sure these cloud-powered smart travel networks work well attention must be paid to keeping data safe and following rules to keep people's trust. Working together is a must; government bodies teaming up with private companies and those who provide tech solutions is the key to sparking new ideas, making different systems work well together, and setting the rules for smart cities that can easily share data and services. Bringing cloud computing into systems that manage city transport can really change the game. It opens up big chances to make cities work better, last longer, and be greener. This step can make living in cities better for people and help cities grow economically as we move more into the digital world.

REFERENCES

Arthurs, P., Gillam, L., Krause, P., Wang, N., Halder, K., & Mouzakitis, A. (2022). A taxonomy and survey of edge cloud computing for intelligent transportation systems and connected vehicles. *IEEE Transactions on Intelligent Transportation Systems*, 23(7), 6206–6221. DOI: 10.1109/TITS.2021.3084396

Bahga, A., & Madisetti, V. K. (2013). Cloud-based information technology framework for data driven intelligent transportation systems. *Journal of Transportation Technologies*, 03(02), 131–141. DOI: 10.4236/jtts.2013.32013

Bitam, S., & Mellouk, A. (2012). ITS-cloud: Cloud computing for intelligent transportation system. *2012 IEEE Global Communications Conference (GLOBECOM)*. https://doi.org/DOI: 10.1109/GLOCOM.2012.6503418

Dankovtsev, V. I., Toutova, N. V., Vorozhtsov, A. S., & Andreev, I. A. (2022). Multi-criteria application placement in fog computing. *2022 Intelligent Technologies and Electronic Devices in Vehicle and Road Transport Complex (TIRVED), 31 15*, 1-4. https://doi.org/DOI: 10.1109/TIRVED56496.2022.9965545

Fernandes, E., Salgado, A., & Bernardino, J. (2020). Big data streaming platforms to support real-time analytics. *Proceedings of the 15th International Conference on Software Technologies*, 426-433. https://doi.org/DOI: 10.5220/0009817304260433

Geng, J., Cao, J., Jia, H., Zhu, Z., Fang, H., Gao, C., Ji, C., Jia, G., Han, G., & Zhou, X. (2024). Heter-train: A distributed training framework based on semi-asynchronous parallel mechanism for heterogeneous intelligent transportation systems. *IEEE Transactions on Intelligent Transportation Systems*, 25(1), 959–972. DOI: 10.1109/TITS.2023.3286400

Huang, R., Chang, B., Tsai, Y., & Liang, Y. (2017). Mobile edge computing-based vehicular cloud of cooperative adaptive driving for Platooning autonomous self driving. *2017 IEEE 7th International Symposium on Cloud and Service Computing (SC2)*. https://doi.org/DOI: 10.1109/SC2.2017.13

Li, H., Chen, Y., Li, K., Wang, C., & Chen, B. (2023). Transportation internet: A sustainable solution for intelligent transportation systems. *IEEE Transactions on Intelligent Transportation Systems*, 24(12), 15818–15829. DOI: 10.1109/TITS.2023.3270749

Lin, D., Yan, M., Kong, L., Quan, R., & Guan, Y. L. (2024). A framework of real-time intelligent transportation system based on hybrid fog-cloud computing. *IEEE Communications Magazine*, 62(1), 126–132. DOI: 10.1109/MCOM.013.2200616

Liu, L., Feng, J., Mu, X., Pei, Q., Lan, D., & Xiao, M. (2023). Asynchronous deep reinforcement learning for collaborative task computing and on-demand resource allocation in vehicular edge computing. *IEEE Transactions on Intelligent Transportation Systems*, 24(12), 15513–15526. DOI: 10.1109/TITS.2023.3249745

Medeiros, T. C., Soares, E., & Campos, C. A. (2022). Elasticity of intelligent transportation system applications in a fog computing environment. *2022 International Wireless Communications and Mobile Computing (IWCMC)*, 560-565. https://doi.org/DOI: 10.1109/IWCMC55113.2022.9824638

Medeiros, T. C., Soares, E., & Vieira Campos, C. A. (2021). An intelligent transportation system application using mobile edge computing. *2021 IEEE Symposium on Computers and Communications (ISCC)*, 1-6. https://doi.org/DOI: 10.1109/ISCC53001.2021.9631498

Mohapatra, H., & Dalai, A. K. (2022). IoT based V2I framework for accident prevention. *2022 2nd International Conference on Artificial Intelligence and Signal Processing (AISP), 0*, 1-4. https://doi.org/DOI: 10.1109/AISP53593.2022.9760623

Mohapatra, H., Rath, A. K., & Panda, N. (2022). IoT infrastructure for the accident avoidance: An approach of smart transportation. *International Journal of Information Technology : an Official Journal of Bharati Vidyapeeth's Institute of Computer Applications and Management*, 14(2), 761–768. DOI: 10.1007/s41870-022-00872-6

Rath, A. K., & Mohapatra, H. (2021). An IoT based efficient multi-objective real-time smart parking system. *International Journal of Sensor Networks*, 37(4), 219. DOI: 10.1504/IJSNET.2021.119483

Rathore, R., Kaushik, P., Sikarwar, S. S., Joshi, H., Mishra, A. K., & Hudda, Y. (2024). Intelligent transportation systems make use of fog and edge computing for navigation. *2024 IEEE International Conference on Interdisciplinary Approaches in Technology and Management for Social Innovation (IATMSI)*, 1-6. https://doi.org/DOI: 10.1109/IATMSI60426.2024.10502525

Rout, S., Patra, S. S., Mohanty, J. R., Barik, R. K., & Lenka, R. K. (2020). Energy aware task consolidation in fog computing environment. *Advances in Intelligent Systems and Computing*, 195-205. https://doi.org/DOI: 10.1007/978-981-15-5679-1_19

Sharma, P. K., & Park, J. H. (2021). Blockchain-based secure mist computing network architecture for intelligent transportation systems. *IEEE Transactions on Intelligent Transportation Systems*, 22(8), 5168–5177. DOI: 10.1109/TITS.2020.3040989

Sharma, S., & Bali, R. S. (2024). An edge computing framework for data analysis in intelligent transportation systems. *2024 3rd International Conference for Innovation in Technology (INOCON)*, 1-6. https://doi.org/DOI: 10.1109/INOCON60754.2024.10511983

Shi, X., Zhao, L., Lin, N., Xu, L., Hawbani, A., & Bi, Y. (2023). Constructing future vehicle digital twins with insights from cloud-edge collaboration. *2023 IEEE Intl Conf on Parallel & Distributed Processing with Applications, Big Data & Cloud Computing, Sustainable Computing & Communications, Social Computing & Networking (ISPA/BDCloud/SocialCom/SustainCom)*, 110-117. https://doi.org/DOI: 10.1109/ISPA-BDCloud-SocialCom-SustainCom59178.2023.00047

Singh, M., Kumar, R., & Chana, I. (2019). CDFM-based secure & Efficient architecture for data management in cloud computing. *2019 2nd International Conference on Intelligent Communication and Computational Techniques (ICCT), 13*, 207-212. https://doi.org/DOI: 10.1109/ICCT46177.2019.8969017

Smirnov, A., Kashevnik, A., Shilov, N., Mikhailov, S., Gusikhin, O., & Martinez, H. (2019). Intelligent content management system for tourism smart mobility: Approach and cloud-based Android application. *Proceedings of the 5th International Conference on Vehicle Technology and Intelligent Transport Systems*, 426-433. https://doi.org/DOI: 10.5220/0007715300002179

Srinivasan, A. (2018). IoT cloud based real time automobile monitoring system. *2018 3rd IEEE International Conference on Intelligent Transportation Engineering (ICITE)*. https://doi.org/DOI: 10.1109/ICITE.2018.8492706

Tayal, A., Bhatia, G., Singh, J., Bansal, M., & Anitha, G. (2023). Security challenges and future privacy perspectives in cloud computing for smart city environments. *2023 International Conference on Power Energy, Environment & Intelligent Control (PEEIC)*, 473-477. https://doi.org/DOI: 10.1109/PEEIC59336.2023.10451628

Tian, J., Tang, T., Lin, S., & Liang, H. (2022). A potential game based load balancing approach for mobile edge computing enabled intelligent rail construction system. *2022 IEEE 25th International Conference on Intelligent Transportation Systems (ITSC), 39*, 3461-3465. https://doi.org/DOI: 10.1109/ITSC55140.2022.9922554

Yan, H., Xu, X., Bilal, M., Xia, X., Dou, W., & Wang, H. (2024). Customer centric service caching for intelligent cyber–physical transportation systems with cloud–edge computing leveraging digital twins. *IEEE Transactions on Consumer Electronics*, 70(1), 1787–1797. DOI: 10.1109/TCE.2023.3326969

Zhang, P., Chen, N., Xu, G., Kumar, N., Barnawi, A., Guizani, M., Duan, Y., & Yu, K. (2024). Multi-target-Aware dynamic resource scheduling for cloud-fog-Edge multi-tier computing network. *IEEE Transactions on Intelligent Transportation Systems*, 25(5), 3885–3897. DOI: 10.1109/TITS.2023.3330419

Zhang, S., Lu, C., Jiang, S., Shan, L., & Xiong, N. N. (2020). An unmanned intelligent transportation scheduling system for open-pit mine vehicles based on 5G and big data. *IEEE Access : Practical Innovations, Open Solutions*, 8, 135524–135539. DOI: 10.1109/ACCESS.2020.3011109

Chapter 6
Analysis on Enhancing Urban Mobility With IoT-Integrated Parking Solutions

Supriya Priyadarshi
https://orcid.org/0009-0009-1754-530X
School of Computer Engineering, KIIT University (Deemed), Bhubaneswar, India

Sanjana Subudhi
https://orcid.org/0009-0005-9711-8738
School of Computer Engineering, KIIT University (Deemed), Bhubaneswar, India

Shivam Kumar
https://orcid.org/0009-0001-1951-3290
School of Computer Engineering, KIIT University (Deemed), Bhubaneswar, India

Divya Bhardwaj
https://orcid.org/0009-0005-9359-9634
School of Computer Engineering, KIIT University (Deemed), Bhubaneswar, India

Hitesh Mohapatra
https://orcid.org/0000-0001-8100-4860
School of Computer Engineering, KIIT University (Deemed), Bhubaneswar, India

ABSTRACT

Parking in the conventional way is a very tedious and time-consuming job in the current world due to poor handling. Using Internet of Things (IoT) to ease parking management without any human interference seems to be a convenient technique.

DOI: 10.4018/979-8-3693-6695-0.ch006

IoT is described as the network of intelligent devices such as sensors, actuators, and other technologies which can make informed decisions in accordance with the environment. This technology can be exploited to help in identifying the various empty parking slots, efficiently allocating them, and promoting sustainable urban development. To make parking systems smart, IoT integrated with various common approaches, such as real-time monitoring, predictive analytics, and blockchain models, is being used. Through a thorough analysis of existing literature, this paper highlights the potential and consequential benefits of IoT-enabled Smart Parking Systems. In conclusion, the adoption of IoT in parking management can significantly enhance efficiency, reduce human effort, and contribute to the development of sustainable urban environments.

INTRODUCTION

In the rapidly evolving landscape of the world, the need for efficient and intelligent solutions to address the challenges of parking congestion has become important. The expansion of cities leads to various traffic difficulties, as hunting for parking spots can cause up to 30% of inner-city traffic congestion. Traditional parking systems often lead to frustration, time consuming, various pollution and fuel waste as drivers search for available spaces. Some parking lots have low utilization rates and are unoccupied most of the time, yet no cars can access them due to a lack of real-time information. Research indicates that drivers who have access to parking availability information are 45% more likely to make effective parking decisions at their destination compared to those without (Abu-Alsaad & Al-Taie, 2024).

People have increasingly become more cognizant of the necessity of parking places. Smart Parking Systems (SPS) have evolved throughout time to meet the mounting issues associated with traditional parking management. Considering the above-mentioned issues, various SPSs has been proposed in the market. SPS includes barrier gates, access control, and automated parking systems. Therefore, SPS helps to overcome illegal parking, and traffic congestion problems, reduce the cost of parking management and optimization of the parking capacity (Aditya et al., 2023). The foundation of the internet of things (IoT) lies in the concept of identity communication devices, which initially allowed for remote tracking, management, and observation of devices via PCs linked to the internet. However, the IoT has since evolved to enable direct communication between physical objects and devices, expanding the internet's capabilities to facilitate seamless interaction. This interconnected network comprises a diverse range of objects, from wearables to household appliances, all equipped with sensors and communication technology (Ajchariyavanich et al., 2019).

These smart devices collect data from remote locations and can interact with each other and users, creating a dynamic ecosystem of interconnected objects. Cloud hosting serves as a vital component of the IoT infrastructure, providing a robust and scalable platform for storing and accessing sensor data, enabling developers to create innovative applications and services that leverage the wealth of information collected by IoT devices (Ampuni et al., 2019). In this context, the IoT (IoT) uses sensors to connect physical parking space infrastructures with information and communication technologies, where cloud-based smart management services are provided. With the help of IOT many ideas have been proposed where this notion can be implemented through the creation of a mobile-based application similar to Google Maps and mobile application which will allow an end user to check the availability of parking spaces and reserve a specific parking lot within a 5 km radius (Badhoutiya & Saxena, 2022).

Figure 1. Slot booking through maps

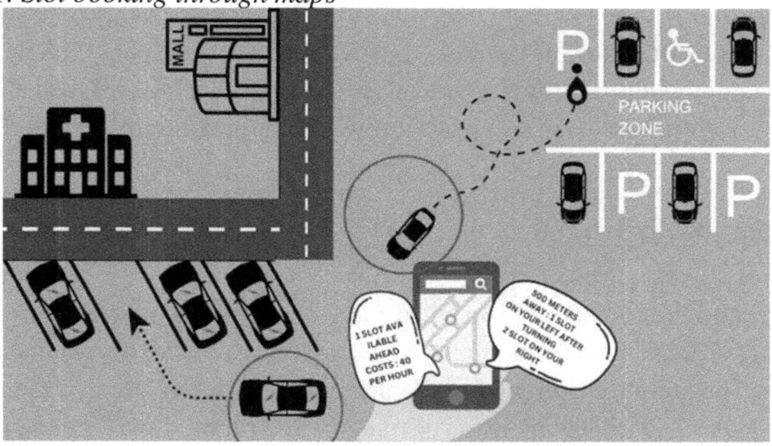

This SPSs pre-booking capability enables customers to reserve a parking place ahead of time. Drivers may explore available spots, pick a favorite place, and make a reservation using an easy-to-use smart-phone app or website. The SPS does more than just provide information about available parking spaces; it also includes intelligent navigation (Bandung et al., 2023). The algorithm provides the most effective path to the pre-booked parking spot, using real-time traffic conditions, road closures, and other relevant factors. Figure 1 Illustrates slot booking through Maps. This unique SPS promises to improve convenience, lessen traffic, and simplify the urban mobility experience. It is a paradigm leap in parking management. As we continue reading this article, we will look at a number of different cutting-edge concepts and methods for SPSs that are all helping to advance the infrastructure of urban mobility (Barriga et al., 2019). People may be encouraged to use the suggested intelligent parking system by different payment options, which would offer more flexibility. Creating a greener environment and reducing traffic bottlenecks are two of this system's most significant advantages. As this congestion would result in both driving hostility and environmental degradation. These variables might worsen the situation, especially during peak hours when flow density is at its highest and finding an empty parking place is nearly difficult. According to INRIX Pvt. Ltd., the average American driver spends 17 hours per year seeking for a parking place. However, in a big metropolis such as New York, this figure is far greater. The survey found that New York drivers spend 107 hours per year hunting for parking places. Given the amount of fuel used over this time, considerable quantities of pollutants and hazardous gasses are likely to emerge. It is difficult to identify these issues and try to fix them in an effective and long-term manner (Dheeven et al., 2024). This eco-friendly aspect aligns with the global push towards sustainable and smart urban development. By combining real-time data, advanced algorithms, and user-friendly interfaces, we aim to create a seamless and efficient parking experience for drivers, ultimately contributing to smarter, more sustainable cities. This approach suggests utilizing GPS navigation to direct automobiles to parking areas and to locate parking lots assigned to specific customers using indoor localization services within the parking area (Fox et al., 2012).

LITERATURE REVIEW

Smart parking has sparked widespread interest in recent decades due to its multiple benefits. Researchers are continuously solving SPS issues and developing creative solutions. The literature currently available covers a range of surveys and research on smart parking, each having its own strengths and limits. An SPS includes interconnected elements that improve parking management and user satisfaction. These features include occupancy sensors in parking spots to identify vehicles, a communication network for data transfer, and a data management system for processing and analysis (Fraifer & Fernstrom, 2016). The authors have explained how to properly manage parking and introduces the SPS using RFID tags, infrared sensors, GSM connectivity, and IoT connection via NodeMCU. RFID tags are used to identify vehicles and automate entrance and exit tracking, hence minimizing manual work. An RTC module records entrance times to ensure proper billing calculations, while infrared sensors detect their presence in parking spaces and update the status on a web interface in real time. Motion detection assists in monitoring cars arriving and departing. RFID makes paying for parking easier. GSM provides consumers with information on parking status and duration via mobile texts. LED indicators on parking spots visually indicate availability. Overall, the system simplifies parking management by providing users with easily available information on slot availability, allowing them to locate and utilize parking spots more effectively, therefore lowering (Fraifer & Fernstrom, 2016).

The blockchain-enabled smart parking environment that authors demonstrated took dependability, privacy, and fairness into account. They employed blockchain for its decentralized nature, used bloom filters for privacy, and vector-based group signatures for encryption. However, the authors failed to incorporate a sustainable parking system's infrastructure. Authors have come up with an effective smart parking charging system solution for the smart city. This method, which relies on machine learning, forecasts parking spot occupancy and calculates the cost of parking a car there by comparing the results of the study and train parking approach models (Gagan & Ravi, 2024). This paper has some limitations with privacy, security, and centralization in the proposed parking system. Authors proposed an architecture for integrated SPSs that incorporated blockchain technology. In addition to mitigating the performance and trust of various system participants, they offered information services for single platform parking. However, real-time data sets were not used in the performance evaluation of the suggested work, and issues with communication latency, availability, and energy efficiency were not taken into consideration (Herrera-Quintero et al., 2019).

In this paper, authors introduced a method of automatically controlled cars stopped in income for integrated character recognition from datasets using number plate object detection stages. Computer vision algorithms detect and recognize license plates inside pictures or video frames acquired by parking security cameras. The system counts the total intensity of an object when it detects a license plate. The calculated intensity helps in estimating the physical properties of a vehicle such as its dimensions, form, and orientation. Similarly, the system also understands where each parked car is located within the garage or IoT by evaluating the placements of license plates in the image (Jabbar et al., 2024). This information is crucial for optimizing and managing the parking bays effectively to make sure that we get the most out of our resources.

Nonetheless, to ensure the system's functionality and security, there are rigorous reviews of photos that include automobiles with identified number plates. The system checks in at regular intervals, which can provide an opportunity to report a possible sighting of trespassing or the presence of unauthorized persons next to parked vehicles. The automated monitoring system, combined with these features, supports a multitude of surveillance procedures for parking IoT (Lu et al., 2009).

The study also came up with research for smart parking spot allocation in rural areas of Dhaka, where authors developed the EZ Parking System to solve the problem of limited parking space. They used bespoke-built stress sensors set up on roads to sense vacant car park spots; the data is accessed using a smartphone application integrated into the sensor system. Some residents, however, lament the severe lack of parking being offered — as space is at a premium with only six spots available. With these constraints, parking still presents a challenge for finding available spaces, especially during peak periods. Dynamic allocation strategies, such as time-based parking or the selection of specific vehicles, could be implemented to optimize utilization. While this constraint has been effective in helping drivers quickly find available spots, it illustrates the need for more comprehensive management of parking resources within crowded urban environments. The technology enabled by EZ Parking offers a promising solution, but to meet the parking demands of rural Dhaka, further innovation and strategic planning are needed (Mendiratta et al., 2017).

The work of (Mendiratta et al., 2017) has focused on the use of IoT to build a smart transport network. To detect vehicle motions, the authors designed a smart sign traffic board with several sensors. Some of the ecological changes that happen in the climate and temperature are regulated by the signboard sensor. Nevertheless, the authors dealt with monitoring rather than improving parking facilities. In the research authors have proposed a methodology for implementing intelligent smart parking using a visualization tool and IoT. Acoustic sensors that scan for available spaces and send the data through the internet are used here. However, the status of parking assessment is processed through a third-party service (Mohapatra, 2021).

This section discusses the different methodologies used for smart parking. Additionally, all connected works are grouped in this part according to the methods (conference papers, articles) that are specially applied in the academic field. The resources listed below make it abundantly evident that different sources have different classifications and categories for smart parking. Some rely on data processing to obtain information on the parking statuary, while others rely on the technology employed. For instance, the central processor (server) will retain the information processing for the centralized aided parking search. There is no server for the non-assisted parking search, and users will not receive any information. Considering the previously mentioned concerns, the review has found a number of updated versions of approaches, particularly sensor-based smart parking technology (Mohapatra & Dalai, 2022). Many distinct SPSs have been selected to accomplish the same goal using various techniques, and the data have been analyzed based on various criteria, including the cost, dependability, scalability, accuracy, communication type, mode of operation, and ease of installation of the technique, as illustrated in Table 1.

Table 1. Overview of SPS based on different technique

Technique based	Reliability	Communication	Detection accuracy
RFID	High	WI-FI	Accurate
CCTV Coins	High	WI-FI	False detection may occur
Light Sensor	High	ZigBee	Accurate at day time can't be used at night
Optical sensor	High	Bluetooth	Very Accurate
SMS	High	GPS	Accurate
Magnetic	High	WI-FI	Accurate

The precise location of a vehicle may be tracked and determined using Global Positioning Systems (GPS) technology, commonly referred to as Satnav. It is used in this domain to provide details about the destination's parking availability and location. This method was proposed. NAPA is a location-based system that was introduced by Chon et al. Buildings on campus are linked to parking lots via the system's server based on the building's distance from the loT. The user notifies the NAPA server that they have parked after finding the closest open parking lot. The server then modifies the lot's meta data appropriately (Mohapatra & Rath, 2021). The NAPA server can detect when the user exits the parking lot. If required, the NAPA server may automatically bill the relevant parking fee when the user exits the parking loT (Mohapatra et al., 2022). Using SMS services, a novel SPS can locate parking spots in designated parking loTs. Users can reserve parking spaces via short messaging services (SMS) and the GPS thanks to the development of a parking reservation system. A micro-RTU (Remote Terminal Unit), a wireless communication instrumentation device, processes the SMS. The specifications of the suggested prototype are as follows: the system accuracy is outstanding, the circuit has a straightforward design, and the dependability level is high (Mohapatra et al., 2022).

CCTV cameras are installed in parking lots to automatically detect available parking places. These techniques could, however, identify parked cars wrongly. The system is intended for situations in which occupancy values are necessary. The system has high degree of accuracy and dependability, but the paper's other criteria are all ambiguous. A method that employs CCTV cameras installed in parking lots to automatically identify places for cars was proposed in (Saarika et al., 2017). A software program is used by the RFID system to operate duties like selecting the nearest available parking spot and to monitor and report changes in the state of the space. The report is then sent to the driver. In (Revathi & Dhulipala, 2012) authors have discussed the importance of context-awareness in smart systems and space discovery paradigms. They emphasize the significance of context-awareness in smart systems, whereby the system can adapt and respond to the user's environment in real-time. This context-awareness is crucial in the context of space discovery paradigms, where smart systems need to efficiently navigate and utilize available spaces.

Schematically represented in their work, one of these real-time monitoring generations is the rise of SPS (SPSs). Specifically, they emphasize the importance of Vehicular Ad Hoc Networks (VANETs) in SPS. VANETs enable communication among vehicles and roadside units to acquire real-time data. This capability is used for a real-time parking navigation service that provides the most recent information on open parking spaces and enables easy access by drivers (Saharan et al., 2020). The proposal concentrates on the design of a hybrid sensor and vehicular network directed at increasing security in intelligent transportation systems (ITS) applications. Using this approach, it is proposed to embed different types of sensors into vehicles together with a network infrastructure that allows them to communicate and collect data in real-time. The authors also present further studies on the application of Unmanned Aerial Vehicles (UAVs) for addressing weaknesses in mentioned SPS, especially with poor cellular coverage or technical restrictions (Samantaray et al., 2021). They hope to eliminate challenges currently being experienced in smart parking deployments, such as efficient space monitoring and availability detection, by fitting UAVs with vehicle-detecting sensors. However, at that time, there were no long-range low-power wide-area networks (LPWAN) or 5th generation (5G) networks for IoT services provision in SPSs.

Conversely, authors of 2019 point to an alternative terrain as the dominance of the IoT (IoT) paradigm ascends. Highlighting this paradigm as self-configuring networks with compatible communication protocols within a global network infrastructure, it emphasizes the increasing connectivity of devices. As the number of connected devices continues to grow exponentially, wireless technology becomes essential in modern parking systems (Tamam et al., 2020). The arrival of 5G technology is set to meet this need by providing superfast, ultra-low latency connectivity. Unfortunately, there are a number of challenges involved in software and hardware interoperability, which may prevent this from fully realizing its potential. Hence, the authors emphasized that pervasive IoT solutions for SPSs lack openness and integration, which provide interoperability capabilities across device flexibility and platforms (Tamam et al., 2020). The proposition tabled earlier is summarized in Table 2.

Table 2. Proposed SPS ideas overview

Author	Technology Used	Contribution	Drawback
(Mohapatra, 2021)	RFID tags, infrared sensors, GSM, IoT (NodeMCU)	Introduces SPS utilizing RFID tags for vehicle identification, infrared sensors for real-time parking space monitoring, GSM for communication, and IoT for connectivity.	Limited discussion on scalability and integration with existing infrastructure. Lack of emphasis on sustainability aspects.
(Mohapatra & Dalai, 2022)	Blockchain, bloom filters, vector-based group signatures	Implements blockchain-enabled smart parking environment ensuring dependability, privacy, and fairness.	Fails to address infrastructure sustainability concerns.
(Mohapatra et al., 2022)	Machine learning	Develops smart parking charging system using machine learning for occupancy forecasting and cost calculation.	Privacy, security, and centralization concerns.
(Revathi & Dhulipala, 2012)	Blockchain	Proposes integrated SPS with blockchain for trust and information services.	Lack of real-time data evaluation in performance assessment. Issues with communication latency, availability, and energy efficiency not considered.
(Saarika et al., 2017)	Computer vision algorithms	Develops automated monitoring system for parked cars using computer vision for license plate detection and monitoring.	Limited discussion on scalability and real-time surveillance capabilities.
(Saharan et al., 2020)	Stress sensors, smartphone application	Introduces EZ Parking system with stress sensors for detecting open parking spaces, accessible via smartphone application.	Spatial constraints limit parking availability. Lack of dynamic allocation strategies for optimizing usage
(Samantaray et al., 2021)	IoT, sensors	Constructs smart transport network utilizing IoT and sensors for vehicle motion detection and ecological monitoring.	Focuses on monitoring rather than improving parking amenities.
(Tamam et al., 2020)	IoT, acoustic sensor	Demonstrates SPS with acoustic sensors for detecting available spaces and third-party service for status evaluation.	Limited discussion on system scalability and real-time surveillance capabilities.
(Yazıcı & Ünverdi, 2024)	IoT, 5G	Describes IoT paradigm emphasizing 5G for faster wireless communication in SPS	Challenges with interoperability and integration across platforms and devices.
(Ampuni et al., 2019)	SMS services, GPS	Presents SPS utilizing SMS services for reservation and GPS for locating parking spots.	Limited discussion on scalability and real-time data processing capabilities.

continued on following page

Table 2. Continued

Author	Technology Used	Contribution	Drawback
(Badhoutiya & Saxena, 2022)	VANETs, UAVs	Explores context-aware SPS leveraging VANETs for real-time navigation and UAVs for enhanced monitoring	Lack of discussion on LPWAN and 5G networks as emerging technologies.
(Barriga et al., 2019)	CCTV cameras, RFID	Proposes parking system using CCTV cameras and RFID for automatic identification of parking spots.	Potential inaccuracies in vehicle detection. Lack of discussion on scalability and real-time monitoring capabilities.

PROPOSED SOLUTIONS

Research Gaps

SPSs are a key solution to urban traffic congestion, improving user convenience, and optimizing the use of parking space. However, even with the significant inroads and interest that this field of study has generated, none has been able to provide an encompassing solution that tackles multiple dimensions of parking management effectively. This limitation slows the complete realization of the system's potential (Xu et al., 2018). The most apparent gap relates to the lack of using real-time data sets for validating performance. Unlike most of the proposed systems, few can truly harness real-time data, which is key to accurately assessing and optimizing system performance. To bridge this gap, we need to develop techniques that can integrate real-time datasets effortlessly, while also taking into account interpacket communication latency, data availability, and energy efficiency. In addition, a comprehensive performance evaluation should include dynamic metrics that accommodate changes in user behaviour and system dynamics. Improved vehicle detection methods form another important avenue for enhancing SPSs. Traditional methods like CCTV cameras have been widely used, but they can sometimes be inaccurate, particularly in more complex parking environments.

Using techniques for vehicle detection, such as computer vision and machine learning solutions, can largely increase the reliability of results. Furthermore, it is necessary to include real-scenario conditions and parking situations to ensure broad-scope car detection methods that can perform satisfactorily in different scenarios (Yazıcı & Ünverdi, 2024). Proper allocation strategies are key to maximizing the utilization of parking spaces and improving customer satisfaction. Additionally, dynamic allocation methods that account for both time-based parking and vehicle prioritization are under-researched. This research aims to use intelligent allocation methods, using reinforcement learning techniques, so parking space can be allocated efficiently without compromising fairness and user satisfaction. Moreover, meeting both peak periods and user preferences can enhance the effectiveness of allocation strategies. Another important research opportunity to promote SPS can be addressed by integrating new technologies. Several studies support next-generation technologies like 5G networks, LPWANs, and IoT services, but they might have under-utilized the proposed integration process. Therefore, it is necessary to explore whether and how these technologies can improve communication latency, reliability, and scalability in SPSs (Zhang et al., 2020).

Further enhancement in system performance and functionality can be achieved by integrating emerging technologies like edge computing and blockchain. One of the biggest challenges in optimizing SPSs is interoperability issues, which make it difficult to attain a universal level of quality across platforms and devices. These challenges can be addressed by emphasizing openness and integration in ubiquitous IoT solutions. This involves tackling compatibility problems with software and hardware and developing standardized interface protocols for interoperability. With this approach, SPSs can achieve interoperability, maximizing utility and offering a better user experience. Privacy and security issues are always crucial in SPSs. These systems require a range of privacy-preserving tools to handle sensitive datasets. Privacy concerns can be significantly addressed by using encryption methods to secure communication channels and maintain user anonymity, thereby providing higher levels of trust. However, it is essential to balance the need for privacy protection with system performance and usability considerations. More broadly, investigating the long-term effects of SPSs is important for scaling up such systems. Factors like scalability, maintenance costs, and community acceptance need to be carefully studied. While early deployment results may appear promising, it is uncertain whether the project will remain viable months after its initial launch. This includes considering economic, social, and environmental impacts and making system adjustments to align with changing needs and preferences over time.

Critical Analysis

The challenge to achieve seamless integration and operational efficiency in the landscape of SPSs faces certain significant bottlenecks. These challenges can range from maximizing the use of parking lots to guaranteeing top-notch security and prompt emergency assistance. At the same time, concerns about future scalability and sustainability are looming, prompting the need for solutions with fresh ideas. Overcoming these barriers is essential to realizing the wider utilization of SPSs and improving the urban mobility experience for all parties involved. Some of the critical challenges gathered from the research gap and their possible solutions have been elaborated as follows:

Optimizing Parking Space Utilization with IoT in SPSs

Another key challenge is insufficient space utilization, as areas of SPSs become over-congested, parking times increase, and the efficient use of urban space is undermined. However, the comprehensive nature of the IoT makes it a solution to this challenge as well. Traditional parking systems are often incapable of dynamically changing as demand shifts and cannot provide real-time data on space occupancy. In contrast, IoT-based solutions install sensors in each space that continuously monitor whether it is occupied and send information to a central database. This database sends information to user-friendly applications or signs, reducing the time spent searching for a spot and ensuring efficient space utilization. Moreover, IoT sensors allow for predictive analytics to optimize space utilization. By feeding information into machine learning algorithms that analyze historical data on space occupancy, operators can predict when demands are likely to peak. Such predictions could stimulate special offers or dynamic pricing, incentivizing users to use their spaces faster or choose other locations. IoT also enables parking reservations, allowing users to book a place in advance through a mobile application. This minimizes space waste as two or more reservations cannot be made for the same spot, ensuring efficient space allocation across a parking area. Lastly, IoT enables operators to plan parking facility expansion or new construction based on historical data on space usage. Thus, they will know where to add or remove spaces based on high or low occupancy rates.

Streamlining Payment and Ticketing Processes in SPSs Through IoT

Some of the recurring issues in SPSs are associated with issues of payment and ticketing, which has remained a concern to users as these lead to inconveniences and may take time. Payment methods and ticketing services that are used in the traditional setting may sometimes be rigid and may cause inconveniences which defeats the purpose of using SPSs. The possibility of applying IoT introduces the algorithms allowing to solve these issues and improve the general effectiveness and convenience in car parking areas. These systems make use of sensors for people and other devices for communication, hence allowing for automated and non-contact exchanges. Customers can even pay for parking through their mobile phones by linking their payment cards hence no use for tickets or money. This not only cuts down the number of transactions to be done but also increases the level of convenience for users in terms of time, which thoroughly supports the fast-evolving lifestyle of today's urban society. Further, IoT makes the application of the dynamic model in parking systems possible. Traditional fixed system may also not be able to produce accurate prices for the space or encourage efficiency of the spaces available.

This means that because of IoT sensors that are installed and used to detect the availability or otherwise of parking spaces, dynamic pricing models can be adopted. Prices can be varied depending on the demand, time or day, special occasion and hence, this will maximize the use of the parking lot and encourage the users to book wisely. This dynamic approach not only helps to improve the organization of the space, but offers another level of usability for them. When IoT is integrated with machine learning algorithms, user preferences can be anticipated such as recommending parking lot or offering best offers and deals from the nearest store. Such personalization is beneficial both in terms of increasing users' satisfaction and optimizing the use of parking assets. Smart sensors that are installed at the entry and exit points are capable of sensing the presence of vehicles as they enter and exit; this can make ticketing to be done programmatically thus not involving the users in that process. The discrete digital tickets people have access to through the IoT can be easily validated during the exit phase such that no enterprise physical tickets are needed. It gets rid of many human errors and at the same time increases the efficiency of parking lot businesses.

Enhancing Security in SPSs Through IoT Integration

Security is a complex issue in SPSs utilizing the IoT, including privacy of the data, infrastructure and users. Among the most important topics within the sphere of security there is one of the most important, and it is the protection of individual data. Internet of things (IoT) connected or otherwise, include sensors, cameras, and other hardware devices that are used to gather specific information about the movements of the vehicle, payment particulars and other information on the preferences of the users. As such, it is significant to guarantee the data's security at both the transfer and storage levels. Use of strong encryption mechanisms, undertaking security assessments and scanning frequently together with adherence to data protection laws are measures that would help to reduce the risk of acts of invasion and compromise in SPSs, thus protecting the identity of the users. However, the physical structure of smart parking facility must be protected against cyber- physical threats that may occur. Based on this, IoT devices remain vulnerable for attackers to gain an initial foothold within a user's system and expand their control from that point (Zhang et al., 2020). Cybersecurity has to become an integral part of the IoT-connected devices' design and management. This includes frequent firmware update, use of intrusion detection system and firewalls which makes form a strong barrier against cyber-crimes.

Further, access controls like, biometric mechanisms, secure tokens or key cards help to enhance security of parking facilities against intruders. Concerning user safety issues that may include cases involving smart parking lots or structures, real-time video surveillance assumes a very important place. Microphones and cameras connected to IoT can remain vigilant to monitor the parking lots or zones for any alert or possible intrusions. The operational efficiency of the IoT surveillance information is improved by the combination of the AI algorithms with security systems ability to identify threats and to react accordingly. It also makes the parking area safer for users as well as reducing cases of criminal activities within parking premises. Moreover, the incorporation of block chain into SPSs has provided a new technique which can improve the security especially in finances and data. Blockchain is a decentralized system to execute secure sales, and there is no chance of forgery or alteration of records. Implementing the use of blockchain when it comes to payment processes and the records of transactions make the smart parking system more secure and transparent while at the same time ensuring the users about safety of their transactions. SPSs can optimize security features for their improvement and at the same time to promote the general idea of the development of networks with more security and intelligence as cities.

Solution to Existing Problems

The proposed IoT integrated SPS has several ideal modules that are intended to improve the architecture of parking for users. Embedded in the center of the system is the sensor or detection module that is placed in each car parking space to measure the distance between two automobiles and an embedded camera mounted at the parking station entrance. This setup facilitates capturing of the vehicle reg number which is very important for parking management. They include; It should be noted that the sensor module works hand in hand with the Raspberry Pi Module which is the connection hub. It provisions Internet connection and is responsible for the management of all communication protocols used by individual sensors installed in the parking implementation. Moreover, the Raspberry Pi-module organizes the data processing and transfer so that the data can be updated and transferred in real time. Enhancing these components is the android module, ULINK, which the ULINK is accessible to the end-users through the mobile application. In this module, the users are provided with capabilities of easily search for parking stations around with available parking spaces. The android module, therefore, brings in more convenience to the parking process coupled with easy interaction with the IoT structures to improve the overall experience of parking. Altogether, these interrelated modules make up a reliable IoT environment fit for changing parking management, providing comfort, speed and real-time availability for operators and the users. Figure 2 below shows the general approach used in the development of a SPS.

Figure 2. Basic methodology for SPS

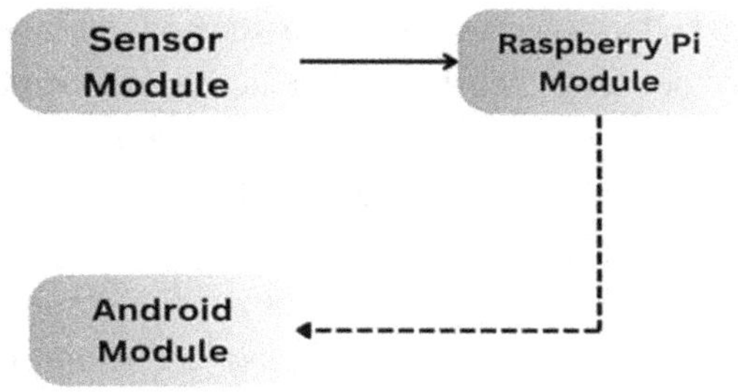

Cutting-Edge Technologies

The envisioned SPS, as illustrated in this paper, harnesses advanced technology aimed at maximizing the system's precision. Leading its list is OCR, an intricate technology designed to recognize characters from the license plate images with a high degree of accuracy. This OCR solution works effectively under open conditions, which means that when it comes to detecting license plate information it performs well despite the differing conditions (Saarika et al., 2017). In this capability is an Ultrasonic Sensor network, properly architected for stability and rarely making errors during long periods of operation. These sensors are therefore designed and tested based on related work to be reliable in their operations, thus improving the reliability of the developed system. Integrating all the parts efficiently and handling the data flow is an embedded controller which is the physical heart of the system, and it connects to cloud services.

This controller allows real-time processing and transfer of data throughout the system in order to promote higher system usability (Saharan & Bawa, 2020). Also, a high-quality camera incorporating advanced pixel technology facilitates easy and clear clicking of images making the system even more efficient. Through the utilization of these advance technologies the SPS is benefited with enhanced accuracy, reliability and efficiency thus revolutionizing the parking experience both from the service provider as well as the users' perspective. Besides focusing on the functional factors, the SPS depicts other non-functional factors that will make the output efficient and secure. The system aims to reduce the time taken by the users to look for the available parking space and the time they spend to look for their cars. It brings order in these procedures thereby improving user experience and eliminating the frustrations likely to be encountered while parking. Also, the system employs technological measures include capturing license plate image and extracting test in the instance of theft of car from parking slots. This goes a long way in safeguarding both drivers and parking operators especially with legal action and conflict-solving. Add to this is the impact it has towards improving the general security of the parking area through enhanced security measures such as surveillance and access control measures. Through efficiency, proper provision of evidence, and different security measures put in place, the SPS will guarantee efficiency and security to all the automobile owners and users of the parking lot.

The system algorithm controls and synchronizes the components of the hardware and incorporates cloud-related technologies into its work. First, the ultrasonic distance sensor serves as a primary one which helps in identifying distances and free parking slots with the help of sound pulse. This process entails setting parking slot status to mean when they are available, they are designated 1 otherwise 0. If an available slot is identified, the sensor checks the limit range for car identification and instructs the camera to take a picture in order for further analysis. After that the slot status is changed indicating that it could not be taken. In case the slot is still occupied, the system provides for the minimum distance needed for cars to exit and hence provides the necessary slot space for subsequent vehicles making entries. Moreover, as seen in Figure 3, through the integration of cloud vision API technology the capabilities of the system are boosted greatly. This API utilizes other features such as OCR engine recognition techniques and classification methods which are similar with those applied in training data-set. The system established in the cloud domain utilizes the cloud level processing to allow fast and precise assessment of image of vehicle. It entails making a key in the clouds API and then installing it into the Raspberry Pi system using a JSON address. The images that are taken by the camera are stored in cloud so as to get the text characters and numbers with extreme accuracy and without any delay. This has reinforced the system complexity hence efficient parking management and value addition through the use of technology among the users.

Figure 3. Integration of cloud-based API technology

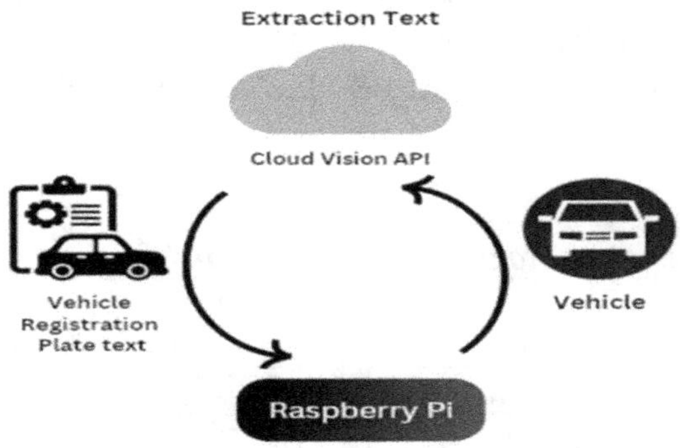

With regard to the security, transparency and availability of the parking system, SPS by utilizing the technology of blockchain can be employed to make the system secure. The correct measures must be taken in order to preserve the identity of the drivers' location. To conceal the drivers' places we employ the strategies of cloaking. The validators based on the blockchain send back available parking offer within the cloaked area. Last of all, the driver chooses the most favorable offer and immediately book with the particular parking place. A set of new analyses is performed to assess the proposed scheme; the conclusions show the feasibility of our scheme. Through the use of blockchain technology, the SPS scheme seeks to optimize parking and showcase ways of improving the same to users. In terms of system design, in the consortium blockchain network, which is under the control of authorized parking lots, parking offers transactions are recorded in the blockchain. Parking lots create their offers, which are then placed on the block chain, which contains such information as the location of a parking lot, the available parking spaces, the services that are offered along with the price rate.

All these offers are stored and safely indexed in the blockchain to make them very transparent and difficult to alter. Due to its decentralized structure and the fact that it is encrypted in cryptographic form, the data in the blockchain cannot be altered in any way. Furthermore, consensus = protocols like Raft guarantees agreement in the content to be stored in the ledger depending on the validators to provide reliability and trust in the system. For drivers, blockchain enables the access of available parking offers while at the same time keeping their identity concealed. Drivers need to ask information from the blockchain network to get parking options in the area of their choice while they wouldn't disclose their personal details. Once a specific parking offer is chosen, it is possible to book a parking space online, with the help of the blockchain-based platform of a high level of security and transparency. Moreover, parking payments in the cryptocurrency like Bitcoin will also increase privacy because users will avoid using their regular payment cards. In essence, blockchain technology is crucial to the SPS by facilitating effective parking services, improving the overall expedition of transactions, and providing increased anonymity to the drivers (Samantaray et al., 2021). Figure 4. Block chain in SPS works in as manner as:

Figure 4. Blockchain technology implementation in SPS

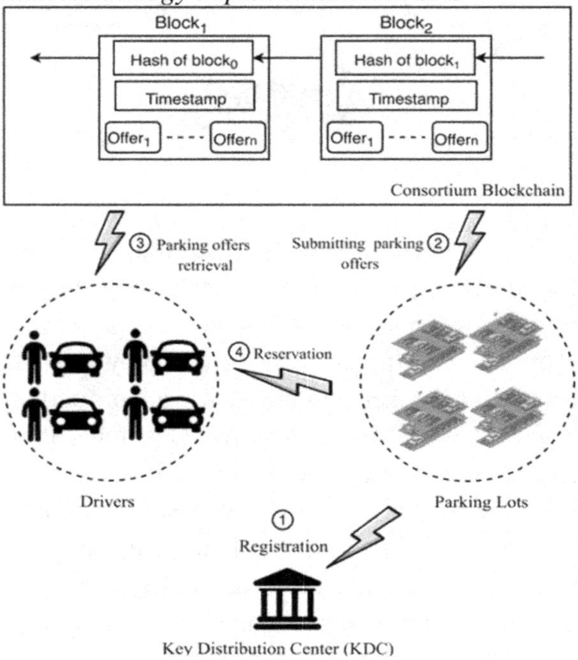

Application Operation

In Application Operation, a user's user name, user ID, password, and vehicle registration plate are required during the registration process before they may use the application. Apps deliver the features outlined in the system application architecture and function requirements when a user registers in. Prior to selecting a parking location, the user has the option to select a specific parking space and confirm whether it is available. Following the technique outlined, the application should notify the user's smartphone when they park, matching the automobile registration plate in the user's database with the retrieved text. The count of people who have parked in the allotted spot is being monitored by the timer system. When the user departed the parking space, they received another notice with the overall timer duration as shown in the following flow-chart of Application Operation in Figure 5.

Figure 5. Application operation basic fundamentals

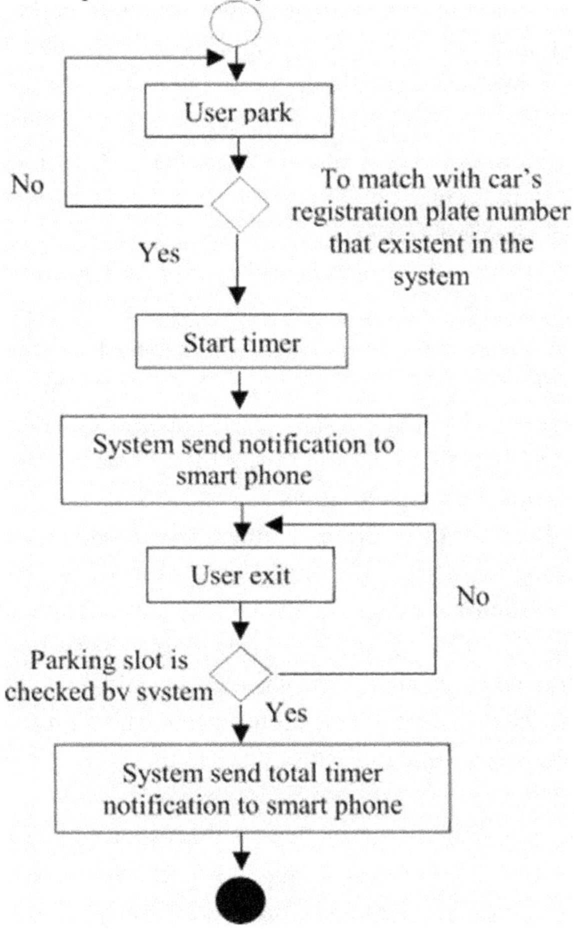

The application operation happens in a user experience process including registration, timely notification and effective tracking process of the parking. The different states in this process includes: The different states in this process includes:

a) User Registration and Login:

For one to be able to use the application, they are required to create an account on the application and fill a few basic details like the user's name, user identification number, password and the registration number of the car they own. It should be understood that this information is safely stored in the user database of the application in question. When registered, the users use their credentials to login and access all the functionalities that are present in the system architecture discussed above.

b) Parking Spot Selection and Verification:

Owners here will be able to see other available parking lots in the application and estimate their availability in real-time. This leads to the ability of the user to choose the most appropriate parking lot to park his/her car. Finally, before they decide to park their car, they can confirm the availability of the selected parking zone and avoid all the inconveniences relating to parking.

c) Notification Mechanism - Parking:

When the driver performs the parking successfully and in any given parking space, there is feedback mechanism that is activated. An appropriate message is sent to the users smart-phone based on the registration plate number in the database associated with the user. This message basically aims at acknowledging the started parking session and such may contain relevant information regarding the parking plan.

d) Timer Mechanism - Parking Duration:

The application incorporates a timer system once the users begin with the parking sessions. This mechanism makes it its business to ensure that time taken by each parked vehicle is well recorded in a manner that leaves no doubt on the number of users that have found their way into the designated area at any given time. Not only does the timer help with fastest space use, moreover it supports the collection and analysis of data for future improvements.

e) Departure Notification:

This is another notification which the users get once they leave the parking space. This covered message pertains to the general information concerning the parking session, principally on the time stayed. The departure notification concludes the parking experience of the user, and informs the user about the time spent which is an important feature for such applications.

f) User Database Maintenance:

The application focuses a lot on the user list specifically the information that is stored about a user. Users' data such as registration information and parking history are recorded meticulously. This repository is also used not only the current functionalities but is also used to build future things that relate to the interaction of the parking system, so that every parking session should provide an efficient and personalized user interface for the user.

FINDINGS AND DISCUSSION

The proposed SPS incorporates components of advanced technologies to improve the experience of the users as well as the overall management of parking. These are the sensor module, Raspberry pi module and android module to give real time information on space available for parking, control communication between all sub-systems and System control and mobile application for easy access to parking facilities. Therefore, it shows a great potential of automating parking management, thus providing more convenience, efficiency and security to operators and users of the parking lots. These aspects include its technology deployment part as well as the user orientation in facilitating a safe and effective parking area that complies with smart city goals and urban growth policies. Some of the improvements brought in by this system are as follows: Some of the improvements brought in by this system are as follows:

- Efficiency Enhancement: The mentioned proposed SPS has the added advantage of increasing efficiency owing to the fact that search times for parking spaces will be brought down, parking processes will be made efficient, and users will be able to access real time updates. This can lead to decrease in traffic flow and thus the number of times a user has to interact with the interface, which in turn increases user satisfaction.
- Security Enhancement: Having strict security measures along with providing strong evidence in case of any event like theft of a car, the system provides better security in the parking area. This can assist in discouraging criminals to perpetrate their acts thus enhancing safety for vehicles and users.
- User Experience Improvement: It can also include such functions as real time notifications, easy selection of the parking spot, and accurate parking time track. This may result in a greater number of users using the parking facility hence increased satisfaction among the users.
- Data-Driven Insights: From the durations the system will record and track the parking durations as well as interactions with the users, data-driven recommendations for further optimization can be made. This can in turn assist parking operators in making right decisions that would further enhance ROI for parking services as well as the general user satisfaction in the long run.

With the growing IoT technology, the governments and researchers have adopted the use of new solutions in managing the parking system. Policy initiatives also highlight the relevance of IoT based policy schemes, regulations and public-private partnerships to promote the use of IoT solutions. In the smart city infrastructure projects and smart city programs Governments want to renovate and improve public areas and upgrade services such as parking lot services as well. In the same regard, funding programs and policy incentives encourage IoT adoption and bring behavioral changes among the drivers and the parking operators. In research, scholars have focused on the technical aspects as well as IoT to enhance parking operation and management. Proposals of their solutions can be based on real-time tracking, risk analysis, and features with focus on the end-users to improve its operations. With the help of IoT sensors for occupancy measurement with the use of ML and AI for predictive analytics for better space management and improved users' experiences. In addition, they discuss the security and privacy issues and suggest the applied encryption approaches and safe communication channels. In conjunction with this support system and technical watts, interdisciplinary cooperation opens the door to improved and balanced perspectives on management of parking facilities and the provision of positive impacts to the operation and sustainability of cities.

Implications and Considerations

Integration of an SPS with IoT technology affects the planning process and policy formulation of many cities. Due to high traffic density and acute parking shortage in urban settings, SPS can apply creativity to help manage the cities more efficiently. Implementation of such systems entails a reformulation of how parking lots are planned and used for spaces in urban centers, affecting elements of urban design, transport legislation, and smart city.

Urban Planning Implications

Optimized Land Use: Thus, through the implementation of SPS, cities are in a position to properly use available space in that the need for vast parking areas is removed. Proper design of parking lots enables the modification of conventional parking lots into versatile infrastructures for various urban requirements like vegetation space or business and public uses. Realtime analysis and forecasting technologies can help the city engineers understand the existing parking demand and also redesign the parking space along with other competing land uses.

Reduced Traffic Congestion: It is a well-known fact that one of the major sources of traffic congestion within urban areas is drivers in the process of seeking for parking space. SPS informs drivers instantly about parking spaces within proximity and can decrease a time parking hunters spend on intersecting roadways, and thus, traffic jams. This has a direct bearing on the on the traffic circulation management and will aid in controlling the emission of vehicles, an area of importance in mayor city solutions to sustainable actions with lesser carbon impact.

Enhanced Public Safety and Security: The use of surveillance technologies and systems including high quality camera systems and OCR systems for LPR improve security in parking zones. This may reduce incidences of crimes, acts as a proof in situations that require it such as theft and disputes and generally, enhance safety of people. Other policies in urban areas could adopt SPS coupled with smart city strategies that intends to uses technology to increase security within cities.

Policy Considerations

Data Privacy and Security Regulations: Some of the key requirements with the implementation of SPS include gathering and interpreting massive amounts of data such as vehicle as well as the users' information. It therefore becomes the responsibility of the policy makers to ensure that data privacy and security measure are put in place that will help to prevent access and use of the user data by unauthorized persons. This includes the application of the law on the protection of personal data such as CCPA in California or GDPR in Europe.

Public-Private Partnerships: SPS that is successfully implemented tend to involve partnerships between public authorities and other players in private sector. Regulations should support such relationships by providing such entities with bonuses, such as tax exemptions or provision of subsidies for smart parking solutions. This approach maybe useful in extending the deployment of SPS across the regions within the urban areas.

Standardization and Interoperability: Based on these arguments, the omission of these aspects, which was identified as a problem when developing SPS for different areas and systems, means that policymakers need to pay attention to the creation of standards for interoperability. That will enable different systems of parking management to share information and work in harmony and enable the users an integrated and effective parking.

Real-World Case Studies

Barcelona, Spain: Barcelona stands out as one of the most advanced smart cities and its smart parking is a proof. The city apply IoT sensors installed in the parking area that can indicate availability of parking space in real time. This data is then used in an application that assists drivers in where to find parking space and this significantly reduces traffic jam and pollution. Hence the system has improved the utilization of available parking space and at the same time extensively improved upon the transportation network of the city.

San Francisco, USA: Another example of the use of SPS is San Francisco's ''SFpark'' program. Some of these features include the use of sensors as well as the ability to vary the prices based on the demand for the parking lots. By adopting the policy of varying parking prices throughout the day the city has been able to solve traffic congestion problem and increase the uptake of public transport. Information acquired from SFpark is also used to guide planning of city's parking spaces, for instance by identifying areas where there is a need to increase or reduce the available spaces.

Amsterdam, Netherlands: The smart parking management system in Amsterdam is connected with the other smart city systems in place. Technologically, the parking operates in real-time with the help of ANPR and sensors for the handling of lots and car park spaces. This information also used by the city as a tool to enforce the various parking policies as well as to manage and organize parking space effectively. Amsterdam's approach also shows how each ST element can be linked with other smart city subdomains such as smart parking with traffic control and public safety subdomains.

Practical Applications

However, a befitting sarcasm based on the implementation of SPS can easily go beyond just about having the parking lots identified. They include:

Dynamic Pricing Models: DNT feature permits price fix as per the need, which would enable the cities to regulate the park space more efficiently and compel people to adopt other means of transport other than parking.

Enhanced User Experience: The mobile apps help the users get the real time updated information about parking availability, payment method and even the directions to the nearest open parking space making the parking experience a more improved one.

Integration with Smart City Infrastructure: SPS can be integrated with other smart city systems such as traffic management, public transport and emergency services, and as a result, made a smart city system that is efficient.

With these implications of urban planning, policy concerns, and real-world scenarios in mind, cities can purposefully incorporate schematic power studies in order to promote urban mobility, mitigate traffic, strengthen security and promote the creation of a more effective urban fabric.

LIMITATION AND FUTURE SCOPE

The introduction of SPSs presents an appealing solution to parking problems but comes with limitations. A substantial barrier to the widespread adoption of SPSs is the high cost and expensive configuration required for implementation. With rapid technology turnover, significant system upgrades are required relatively often, quickly driving up what is typically a very high total cost of implementation. Moreover, combining components such as sensors, RFID, and related devices into a single system raises consumer costs. To address these problems, researchers have proposed solutions to reduce costs and ease deployment. One solution could involve using IoT technology as a means to reduce or even eliminate deployment overheads. Researchers aim to reduce the financial hurdle by developing affordable sensors and communication devices. Additionally, investing in the creation of open-source software platforms for SPSs should be encouraged, offering cities and parking operators more tailored solutions while reducing dependence on costly proprietary programs.

Finally, wireless communication technologies such as LoRaWAN and NB-IoT could bridge the gap by allowing for widespread coverage without the need for expensive infrastructure. In addition, the use of privacy-preserving data management strategies, like encryption and decentralized storage, addresses concerns about security while also protecting user information. Moreover, the incorporation of offline capabilities and redundancy mechanisms in SPSs mitigates connectivity failures, allowing for continued operation even under limited internet availability conditions. Through community outreach and educational programs, researchers aim to educate people about smart parking technologies, increasing awareness and reducing resistance to change, which in turn promotes acceptance and adoption of these new technologies. A standout feature of SPS is its capacity to optimize parking space usage by informing drivers in real-time about available parking, which eliminates time wasted on searching for a spot. On the other hand, SPS offers features like online payment for parking, notifications to drivers after their parking time has elapsed, and a car searching functionality for large parking lots. Additionally, SPS contributes to a cleaner environment by reducing greenhouse gas emissions and promoting the use of clean energy, such as solar power, which reduces the carbon footprint and serves as a low-cost energy source. SPS also addresses security challenges by incorporating surveillance and access control, increasing safety in parking areas. SPS is headquartered in the US and is currently operating across different industry segments (commercial, government) and territories globally, providing opportunities not only for product innovation but also for market expansion in smart parking.

CONCLUSION

In conclusion, SPS powered by the IoT is poised to revolutionize urban transportation and make a dent in old age parking problems. These solutions leverage IoT-enabled tools like sensors, live monitoring, predictive analysis, and blockchain architectures to optimize the demand-supply gap for parking spaces while improving overall end-user satisfaction levels, even as they align with fostering sustainable urban developments. SPS is very flexible, and it promotes benefits beyond just convenience. They are consistent with larger aims, such as reducing urban traffic and minimizing environmental effects. Nevertheless, there is still much work that must be done in order to address some challenges for the next generation of research and innovation. These concerns include security, accuracy, scalability, and sustainability. Addressing these obstacles will be crucial for realizing the promise of SPSs and that they can effectively become part of an urban landscape.

In essence, SPSs pave the way for smarter and more efficient eco-cities. As the enablers continue, they pave the way for a future of improved transportation systems in cities. Essentially, IoT-powered SPSs transcend mere parking management. About using technology to make our cities more livable. They are the real-time applications that help to take decisions based on realistic data, which ultimately focus on improving end-user satisfaction. And they are in favor of sustainable urban development, reducing congestion and the environment alike. Still, this ascension to social-safety-net holy grail is not easy. Common concerns grow without security, accuracy, or sustainability. They represent opportunities for more research and innovation.

However, it is not all doom and gloom for the future of SPSs. Collaboration on and commitment to the right measures, after all, could lead us towards smarter cities that are less energy-hungry. There will be different stakeholders — the policymakers, the urban planners, and then of course, the technology companies. Together, they can deliver comprehensive answers to the adaptations of urban communities. To sum up, SPSs powered with IoT are the next biggest leap we must take to make our cities better to live in. They demonstrate how technology can help solve everyday issues and improve the lives of all urban residents. And as we press on to learn more and improve; there is no question that these systems will play a vital part in shaping the future of city mobility.

REFERENCES

Abu-Alsaad, H. A., & Al-Taie, R. R. K. (2024). Smart Parking System Using IoT. 2024 16th International Conference on Electronics, Computers and Artificial Intelligence (ECAI), 1–7. DOI: 10.1109/ECAI61503.2024.10607419

Aditya, A., Anwarul, S., Tanwar, R., & Koneru, S. K. V. (2023). An IoT assisted Intelligent Parking System (IPS) for Smart Cities. *Procedia Computer Science*, 218, 1045–1054. DOI: 10.1016/j.procs.2023.01.084

Ajchariyavanich, C., Limpisthira, T., Chanjarasvichai, N., Jareonwatanan, T., Phongphanpanya, W., Wareechuensuk, S., Srichareonkul, S., Tachatanitanont, S., Ratanamahatana, C., Prompoon, N., & Pipattanasomporn, M. (2019). Park King: An IoT-based Smart Parking System. 2019 IEEE International Smart Cities Conference (ISC2), 729–734. DOI: 10.1109/ISC246665.2019.9071721

Ampuni, A., Fonataba, S., Fitrianto, A., & Wang, G. (2019). Smart Parking System With Automatic Cashier Machine Utilize the IoT Technology. 2019 International Conference on ICT for Smart Society (ICISS), 1–4. DOI: 10.1109/ICISS48059.2019.8969793

Badhoutiya, A., & Saxena, A. (2022). SPS: A framework Of Smart Parking System Using IoT Technology. 2022 6th International Conference On Computing, Communication, Control And Automation (ICCUBEA, 1–4. DOI: 10.1109/ICCUBEA54992.2022.10011086

Bandung, Y., Aldiansyah, M. F. N., Dwi Putra, M. R., & Syiraaj, M. I. I. (2023). Design and Implementation of IoT-Based Smart Parking System in Campus Area. 2023 10th International Conference on ICT for Smart Society (ICISS), 1–6. DOI: 10.1109/ICISS59129.2023.10292040

Barriga, J. J., Sulca, J., León, J. L., Ulloa, A., Portero, D., Andrade, R., & Yoo, S. G. (2019). Smart Parking: A Literature Review from the Technological Perspective. *Applied Sciences (Basel, Switzerland)*, 21(21), 4569. DOI: 10.3390/app9214569

Dheeven, T. A., Kumar, P. M., Venkatesh, V., & Sailaja, K. A. I. (2024). IoT based sensor enabled vehicle parking system. *Measurement. Sensors*, 100953, 100953. Advance online publication. DOI: 10.1016/j.measen.2023.100953

Fox, G. C., Kamburugamuve, S., & Hartman, R. D. (2012). Architecture and measured characteristics of a cloud based internet of things. *2012 International Conference on Collaboration Technologies and Systems (CTS)*, 6–12. DOI: 10.1109/CTS.2012.6261020

Fraifer, M., & Fernstrom, M. (2016). Smart car parking system prototype utilizing CCTV nodes: A proof of concept prototype of a novel approach towards IoT-concept based smart parking. 2016 IEEE 3rd World Forum on Internet of Things (WF-IoT), 649–654. DOI: 10.1109/WF-IoT.2016.7845458

Gagan, S., & Ravi, H. K. (2024). ParkNest - The Smart Parking System Using IoT. *2024 International Conference on Smart Systems for Applications in Electrical Sciences (ICSSES)*, 1–6. DOI: 10.1109/ICSSES62373.2024.10561362

Herrera-Quintero, L. F., Vega-Alfonso, J., Bermudez, D., Marentes, L. A., & Banse, K. (2019). ITS for Smart Parking Systems, towards the creation of smart city services using IoT and cloud approaches. *2019 Smart City Symposium Prague (SCSP)*, 1–7. DOI: 10.1109/SCSP.2019.8805705

Jabbar, W. A., Tiew, L. Y., & Ali Shah, N. Y. (2024). Internet of things enabled parking management system using long range wide area network for smart city. Internet of Things and Cyber-Physical Systems, 82–98. DOI: 10.1016/j.iotcps.2023.09.001

Lu, R., Lin, X., Zhu, H., & Shen, X. (2009). SPARK: A New VANET-Based Smart Parking Scheme for Large Parking Lots. *IEEE INFOCOM*, 2009, 1413–1421. Advance online publication. DOI: 10.1109/INFCOM.2009.5062057

Mendiratta, S., Dey, D., & Rani Sona, D. (2017). Automatic car parking system with visual indicator along with IoT. *2017 International Conference on Microelectronic Devices, Circuits and Systems (ICMDCS)*, 1–3. DOI: 10.1109/ICMDCS.2017.8211549

Mohapatra, H. (2021). Socio-technical Challenges in the Implementation of Smart City. 2021 International Conference on Innovation and Intelligence for Informatics, Computing, and Technologies (3ICT), 57–62. DOI: 10.1109/3ICT53449.2021.9581905

Mohapatra, H., & Dalai, A. K. (2022). IoT Based V2I Framework For Accident Prevention. 2022 2nd International Conference on Artificial Intelligence and Signal Processing (AISP), 1–4. DOI: 10.1109/AISP53593.2022.9760623

Mohapatra, H., & Rath, A. K. (2021). An IoT based efficient multi-objective real-time smart parking system. *International Journal of Sensor Networks*, 4(4), 219. DOI: 10.1504/IJSNET.2021.119483

Mohapatra, H., Rath, A. K., & Panda, N. (2022). IoT infrastructure for the accident avoidance: An approach of smart transportation. *International Journal of Information Technology : an Official Journal of Bharati Vidyapeeth's Institute of Computer Applications and Management*, 2(2), 761–768. DOI: 10.1007/s41870-022-00872-6

Revathi, G., & Dhulipala, V. R. S. (2012). Smart parking systems and sensors: A survey. *2012 International Conference on Computing, Communication and Applications*, 1–5. DOI: 10.1109/ICCCA.2012.6179195

Saarika, P. S., Sandhya, K., & Sudha, T. (2017). Smart transportation system using IoT. *2017 International Conference On Smart Technologies For Smart Nation (SmartTechCon)*. DOI: 10.1109/SmartTechCon.2017.8358540

Saharan, S., Kumar, N., & Bawa, S. (2020). An efficient smart parking pricing system for smart city environment: A machine-learning based approach. *Future Generation Computer Systems*, 106, 622–640. DOI: 10.1016/j.future.2020.01.031

Samantaray, M., Biswal, A. K., Singh, D., Samanta, D., Karuppiah, M., & Joseph, N. P. (2021). Optical Character Recognition (OCR) based Vehicle's License Plate Recognition System Using Python and OpenCV. 2021 5th International Conference on Electronics, Communication and Aerospace Technology (ICECA), 849–853. DOI: 10.1109/ICECA52323.2021.9676015

Tamam, I., Wang, S., & Djahel, S. (2020). An IoT-based Eco-Parking System for Smart Cities. 2020 IEEE International Smart Cities Conference (ISC2), 1–6. DOI: 10.1109/ISC251055.2020.9239041

Xu, W., Yan, C., Jia, W., Ji, X., & Liu, J. (2018). Analyzing and Enhancing the Security of Ultrasonic Sensors for Autonomous Vehicles. *IEEE Internet of Things Journal*, 6(6), 5015–5029. DOI: 10.1109/JIOT.2018.2867917

Yazıcı, A. B., & Ünverdi, N. Ö. (2024). LoRa Technology Overview and Smart Parking System Design. 2024 32nd Signal Processing and Communications Applications Conference (SIU), 1–4. DOI: 10.1109/SIU61531.2024.10600948

Zhang, C., Zhu, L., Xu, C., Zhang, C., Sharif, K., Wu, H., & Westermann, H. (2020). BSFP: Blockchain-Enabled Smart Parking With Fairness, Reliability and Privacy Protection. *IEEE Transactions on Vehicular Technology*, 6(6), 6578–6591. DOI: 10.1109/TVT.2020.2984621

Chapter 7

Development of Analytical Approaches Using Smart Card Data While Investigating Travel Behaviors in Terms of Space and Planning: Izmir Case

Hakan Uzun
https://orcid.org/0009-0005-8465-5016
Dokuz Eylul University, Turkey

K. Mert Cubukcu
https://orcid.org/0000-0003-3604-7014
Dokuz Eylul University, Turkey

ABSTRACT

The objective of this study is to ascertain the distinctions in the mobility patterns of elderly individuals and to formulate recommendations that will enhance the accessibility and suitability of public transportation systems for this demographic. To achieve this objective, data from the electronic fare collection system used in public transportation in Izmir city for elderly individuals with smart cards has been utilized. To investigate the variables that differentiate the mobility of elderly individuals, one-way variance analysis (ANOVA) and Student's T-test analysis were employed. The results demonstrate that elderly individuals are not a homogeneous group;

DOI: 10.4018/979-8-3693-6695-0.ch007

gender, age group, and distance from the city center have an impact on mobility. The issue of increasing the mobility of elderly individuals will have an impact on spatial planning and transportation planning.

BACKGROUND

The issue of transportation is a pervasive challenge faced by urban centers across the globe. The exponential growth in the number of vehicles on the road is causing significant issues in urban areas, particularly in traditional city centers. Concurrently with the growing economic significance of expedient human and material mobility, governmental and municipal authorities are allocating substantial financial resources to transportation infrastructure development. The most traditional approach to addressing transportation issues is to enhance public transportation while reducing reliance on private vehicles. In this context, it is inevitable that developed and developing countries with aging populations will seek to comprehend the transportation behaviors of elderly individuals with a high level of vehicle dependence.

The average age in the world, especially in industrialized countries, is gradually increasing (Smith & Sylvestre, 2001). According to projections for 2025, it is predicted that these countries will reach twice the rate of population over 65 years (Alsnih & Hensher, 2003; Boerema, van Velsen, Vollenbroek-Hutten, & Hermens, 2016; Rosenbloom, 2001). According to World Health Organization data, it is estimated that by 2050 there will be over 1.5 billion people over 65 years of age worldwide (World Health Organization [WHO], 2011).

It can be stated that there is currently a predominant time-geography perspective, which has placed the human being at the center of analysis and planning for transportation systems since the 1970s. In conjunction with the quantitative revolution, some urban and regional planners have discussed the optimal physical plan concept that can be achieved through the utilization of diverse mathematical methodologies.

With the aim of sustaining their lives, people move in space and time. Restrictions on space and time are determinative of people's transportation behavior. Each point that people can reach from the starting point to the ending point at maximum speed is called the "potential space-time path." The cluster formed by all these paths constitutes the potential space-time prism of the person. Space-time paths taken by an individual are controlled by constraints on these space-time prisms (Hagerstrand, 1970; Steven, 1997).

The distance traversed constitutes a significant element in the evaluation of this potential route. The most crucial aspect of travel behavior research is the active engagement with time and space. This enables a tangible link to be made between individual decisions and the contextual constraints that inform them. In considering the potential space-time prisms of individuals, three principal constraint groups may be identified. Capacity constraints encompass the psychological and biological factors that impede human mobility. Relational constraints, given that individuals' space-time paths intersect, exert an influence on one another. Authorization restrictions pertain to the rules, laws, or economic barriers that govern access.

In parallel with the adoption of individuals at the center of human behavioral analyses (transport-oriented), the space-time approach described above has begun to be used to examine the behavior of elderly individuals (Pred, 1977; Mercado & Páez, 2009; Cullen, 1976).

It is of significant importance at this stage to examine the impact of an aging population on transportation and urban use systems, and vice versa, in order to gain insight into how mobility patterns of the elderly are affected. Elderly people maintaining their active lives necessitates that transportation systems examine the needs of this demographic (Alsnih & Hensher, 2003). Considering studies showing that mobility is directly related to quality of life, the importance of this issue is increasing (Boerema et al., 2016).

As previously stated, transportation and urban use systems have an impact on the mobility of elderly individuals. The lack of access to transportation among the elderly population increases the likelihood of engaging in low-level physical activity, which in turn leads to a decline in independence and an exacerbation of health issues (Hess, 2009).

All in all, individuals over 65 years of age are considered to be elderly. In some studies, those older than 60 years can be considered elderly (Hess, 2009). However, in some studies, this wide range has been divided into two groups: new elderly (65-75) and old elderly (75+) (Alsnih & Hensher, 2003). Many studies have found that the mobility of elderly individuals differs from that of other age groups (Newbold, Scott, Spinney, Kanaroglou, & Páez, 2005; Zhang, Mao, Liu, Chen, & Guo, 2007).

The objective of this study is to ascertain the distinctions in the mobility patterns of elderly individuals and to formulate recommendations that will enhance the accessibility and suitability of public transportation systems for this demographic.

LITERATURE

The concept of behavior can be explained as all kinds of signs caused by numerous stimuli originating from both internal and external sources of the organism (Eroğlu, 2017). Behavioral science, in its effort to explain the areas it focuses on, is in relation with psychology, sociology, anthropology, and other disciplines.

For transportation planners, information about personal travel choices, which is necessary for increasing the mobility and efficiency of countries, developing options, and forecasting demand, becomes crucial (Miller, Hoel, Goswami, & Ulmer, 2006). This is because it is possible to influence people's transportation behavior through various policies (Elker, 1999).

Behavior refers to a person's movements that arise as a result of interactions (motivation) triggered by stimuli from within or outside. In studying the reasons behind people's behavior, it is essential to examine the motivations that drive them (Baydar, 2007). The sources of motivations are given in Figure 1, and these sources have three dimensions: physical, mental, and social.

Figure 1. The origins of motivations (Baydar, 2007)

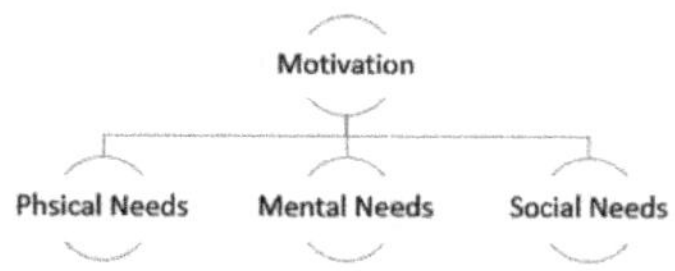

When transportation planning is approached from the perspective of understanding travel behavior, it becomes evident that the foundation of studies is the identification of the needs that form the source of mobility and the efforts to reveal the effects of the parameters that trigger these needs.

When examining definitions of mobility, we can see that it encompasses all types of movements, from waking up in the morning and moving from bed to a chair to walking for errands and leisure, working, exercising, driving a car, and using various forms of public transport (Boerema, van Velsen, Vollenbroek-Hutten, & Hermens, 2016; Rantakokko, Mänty, & Rantanen, 2013). Additionally, mobility must also be regarded as the ability to move (Giuliano, Hu, & Lee, 2003). This definition effectively explains the concept, although the dimension of accessibility is often evaluated in studies examining transportation behaviors related to mobility. Accessibility refers to the ease with which people can reach their destinations. Although mobility and accessibility are similar concepts, they are distinct. For example, two points that are close together may have high accessibility but low mobility value (Levine & Garb,

2002; Giuliano et al., 2003; Metz, 2000). Mobility can be viewed as a space that provides links among places (Tim Schwanen & Páez, 2010). Whether individuals have the necessary knowledge, equipment, infrastructure, and ability to travel where and when they want for any purpose constitutes the dimensions of activity. Mobility refers to the freedom, independence, and compatibility of action (Burns, 1999).

In the contemporary era, the majority of the global population is concentrated within urban settlements. The concept of mobility is increasingly significant in terms of its impact on the economic, social, and geographical development of cities (Albalate & Bel, 2010).

Mobility is closely linked to one's independence, well-being, and quality of life (Spinney, Scott, & Newbold, 2009; Tacken, 1998). Classical travel demand and supply analysis aims to improve mobility by evaluating it as access to people and places. It is also known that mobility has psychological and physiological effects on individuals, as well as impacts on socialization, life satisfaction, and morale enhancement (Metz, 2000; Alsnih & Hensher, 2003; Dickerson et al., 2007; Whelan, Langford, Oxley, Koppel, & Charlton, 2006). Therefore, the mobility of elderly individuals should be studied in greater detail (Ipingbemi, 2010). Human beings are social entities; mobility is not only a personal issue but also related to family and society.

The concept of elderly mobility is multifaceted and encompasses a range of dimensions, as illustrated in Figure 2. The initial factor to consider is the personal characteristics of elderly individuals (Spinney et al., 2009). Personal characteristics include variables such as age, gender, ethnicity, income, educational status, and vehicle ownership. It has been demonstrated that the propensity to travel among the elderly significantly decreases with increases in both travel time and age. The mean journey distance for elderly individuals is lower than that observed for other groups (2.9 km). Most journeys (75.5%) end at a distance of less than 2 km. The most common pattern among the elderly is returning home after shopping, which represents 45-48% of all journeys. The second most common type of trip is home-leisure-home, accounting for 20-26% of all trips (Zhang et al., 2007).

Figure 2. Dimensions of elderly mobility

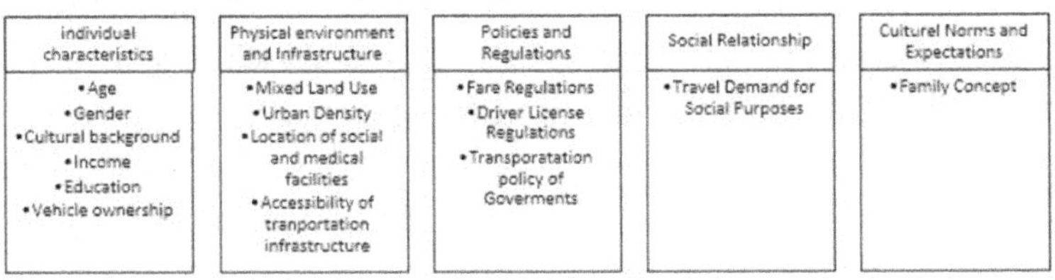

When investigating parameters related to average travel distance, it is evident that the travel behaviors of the elderly vary depending on factors such as gender, ethnicity, and geography. For instance, Black individuals and older men are more likely to use public transportation and travel shorter distances (Mercado & Páez, 2009). Research has found that elderly women travel less frequently than elderly men (Zhang et al., 2007). A study in the Netherlands showed that elderly individuals are more active in social travel than average and are as mobile as younger age groups. Other effective factors include gender, household structure, education level, car ownership, disability, city density, distance, and social activity (Cho, Sohn, & Park, 2016; Zhang et al., 2007; Hu, Wang, & Wang, 2013).

The second dimension of elderly mobility is the physical environment, including transportation and communication infrastructure, as well as other related infrastructures. This dimension can be addressed in two main topics. First, the effects of urban density and diversity of land use are examined. Mixed land use, for example, reduces the average distance traveled by private vehicles (Mercado & Páez, 2009). Factors such as urban intensity, distance, and social activity are also influential. Moreover, the locations of social assistance facilities, sports facilities, and medical facilities are crucial land use variables for elderly mobility (Choo et al., 2016).

Studies focusing on the accessibility of transportation infrastructure have determined that the perceived walking distance to transportation options impacts the frequency of their use. For instance, every five-minute increase in perceived walking distance reduces public transport use by 5% among non-motorized elderly individuals and by 25% among elderly drivers. A survey conducted in London by Su & Bell (2009) found that elderly people use fewer stops and destinations for shopping trips. Srichuae, Nitivattananon, & Perera (2016) discovered that public transport within walking distance significantly affects travel behaviors. It is emphasized in the studies that urban development and planning are crucial to support the public domain hierarchy and the mobility of the elderly. Traditional fixed-route public transport is also questioned for its limited ability to meet the needs of the elderly (Alsnih & Hensher, 2003).

The third dimension of elderly mobility involves policies and regulations. High transportation costs, lack of bus stops, inadequate roads, traffic congestion, insufficient attitudes of commercial transport operators, and lack of state transportation services have been identified as limitations for elderly people (Olawole & Aloba, 2014). In some countries, older drivers' licenses may be confiscated, making pedestrian and public transportation modes more important for elderly people who lose their driver's licenses (Su & Bell, 2009).

The fourth dimension of elderly mobility is social relations, including family, friends, and society. Social characteristics significantly affect mobility (van den Berg et al., 2011).

Cultural norms and expectations represent the fifth dimension of elderly mobility. The cultural differences between societies have a significant impact on the mobility of the elderly. Particularly, the values of Western and Eastern communities differ. The concept of family and the role of elderly individuals within it is crucial in this context (Chang, 2013).

The main components of mobility are the number of journeys, their length, and their duration. Travel length is geo-referenced and considered a measure of sustainable transport, mobility, and an indicator of quality of life in many countries, reflecting surrounding environmental factors (Mercado & Páez, 2009; Tacken, 1998).

Measuring all aspects of mobility is challenging. Studies often focus on the number, duration, and length of journeys rather than their impacts on psychological and physiological health (Metz, 2000). Additionally, research often examines the mode and timing of mobility to understand the character of travel (Choo et al., 2016). Studies examining travel distance as a parameter for measuring mobility show that elderly individuals tend to travel shorter distances compared to younger individuals (Zhang et al., 2007; Rouwendal & Rietveld, 1994). The relationship between travel distance and age is often direct, though other factors may also influence this relationship, highlighting the strong effect of age (Schwanen, Dieleman, & Dijst, 2001).

Active aging is defined as the process of optimizing health, participation, and security opportunities to improve quality of life as individuals age (WHO, 2002). Mobility is crucial for health and daily life participation, with travel distance serving as an important measurable indicator of daily life engagement and mobility. Transport and self-maintenance capabilities are integral to independence (Schaie, Boron, & Willis, 2005). Literature reveals that various factors affect the mobility of older people. Most studies treat the elderly as a single group and compare them to other age groups. This study explores whether factors affecting mobility vary within different age groups designated as elderly or according to the urban areas in which they live.

A review of the literature reveals that it is challenging to comprehensively assess all dimensions of mobility in a single study. Researchers, in previous studies, have only been able to explore this complex structure from specific perspectives, based on the data available to them. The majority of these studies have focused on developed countries, where aging populations are growing rapidly, while research on developing countries, particularly those like Turkey, remains limited. In this context, the aim of our study is to contribute to the understanding of the mobility of elderly individuals in Turkey and other developing nations. Furthermore, while earlier research has emphasized the conflict between private vehicle use and public transportation, there has yet to be a study that specifically examines the mobility of elderly individuals within the framework of public transportation systems.

DATA

Izmir is the third most populous city in Turkey, with a population exceeding 4 million. The Izmir Metropolitan Municipality manages public transportation throughout the city, which includes a comprehensive network of bus, ferry, subway, and light rail systems. All public transport systems operated by the municipality are integrated with the ESHOT General Directorate's "electronic fare collection system." This system allows for the retrieval of data on the temporal use of user cards (based on ID verification) and the points of transfer within the transportation network. The data is geospatial, aligned with the coordinates of the respective stations, and enables the analysis of passenger behavior in each neighborhood and district. Comparing this information with other geographical features significantly contributes to understanding the factors influencing transport behavior.

The electronic fare collection system includes both standard and personalized cards. Personalized cards are used by students, teachers, passengers aged 60 to 65 years, and those aged 65 years and above. In Izmir, there are 399,294 individuals aged 65 years or older. This study focused on the travel movements of passengers over 65 who possessed personalized cards during a three-month period, from January to March 2015. The number of cardholders aged 65 and above was recorded as 163,924. The electronic fare collection system contains data on the age, gender, and place of residence of cardholders with personalized cards.

For each personalized card:

- Card Transaction Information
- Travel Information
- Line and Stop Information

are available.

When card usage was examined, it was found that, although there are 163,924 cards registered in the system throughout the city, the cards of 110,004 individuals were used at least once during January, February, and March 2015.

To determine the sample size, a reliability level of 95% was set for the study. The sample size for the 110,004 individuals was calculated to be 1,057 using the minimum sample size formula. Out of approximately 110,000 people, 1,057 were selected using a simple random sampling method. Instead of using a random number table, functions included in Excel software were employed.

The riders were tracked over a period of three months, and distances were measured using Esri ArcMap (GIS) software and Google Maps algorithms. The mileage information was obtained as one of the key indicators of mobility.

It was also decided that elderly people would be informed about their home location using the addresses they have officially registered, rather than the stops they have boarded. By employing the geocoding method, addresses were identified using latitude and longitude information from available software. Geocoding is a geographic coding process that translates addresses into geographical coordinates using online open-access sources.

In this study, the total distance traveled by elderly individuals has been determined as the dependent variable. It has been observed that travel distance has also been utilized as a variable in previous studies (Mercado and Páez, 2009; Zhang et al., 2007; Schwanen et al., 2001). Within the public transportation system, travel can be accomplished using metro, suburban, bus, and ferry systems. The distances traveled have been calculated by analyzing personalized card data from the electronic fare collection system. Cardholders present their cards when boarding any mode of public transportation within the system. They do not re-scan their cards when exiting the system. Therefore, it is necessary to estimate individuals' disembarkation points based on their subsequent boarding actions. The journey chain method has been employed to predict destination points (Nasiboglu, Kuvvetli, Ozkilcik, and Eliiyi, 2012).

In this context, a new approach has been adopted in the calculation method for the two key parameters accepted in the literature for measuring mobility: the number of boardings and the total kilometers traveled. With this new approach, the distances between the points where individuals move are obtained not as straight-line distances but through routing (Google API) (Google, 2017). Furthermore, a new definition of a trip has been introduced, where the last boarding made within 60 minutes is identified as the endpoint of the journey. This decision is based on findings from the 2015 İzmir transportation master plan revision, which determined that the average travel time in public transportation was 55.7 minutes (İBŞB, 2018).

As a result of the analysis of three months of data, the average distance traveled by elderly individuals has been identified as 276 kilometers, with the longest distance being 1,618 kilometers. When examining studies focused on the behaviors of elderly individuals, it has been shown that gender impacts mobility (Mercado and Páez, 2009). Accordingly, in this thesis study, the gender parameter has been treated as an independent variable and analyzed. When examining the gender distribution of cardholders, it is observed that 62.64% are male while 37.36% are female.

In analyzing the distance statistics of elderly individuals based on gender, it has been found that men travel longer distances than women. Average distances indicate that men cover more kilometers than women.

The second independent variable is age groups. In reviewing past literature, it is evident that individuals over the age of 65 are considered elderly (Hess, 2009). The mobility of elderly individuals has been the subject of many studies, showing differences compared to other age groups (Newbold et al., 2005; Zhang et al., 2007). However, there are limited studies examining whether differences exist among elderly individuals themselves (Alsnih and Hensher, 2003; Mercado and Páez, 2009). In this context, individuals over 65 years of age have been categorized into groups: 65-69, 70-74, 75-79, and 80+.

When examining the distribution of elderly individuals by age group, it is observed that 41.82% of the elderly individuals included in the study fall within the 65-70 age range. According to the distance statistics for elderly individuals by age group, the group that traveled the least distance is those over 80 years of age. The group that traveled the most, in terms of average distance, is those aged 70-74, while the group aged 80 and above traveled the least.

Another variable group pertains to urban areas referenced by location. Since one of the significant research topics of our study is the effects of the urban area where elderly individuals reside on their mobility, it is crucial to accurately determine individuals' residential locations. To achieve this, geographic coding (geocoding) has been utilized. It is possible to estimate individuals' residential addresses based on the stops where they board most frequently. However, geographic coding has been performed to prevent possible discrepancies.

The geographic coding method is based on two parameters. The first is addresses that comply with the national addressing system. Addresses must be in text format and should not contain special identifiers. The second parameter is the API key. With this key, the software can retrieve the geographic coordinates of textual addresses from Google services. In our study, using Google APIs in a web-based geographic coding program, the addresses in the personalized card database have been converted to geographic coordinates. Geocoding refers to the process of converting addresses into geographic coordinates (Google, 2015).

Using geographic information system software (Esri ArcMap 10.3), the central district of İzmir has been designated as the provincial building, and urban areas encompassing distances of 0-10, 10-20, 20-30, 30-40, and 40+ kilometers have been created from this point. The residential coordinates of elderly individuals obtained through geocoding have been matched with urban areas, thus creating a unique urban area code for each elderly individual.

The thirty districts located in the city have been distributed into the five regions designated as "urban areas." In this allocation, the positions of district centers have been taken into account. The greatest number of districts is found in the "40+" sector, located more than forty kilometers from the center. There are twelve districts in the area beyond forty kilometers. In the first urban area (0-10), eight districts are located, where 63.67% of the population resides.

When examining the distance statistics of elderly individuals based on their urban region, it is observed that, on average, the longest distances are attributed to the urban area encompassing "10-20" kilometers.

MODAL

The most important indicators of elderly individuals' mobility are the number and distance of their trips. Various dimensions, including personal characteristics, the physical environment, policies, social relationships, and cultural norms, emerge as factors that influence the number and distance of these trips. However, it is not always possible to find a dataset that allows for the simultaneous analysis of all these factors. Through the use of personalized cards in electronic fare collection systems, it is possible to access a wealth of data, such as the distance between public transportation users' residential locations and urban centers or sub-centers, as well as details regarding age, gender, number of trips, boarding location, and time, over extended time series.

The aim of this study is to assess the mobility of elderly individuals who use public transportation by analyzing the total distance they travel. Many studies assume that all "elderly" individuals exhibit similar mobility characteristics. Therefore, this study examines whether differences exist within the elderly population itself.

When measuring mobility, the length, duration, and frequency of trips are critical factors. In this study, total distance traveled was selected as the dependent variable. In order to explain the dependent variable using the data available from the electronic fare collection system, three independent variables were employed. The first variable is gender. The second variable is age group. The final independent variable is the distance between the individuals' residential locations and the city center, aiming to explore the effect of proximity to the city center on their mobility.

To analyze the variables affecting the travel distances of the elderly, Simple Linear Regression Analysis was initially employed. Regression analysis is one of the most fundamental statistical techniques, used to examine how one variable changes when another changes by a certain amount. A review of the literature shows that various regression analyses have been employed in numerous studies.

The initial analysis using the simple regression technique did not yield significant results indicating a meaningful relationship between the dependent and independent variables. Upon reviewing the randomly selected sample structure, it was observed that, due to the high number of individuals living in the city center, the standard deviation in the data often exceeded the mean. This condition, which indicates high deviations from the average in the total travel distances, complicates the modeling of the data.

Due to the inability to obtain desired results from multiple regression attempts, the measurement aspects of the study were revisited. In this context, a new approach was adopted in the calculation of two key parameters commonly used to measure mobility in the literature: the number of trips and total distance traveled. Unlike the initial approach, which calculated distances as the crow flies, the new method utilized routing to determine the distances between the points traveled by individuals. The resulting data were then analyzed using variance analysis methods.

The dependent variable is the total distance covered by elderly individuals. The travel distance is also used as a variable in previous studies (Mercado & Páez, 2009; Zhang et al., 2007; T. Schwanen et al., 2001). The public transport system operates using various modes, including the metro, light rail system, bus, and ferry. The distance traveled is calculated by analyzing personalized card data from the electronic fare collection system. Cardholders are required to validate their tickets upon boarding any public transportation vehicle, but there is no record of when a passenger exits the system. Therefore, it is necessary to estimate the next stop at which the individual will disembark. Destination points were estimated using the travel chain method (Nasiboglu, Güçlü, Ozkilcik, & Eliiyi, 2012). A novel approach has been adopted for calculating the number of boarding passes and total mileage, which are two fundamental parameters for measuring mobility. This new method calculates distances between movement points using routing (Google API) rather than air distance. Additionally, a revised travel definition is introduced, where the final journey of a ride within a 90-minute period is considered the end of the journey.

The first independent variable is age groups. According to previous literature, individuals over 65 years are considered elderly (Hess, 2009). Some studies also analyze differences among individuals over 65 by grouping them into age categories (Alsnih & Hensher, 2003; Mercado & Páez, 2009). In this context, individuals over 65 are divided into the following groups: 65-69, 70-74, 75-79, and 80+.

Another variable group is urban areas, which refer to the location. Addresses of elderly individuals in the database were converted into geographical coordinates through a geocoding process. Using geographical information system software (Esri ArcMap 10.3), the governor's building in İzmir was accepted as the city center, from which urban areas were defined as follows: 0-10 km, 10-20 km, 20-30 km, 30-40 km, and 50+ km. The residence coordinates of elderly individuals obtained from the geocoding process were matched to these urban areas, creating an urban area code for each individual.

The final variable group is gender. There are studies in the literature examining the effects of personal characteristics on mobility (Spinney et al., 2009), with gender being identified as an important variable.

ANALYSES

T-tests and ANOVA (analysis of variance) are widely employed statistical techniques that are contingent upon specific assumptions to ensure the veracity of the resulting data. A fundamental assumption is that the data follows a normal distribution. This implies that the sample data should exhibit a normal distribution, particularly when the sample size is limited. Furthermore, the t-test is contingent upon the assumption of equal variances among the groups being compared, which is known as homogeneity of variance. This assumption posits that the variability within each group should be analogous to ensure that the results are not biased by unequal variances. A further fundamental assumption is that of random sampling, whereby the samples must be selected at random from the population in order to ensure that the results are generalizable.

However, in the event of a breach of these assumptions, the results of the tests may become unreliable. For instance, in the event that the normality assumption is not satisfied, the test statistics may not adhere to the anticipated distribution, resulting in an inaccurate p-value. Similarly, in the event of a breach of the assumption of equal variances, there is a possibility of an increased incidence of Type I error rates, which may result in the erroneous rejection of the null hypothesis. In such cases, alternative statistical methods, such as non-parametric tests (e.g., Mann-Whitney U test for t-tests or Kruskal-Wallis test for ANOVA), can be employed, as they do not rely on the same assumptions regarding normality and homogeneity of variance.

Moreover, potential biases in the data collection process can significantly impact the validity of the results. These biases may arise from various sources, including selection bias, measurement bias, or response bias. For instance, if the sample is not representative of the population, the findings may not be generalizable. Therefore, it is essential to acknowledge and address these biases to enhance the reliability and validity of the research outcomes. Understanding these limitations and addressing them appropriately is crucial for drawing accurate conclusions

The Effects of Gender Variable on Elderly Mobility

In this study, since gender consists of two distinct groups, an independent samples t-test (Student's t-test) is applied. The t-test is used to compare the means of two independent samples, where observations are obtained from equally spaced or equal ratio scales belonging to one variable. The results are presented in Tables 1 and 2. The formula for the t-test is:

$$t = \frac{|\bar{x} - \bar{y}|}{s_{xy}\sqrt{\frac{2}{n}}} \tag{3.1}$$

(n): observation number of x or y
($\bar{x}$): mean of (x)
($\bar{y}$): mean of (y)
(s_{xy}): standard deviation of (x) and (y)
where:

Null Hypothesis (H_0): two samples were selected randomly from the same or equal group or groups with normal distribution and the mean of two groups is the same.

Alternative Hypothesis (H_A): two samples were selected randomly from equal group or groups with normal disribution and the mean of two groups is different.

The rejection of Null Hypothesis reveals that there is a significant difference between groups (Çubukçu, 2015).

Table 1. Group statistics

Group Statistics					
Gender		**N**	**Mean**	**Std. Deviation**	**Std. Error Mean**
Total Distance	0	395	207,176	197,258	9,925
	1	662	317,156	287,478	11,173

Table 2. Independent sample test results

Independent Samples Test										
		Levene's Test for Equality of Variances		**t-test for Equality of Means**						
		F	**Sig.**	**t**	**df**	**Sig. (2-tailed)**	**Mean Difference**	**Std. Error Difference**	**95% Confidence Interval of the Difference**	
									Lower	**Upper**
Total Distance	Equal variances assumed	47,975	,000	-6,717	1055,000	,000	-109,979	16,372	-142,105	-77,854
	Equal variances not assumed			-7,359	1034,795	,000	-109,979	14,945	-139,305	-80,654

In the analysis conducted, it was observed that the variances are not homogeneous. As a result, the null hypothesis (H_0) is rejected because the significance value is less than 0.05. This indicates a significant difference in total distance traveled between men and women. Specifically, there is a statistically significant difference in mobility, as measured by the distance covered, between men and women among the population over 65 years of age. The findings suggest that men are more mobile than women.

The Impact of Age Variance on Elderly Mobility

One-way analysis of variance (ANOVA) is used to investigate whether there are differences between three or more groups based on a particular variable. ANOVA compares the variance between sample groups to the variance within the groups to determine which is greater. For this method to be applicable, there must be at least three groups, the sample must exhibit a normal distribution, and the variances of the groups must be homogeneous. Additionally, the groups should be formed randomly (Çubukçu, 2015).

$$F = \frac{\frac{\sum_{i=1}^{I} n_i (\overline{x}_i - x)^2}{I - 1}}{\frac{\sum_{i=1}^{I} (n_i - 1) s_i^2}{I - 1}} \tag{3.2}$$

(I): Number of sample
(N): Total number of observation
(n_i): Observation number of (i)
($\overline{x}$): Mean of whole observation
($\overline{x}_i$): Mean of (i)
(s_i): Standard deviation of (i)
where

Null Hypothesis (H_0): Whole sample were selected randomly from the same or equal group or groups with normal disribution and the mean of two groups is the same.

Alternative Hypothesis (H_A): Samples were selected randomly from two groups with normal distribution and the mean of two groups is different. (Çubukçu, 2015).

In the analysis of mobility among the elderly, individuals in the sample were initially evaluated based on the total distance they covered. While existing literature has explored whether the elderly are less mobile compared to other age groups, there is a lack of studies investigating differences in mobility patterns within the elderly population, specifically among those aged 65 and above (Alsnih & Hensher, 2003).

In this context, elderly individuals were divided into age groups for the analysis: (65-69), (70-74), (75-79), and (80+). The hypothesis was tested using one-way analysis of variance (ANOVA). The first step was to determine whether the variances among the groups were homogeneous. A statistically significant difference was identified, with a p-value of 0.038 and a 95% confidence interval. Since the p-value is less than 0.05, the null hypothesis is rejected. The one-way ANOVA was then conducted, and the results are presented in Table 3

Table 3. Age groups - total distance ANOVA analysis results

Descriptives								
	N	Mean	Std. Deviation	Std. Error	95% Confidence Interval for Mean		Minimum	Maximum
					Lower Bound	Upper Bound		
65-69	427	296,517	269,370	13,287	270,398	322,636	0,214	1377,010
70-74	525	272,065	268,582	11,905	248,677	295,454	0,941	1617,965
75-79	71	202,984	217,871	29,378	144,085	261,882	8,483	980,236
80+	34	144,983	178,388	42,047	56,272	233,693	1,730	601,434
Total	1057	276,056	266,175	8,447	259,480	292,632	0,214	1617,965

Test of Homogeneity of Variances			
Levene Statistic	df1	df2	Sig.
2,824	3	1053	,038

ANOVA					
	Sum of Squares	df	Mean Square	F	Sig.
Between Groups	834511,445	3	278170,482	4,062	,007
Within Groups	72116190,426	1053	68486,411		
Total	72950701,871	1056			

The null hypothesis (H_0) was rejected because the significance value is less than 0.05, indicating a significant difference in mobility between the groups. Since the variances among the groups were not homogeneous, the Tamhane method was chosen for the Post Hoc analysis. The results are presented in Table 4. The Tamhane method is preferred in ANOVA tests for several key reasons. Firstly, this method does not require the assumption of homoscedasticity, which is the equality of variances across groups, making it particularly beneficial for groups with differing variances. Additionally, it provides more reliable results in cases of small sample sizes, serving as a suitable alternative when the assumptions of normal ANOVA are not met. The Tamhane method controls the Type I error rate in multiple comparisons, reducing the likelihood of false-positive results, thereby enhancing the reliability of the findings. Its less stringent assumptions compared to other methods allow for a more flexible approach to addressing differences in the data, and it is also suitable for complex experimental designs.

Table 4. Age group –post-hoc test results for total distance ANOVA analysis

Multiple Comparisons						
Dependent Variable:	Total Distance					
Tamhane					95% Confidence Interval	
(I) Age group		Mean Difference (I-J)	Std. Error	Sig.	Lower Bound	Upper Bound
65-69	70-74	24,391	17,085	0,633	-20,655	69,437
	75-79	93,361	30,237	0,016	11,913	174,808
	80+	152,179	41,759	0,009	31,449	272,910
70-74	65-69	-24,391	17,085	0,633	-69,437	20,655
	75-79	68,970	29,716	0,130	-11,208	149,147
	80+	127,789	41,383	0,033	7,713	247,864
75-79	65-69	-93,361	30,237	0,016	-174,808	-11,913
	70-74	-68,970	29,716	0,130	-149,147	11,208
	80+	58,819	48,321	0,794	-75,554	193,192
80+	65-69	-152,179	41,759	0,009	-272,910	-31,449
	70-74	-127,789	41,383	0,033	-247,864	-7,713
	75-79	-58,819	48,321	0,794	-193,192	75,554

*. The mean difference is significant at the 0.05 level.

Table 5. Age group – sum of total distance analysis

Age group	**65-69**	**70-74**	**75-79**	**80+**
65-69			x	x
70-74				x
75-79	x			
80+	x	x		

(x) refers to significant differences between the groups

Significant differences were found between the groups with a significance level lower than 0.05, and the groups with these differences are indicated in Table 5. The analysis reveals a pattern of varying mobility among the age groups. Specifically, elderly individuals aged 65-74 years are more mobile compared to those in other age groups, as they cover a greater distance.

The Impacts of Distance From City Center Variable on Elderly Mobility

Çeviri hatasıOne important dimension of elderly mobility is the physical environment, including transportation and communication infrastructure. In this context, the residence addresses of individuals in the sample were processed using geographic information systems to measure distances from the province center. The results support the hypothesis that elderly mobility decreases as distance from the center increases. Residences were divided into five groups based on their distance from the city center, with each region representing a ten-kilometer increment from the center. A one-way analysis of variance (ANOVA) was performed, which revealed that while the variance among groups was not homogeneous, there was a significant difference in the averages between the groups. The results are presented in Table 6.

Table 6. Urban areas– total distance ANOVA analysis results

Descriptives								
	N	**Mean**	**Std. Deviation**	**Std. Error**	**95% Confidence Interval for Mean**		**Minimum**	**Maximum**
					Lower Bound	**Upper Bound**		
0- 10	791	282,615	259,842	9,239	264,480	300,751	0,214	1550,658
10-20	167	320,843	306,249	23,698	274,054	367,632	10,877	1617,965
20-30	23	236,809	213,799	44,580	144,355	329,262	3,712	740,807
30-40	35	120,272	108,874	18,403	82,872	157,671	5,328	381,946
40+	41	122,098	101,039	15,780	90,206	153,990	7,919	494,914
Total	1057	276,056	262,835	8,084	260,193	291,920	0,214	1617,965
Test of Homogeneity of Variances								
Levene Statistic	df1		df2			Sig.		
10,139	4		1052			,000		
ANOVA								
	Sum of Squares	df	Mean Square	F	Sig.			
Between Groups	2225667,823	4	556416,956	8,28	0,00			
Within Groups	70725034,048	1052	67229,120					
Total	72950701,871	1056						

The null hypothesis (H_0) was rejected because the significance value was less than 0.05. This indicates a significant difference in the total distance traveled between the urban areas where the elderly reside. The analysis reveals that the variance among the groups is not homogeneous. Therefore, the Tamhane method was used for the Post Hoc test. The results are presented in Table 7. The analysis revealed a behavioral pattern indicating a decrease in mobility among the elderly as they move further away from the city center.

Table 7. Urban areas – post hoc test results for total distance ANOVA analysis

Multiple Comparisons						
Dependent Variable:	**Total Distance**					
Tamhane						
(I)Urban areas		Mean Difference (I-J)	Std. Error	Sig.	95% Confidence Interval	
					Lower Bound	Upper Bound
0-10	10- 20	-38,227	25,436	,764	-110,162	33,707
	20- 30	45,807	45,527	,980	-94,502	186,115
	30- 40	162,344	20,592	,000	102,202	222,485
	40+	160,517	18,285	,000	107,701	213,333
10-20	0- 10	38,227	25,436	,764	-33,707	110,162
	20-30	84,034	50,488	,669	-66,556	234,624
	30-40	200,571	30,005	,000	115,346	285,797
	40+	198,744	28,471	,000	118,101	279,388
20-30	0-10	-45,807	45,527	,980	-186,115	94,502
	10-20	-84,034	50,488	,669	-234,624	66,556
	30-40	116,537	48,229	,200	-29,311	262,385
	40+	114,711	47,291	,200	-29,110	258,532
30-40	0-10	-162,344	20,592	,000	-222,485	-102,202
	10-20	-200,571	30,005	,000	-285,797	-115,346
	20-30	-116,537	48,229	,200	-262,385	29,311
	40+	-1,827	24,242	1,000	-71,897	68,243
40+	0-10	-160,517	18,285	,000	-213,333	-107,701
	10-20	-198,744	28,471	,000	-279,388	-118,101
	20-30	-114,711	47,291	,200	-258,532	29,110
	30-40	1,827	24,242	1,000	-68,243	71,897

*. The mean difference is significant at the 0.05 level.

Table 8. Urban Area - sum of total distance analysis

Urban areas	0-10	10-20	20-30	30-40	40+
0-10				x	x
10-20				x	x
20-30				x	x
30-40	x	x			
40+	x	x			

(x) refers to significant differences between the groups

CONCLUSION

This study aims to identify the parameters affecting the mobility of elderly individuals and to formulate policies aimed at enhancing mobility, which significantly impacts their quality of life. The mobility of elderly individuals encompasses several dimensions, including personal characteristics, the physical environment and infrastructure, policies, and social relationships.

A review of the literature reveals that it is not feasible to evaluate all dimensions of mobility within a single study. Prior research indicates that researchers have attempted to explain specific dimensions of the complex relational structure based on the data available to them (Spinney et al., 2009; Mercado and Páez, 2009; Olawole and Aloba, 2014).

Most previous studies pertain to developed countries experiencing a rapid increase in their elderly populations, while there is a scarcity of research concerning developing countries like ours. In this regard, our study aims to contribute to the existing literature on understanding the mobility of elderly individuals living in our country and other developing nations. Furthermore, previous studies have emphasized the conflict between private vehicle use and public transportation, and there is a lack of studies specifically examining the mobility of elderly individuals within the context of public transportation systems.

This research analyzes the behaviors of individuals aged 65 and older who use public transportation across the city of İzmir. A review of prior studies shows that they were conducted using household surveys in terms of sample size and time series. In contrast, the data obtained from electronic fare collection systems provides much more detailed information on mobility compared to household surveys. In this context, all movements of elderly individuals over a three-month period were taken into account in the analyses.

The increase in the number of elderly individuals holding driver's licenses has been shown to have adverse effects on urban traffic loads and safety, highlighting the need for accessible public transportation options.

In relation to the personal characteristics dimension of mobility, results supporting the hypothesis that individuals aged 65 and older do not constitute a homogeneous group have been obtained. In this context, our findings indicate that gender has an effect on mobility, with males being more mobile than females.

Notably, there is a significant decrease in mobility among individuals aged 75 and older compared to those aged 65-74. Some studies in the literature corroborate this finding (Hess, 2009). Consequently, when formulating transportation policies, it is essential for both central and local governments to conduct needs analyses for individuals aged 75 and older in order to enhance the accessibility of public transportation or paratransit systems, thereby improving the quality of life for the elderly.

Another finding of our study relates to the physical environment and infrastructure dimension of elderly mobility. Results supporting the hypothesis that mobility decreases as one moves away from the city center have been obtained. Individuals residing thirty kilometers away from the city center are found to be less mobile than those living closer. To increase the mobility of elderly public transport users, it is necessary to conduct a detailed analysis of the mobility needs of individuals residing on the urban periphery.

Research conducted in developed countries shows that vehicle ownership is associated with mobility freedom for elderly individuals (Newbold et al., 2005). To prevent our country from facing the issues experienced by developed nations, studies should be conducted on how to provide accessible, reliable, and safe public transportation for elderly individuals (Alsnih and Hensher, 2003). The aim should be to enhance the quality of life through active aging strategies (WHO, 2002).

There are various initiatives both nationally and globally aimed at increasing the mobility of the elderly within the public transportation dimension. These initiatives can be grouped into four categories. As illustrated in Figure 3, the first category involves making public transport more economical.

Figure 3. Dimensions of elderly mobility

Applications granting preferential travel rights to encourage elderly individuals to use public transportation are commonly observed. Following recent legal regulations in our country, individuals aged 65 and older have been granted free access to publicly operated public transport. Initiatives like providing free urban public transportation for individuals aged 65 and older, which has been in place since 2014, should continue. Positive discrimination in fare tariffs has beneficial effects on mobility (Metz, 2003).

A similar application is observed for bus services across the United Kingdom. In London, individuals aged 60 and above can use all public transportation systems free of charge. Additionally, taxi card programs, discounts for suburban and intercity trains and buses, and other measures aim to increase interest in public transportation. The positive impact of personalized cards allowing discounted access to services outside of transportation can also be noted.

In addition to providing affordable public transportation services for the elderly, it is suggested that the government incorporate mobility and safety issues related to the elderly into its transportation policy (Olawole and Aloba, 2014).

To enhance accessibility, it is crucial that transportation planning is conducted in a manner that integrates all systems. In this context, ministries, metropolitan municipalities, and district municipalities in our country offer various social services, particularly to ensure that the elderly can access healthcare services. However, these services are often implemented independently and without coordination. By planning all resources centrally, the accessibility of services for elderly individuals can be increased. To ensure integration, it is necessary to develop passenger information, ticketing, and transfer centers. Non-profit organizations can provide demand-driven and flexible transportation services, particularly aiding elderly individuals residing in rural areas where public transportation systems are underdeveloped. These services not only facilitate access to essential services for the elderly but also enhance their relationships with non-governmental organizations. Minibus and taxi services can be utilized for door-to-door transportation. These services, which can be accessed through telephone reservations, can provide pick-up and drop-off services for isolated individuals.

Due to the high investment required for traditional public transportation, there are suggestions for the use of private services (Su and Bell, 2009). Especially with the development of rail systems, door-to-door transportation (e.g., buses, minibuses, and shared taxis) is increasingly being reduced by local authorities. Considering the challenges associated with increasing walking distances and transfer journeys, it is evident that elderly mobility requires diverse needs. Particularly, connecting individuals aged 75 and older with alternative transportation options can enhance their independence and mobility.

To increase accessibility to services, vehicles used in public transportation should be designed to meet the needs of elderly individuals. Although a legal framework has been prepared in our country, it cannot be claimed that it is fully implemented. The presence of elevators, escalators, and other facilities at stations for accessing rail systems is important. Taxis need to be made accessible for individuals with disabilities. Arrangements must be made to allow elderly individuals with mobility impairments to board taxis with their wheelchairs. Adequate lighting on roads is vital for ensuring safe travel. It is known that elderly individuals often conduct many of their journeys on foot. In this regard, easily usable and safe pedestrian pathways can contribute positively to elderly mobility.

It is well-documented that dependency on vehicles increases with aging. Thus, increasing the use of car-sharing and rental systems can enhance travel options for elderly individuals, contributing to a decrease in vehicle usage rates in urban areas.

Elderly individuals are generally more vulnerable to external dangers compared to other groups. Therefore, it is essential to enhance the suitability of various modes of transportation to meet the needs of the elderly. In this context, increasing individual safety in transportation is of paramount importance. It is crucial to enhance the number of security cameras and personnel in public transportation systems at stations and inside buses. Additionally, safe driving training should be provided to drivers to prevent injuries that may occur during transit. Local governments have responsibilities in raising awareness about safe public transportation usage. Providing training for the elderly on safely using public transportation is essential. Transportation planning efforts must consider the demographic structure of the planned area, the proportion of the elderly population, and its characteristics. Our study demonstrates that elderly individuals are not a homogeneous group. Moreover, factors such as gender, age group, and distance from the city center affect the mobility of elderly individuals

ACKNOWLEDGMENT

This study was conducted as part of the PhD Dissertation in City and Regional Planning Department, Dokuz Eylül University, The Graduate School of Natural and Applied Sciences.

REFERENCES

Albalate, D., ve Bel, G. (2010). What shapes local public transportation in Europe? Economics, mobility, institutions, and geography. Transportation Research Part E: Logistics and Transportation Review, 46(5), 775–790. 11 Mayıs 2012 DOI: 10.1016/j.tre.2009.12.010

Alsnih, R., ve Hensher, D. (2003). The mobility and accessibility expectations of seniors in an aging population. Transportation Research Part A: Policy and Practice, 37(10), 903–916. 22 Aralık 2014, DOI: 10.1016/S0965-8564(03)00073-9

Baydar, T. (2007). İnsan ilişkileri ve davranış. 11 Haziran 2013, https://www.istanbul.edu.tr/yuksekokullar/sosyalbilimler/iuwebson/turgutbaydar07guz/davranis.pdf

Boerema, S. T., Van Velsen, L., Vollenbroek-Hutten, M. M. R., ve Hermens, H. J. (2016). Value-based design for the elderly: An application in the field of mobility aids. Assistive Technology, 29(2), 1–9. 25 Haziran 2013, DOI: 10.1080/10400435.2016.1208303

Burns, P. C. (1999). Navigation and the mobility of older drivers. The Journals of Gerontology Series B: Psychological Sciences and Social Sciences, 54B(1), S49–S55. 24 Ocak 2018, DOI: 10.1093/geronb/54B.1.S49

Chang, Y.C. (2013). Factors affecting airport access mode choice for elderly air passengers. Transportation Research Part E: Logistics and Transportation Review, 57, 105–112. 9 Eylül 2014, DOI: 10.1016/j.tre.2013.01.010

Choo, S., Sohn, D., ve Park, M. (2016). Mobility characteristics of the elderly: A case for Seoul Metropolitan Area. KSCE Journal of Civil Engineering, 20(3), 1023–1031. 20 Şubat 2018, DOI: 10.1007/s12205-016-0651-x

Corbett, J. (2001). Torsten Hägerstrand: Time geography. CSISS Classics. UC Santa Barbara: Center for Spatially Integrated Social Science. 24 Haziran 2018, https://escholarship.org/uc/item/2t75b8sj

Çubukçu, K. M. (2015). *Planlama ve coğrafyada temel istatistik ve mekansal istatistik.* Nobel Akademik Yayıncılık Eğitim Danışmanlık Tic. Ltd. Şti.

Cullen, I. (1976). Reviews. *Lund Studies in Geography*, 44, 961–972.

Dickerson, A. E., Molnar, L. J., Eby, D. W., Adler, G., Bedard, M., Berg-Weger, M.,Trujillo, L. (2007). Transportation and aging: A research agenda for advancing safe mobility. The Gerontologist, 47(5), 578–590. 25 Ocak 2018, DOI: 10.1093/ geront/47.5.578

Elker, C. (1999). Çağdaş uaşım politikaları. II. Ulaşım ve Trafik Kongresi- Sergisi, 175–184. 8 Şubat 2016, http://arsiv.mmo.org.tr/pdf/11164.pdf

Eroğlu, F. (2017). *Davranış bilimleri (15.Baskı)*. Beta Yayın Evi.

Giuliano, G., & Hu, H. ve Lee, K. (2003). Travel patterns of the elderly: The role of land use. 24 Ocak 2018, https://www.metrans.org/sites/default/files/research -project/00-08_Final_0_0.pdf

Google. (2015). The Google geocoding API. 1 Mart 2017, https://developers.google .com/maps/documentation/geocoding/

Google. (2017). Distance matrix API. 1 Mart 2017, https://developers.google.com/ maps/documentation/distance-matrix/start

Hagerstrand, T. (1970). What about people in regional science. Papers Of The Regional Science Association, 24, 7–21. 24 Haziran 2014, https://link.springer.com/ content/pdf/10.1007/BF01936872.pdf

Hess, D. B. (2009). Access to public transit and its influence on ridership for older adults in two U. S. Cities. Journal of Transport and LandUse 2, 2(1), 3–27. 2 Mayıs 2013, https://www.jtlu.org/index.php/jtlu/article/view/11

Hu, X., Wang, J., ve Wang, L. (2013). Understanding the travel behavior of elderly people in the developing country: a case study of Changchun, China. Procedia - Social and Behavioral Sciences, 96(Cictp), 873–880. 4 Ocak 2015, DOI: 10.1016/j. sbspro.2013.08.099

Ipingbemi, O. (2010). Travel characteristics and mobility constraints of the elderly in Ibadan, Nigeria. Journal of Transport Geography, 18(2), 285–291. 25 Ocak 2018, https://doi.org/DOI: 10.1016/j.jtrangeo.2009.05.011

İzmir Büyükşehir Belediyesi. (2018). İzmir ulaşım ana planı saha araştırmaları bilgi notu (2). 31 Ocak 2018, https://www.izmir.bel.tr/YuklenenDosyalar/file/ HIZMETREHBERI/idari/Upi_SahaArastirma.pdf

Levine, J., ve Garb, Y. (2002). Congestion pricing's conditional promise: promotion of accessibility or mobility? Transport Policy, 9(3), 179–188. October 1 Ocak 2018, DOI: 10.1016/S0967-070X(02)00007-0

Mercado, R., ve Páez, A. (2009). Determinants of distance traveled with a focus on the elderly: a multilevel analysis in the Hamilton CMA, Canada. Journal of Transport Geography, 17(1), 65–76. 2 Eylül 2012, DOI: 10.1016/j.jtrangeo.2008.04.012

Metz, D. H. (2000). Mobility of older people and their quality of life. Transport Policy, 7(2), 149–152. 5 Eylül 2014, DOI: 10.1016/S0967-070X(00)00004-4

Metz, D. H. (2003). Transport policy for an ageing population. Transport Reviews, 23(4), 375–386. 21 Nisan 2014, DOI: 10.1080/0144164032000048573

Miller, J. S., Hoel, L. a., Goswami, A., ve Ulmer, J. M. (2006). Assessing the utility of private information in transportation planning studies: A case study of trip generation analysis. Socio-Economic Planning Sciences, 40(2), 94–118. 1 Haziran 2012, DOI: 10.1016/j.seps.2004.10.003

Nasiboglu, E., Kuvvetli, U., & Ozkilcik, M. ve Eliiyi, U. (2012). Origin-destination matrix generation using smart card data: Case study for Izmir. 2012 IV International Conference "Problems of Cybernetics and Informatics" (PCI), 1–4. 15 Haziran 2014, DOI: 10.1109/ICPCI.2012.6486315

Newbold, K. B., Scott, D. M., Spinney, J. E. L., Kanaroglou, P., ve Páez, A. (2005). Travel behavior within Canada's older population: A cohort analysis. Journal of Transport Geography, 13(4), 340–351. 10 Mayıs 2015, DOI: 10.1016/j.jtrangeo.2004.07.007

Olawole, M. O., ve Aloba, O. (2014). Mobility characteristics of the elderly and their associated level of satisfaction with transport services in Osogbo, Southwestern Nigeria. Transport Policy, 35, 105–116. 10 Aralık 2017 DOI: 10.1016/j.tranpol.2014.05.018

Pred, A. (1977). The Choreography of existence: comments on Hägerstrand' s time-geography and its usefulness. Economic Geography, 53(2). 7 Mart 2013, https://www.jstor.org/stable/142726?seq=1#page_scan_tab_contents

Rantakokko, M., Mänty, M., ve Rantanen, T. (2013). Mobility decline in old age. Exercise and Sport Sciences Reviews, 41(1), 19–25. 1 Ocak 2018, DOI: 10.1097/JES.0b013e3182556f1e

Rosenbloom, S. (2001). Sustainability and automobility among the elderly: An international assessment. Transportation, 28(4), 375–408. 30 Haziran 2013, DOI: 10.1023/A:1011802707259

Rouwendal, J., ve Rietveld, P. (1994). Changes in commuting distances of Dutch households. Urban Studies, 31(9), 1545–1557. 26 Ocak 2018, DOI: 10.1080/00420989420081421

Schaie, K., & Boron, J. ve Willis, S. (2005). Everyday competence in older adults. Cambridge: Cambridge University Press. DOI: 10.1017/CBO9780511610714

Schwanen, T., ve Páez, A. (2010). The mobility of older people – an introduction. Journal of Transport Geography, 18(5), 591–595. 20 Mayıs 2013, DOI: 10.1016/j.jtrangeo.2010.06.001

Schwanen, T., Dieleman, F. M., ve Dijst, M. (2001). Travel behaviour in Dutch monocentric and policentric urban systems. Journal of Transport Geography, 9(3), 173–186. 20 Mayıs 2013, DOI: 10.1016/S0966-6923(01)00009-6

Smith, G. C., ve Sylvestre, G. M. (2001). Determinants of the travel behavior of the suburban elderly. Growth and Change, 32(3), 395–412. 30 Mayıs 2015, DOI: 10.1111/0017-4815.00165

Spinney, J. E. L., Scott, D. M., ve Newbold, K. B. (2009). Transport mobility benefits and quality of life: A time-use perspective of elderly Canadians. Transport Policy, 16(1), 1–11. 1 Ocak 2018, DOI: 10.1016/j.tranpol.2009.01.002

Srichuae, S., Nitivattananon, V., ve Perera, R. (2016). Aging society in Bangkok and the factors affecting mobility of elderly in urban public spaces and transportation facilities. IATSS Research, 40(1), 26–34. 22 Mayıs 2017, DOI: 10.1016/j.iatssr.2015.12.004

Steven, R. (1997). *Constraints on elderly daily travel behavior, Master Theses*. The University of Westerm Ontario.

Su, F., ve Bell, M. G. H. (2009). Transport for older people: Characteristics and solutions. Research in Transportation Economics, 25(1), 46–55. 26 Ocak 2018, DOI: 10.1016/j.retrec.2009.08.006

Tacken, M. (1998). Mobility of the elderly in time and space in the Netherlands: An analysis of the Dutch National Travel Survey. Transportation, 25(4), 379–393. 30 Haziran 2013, DOI: 10.1023/A:1005042614848

Van Den Berg, P., Arentze, T., ve Timmermans, H. (2011). Estimating social travel demand of senior citizens in the Netherlands. Journal of Transport Geography, 19(2), 323–331. 30 Haziran 2013, DOI: 10.1016/j.jtrangeo.2010.03.018

Whelan, M., & Langford, J. Oxley, J.; Koppel, S., ve Charlton, J.; (2006). The elderly and mobility: a review of the literatre. 30 Haziran 2013, https://www.monash.edu/muarc/research/our-publications/muarc255

World Health Organization (2002). Active ageing: A policy framework. The Aging Male, 5(1), 1–37. 24 Haziran 2014, DOI: 10.1080/713604647

World Health Organization. (2011). Global health and aging. NIH Publication No 117737, 1(4), 273–277. 10 Haziran 2015, https://doi.org/DOI: 11-7737

Zhang, Z., Mao, B., Liu, M., Chen, J., ve Guo, J. (2007). Analysis of Travel Characteristics of Elders in Beijing. Journal of Transportation Systems Engineering and Information Technology, *7*(6), 11–20. 24 Eylül 2016, .DOI: 10.1016/S1570-6672(08)60001-X

Chapter 8

Microsimulation Analysis of Public Transport: The Case of Bagdat Street, Istanbul

İrem Merve Ulu
https://orcid.org/0000-0002-3853-3860
Istanbul Okan University, Turkey

Burcu Ozdemir
Informatics Institute, Istanbul, Turkey

ABSTRACT

Metropolitan areas around the world are facing traffic congestion and mobility issues. In this context, improving public transportation systems has become a critical requirement for ensuring the sustainability of cities. This study examines the impacts of public transport strategies implemented on Bagdat Street, one of Istanbul's main corridors, on pedestrian and vehicle networks. Using PTV VISSIM for traffic simulation, various scenarios, including changes in bus frequency, capacity, number of stops, and dedicated lanes, were analyzed. Findings reveal that increasing the frequency and reducing the capacity of the existing public transport buses not only improves traffic flow but also enhances passenger satisfaction. These insights provide valuable guidance for urban transportation planning and policy making, which can be adapted to similar contexts in other urban environments.

DOI: 10.4018/979-8-3693-6695-0.ch008

INTRODUCTION

In contemporary metropolitan areas, transportation-related challenges such as traffic congestion, air and noise pollution, traffic accidents, and the inefficient use of energy and time are prevalent. To address these issues, a range of strategies has been employed, including promoting the use of public transportation (PT), improving infrastructure for bicycles and pedestrians, incentivizing shared and flexible mobility, incorporating environmentally sustainable transportation solutions, integrating different modes of transport, and implementing policies for managing travel demand and pricing. Furthermore, the development of Intelligent Transportation Systems (ITS) and Information and Communication Technology has facilitated innovative approaches such as the prediction of traffic congestion, traffic flow, travel time, and accidents (Kaffash et al., 2021). To mitigate the negative impacts of traffic congestion, policies such as traffic diversion, fuel tax, road taxes, and strengthening PT have been put into action (Abroshan et al., 2021; Bjertnæs, 2019; Isa et al., 2015; Mathew, et al., 2020; Li et al., 2009; Paricio & Lopez-Carmona 2019; Stefanello, et al., 2017; Storchmann, 2001).

The current transportation policies aim to create more socially, environmentally, and economically sustainable cities by promoting the use of PT over private cars (European Commission, 2017). Challenges such as social growth, environmental issues, long travel distances, space competition, and the need for social equality/ justice underscore the importance of increasing PT usage today (Abenoza et al., 2017). PT offers benefits such as energy savings, environmental protection, high passenger capacity, and efficiency. Promoting PT can alleviate traffic congestion caused by private car use (Mokonyama and Venter, 2013), consume less energy, produce fewer emissions per passenger (Waygood et al., 2014), and save approximately 20 times the space for vehicles traveling at 50 km/h (Héran and Ravalet, 2008). Therefore, promoting PT is crucial in reducing traffic congestion, enhancing people's living standards, and improving government public services (Ni et al., 2020).

BACKGROUND

Cities worldwide, particularly in developing countries, are promoting PT services to reduce reliance on vehicles (Hasan et al., 2021). To improve PT services, various measures are being implemented, including special PT lanes (dedicated lanes), limited stops, adjustments to vehicle capacity and frequency, integration with other modes, improved pricing policies, and integration with ITS (Conlon et al., 2001; El-Geneidy & Surprenant-Legault, 2010; Ercolano, 2011; Hasan et al., 2021; Jara-Diaz & Gschwender, 2009; Jara-Diaz et al., 2017; Renne et al., 2016;

Zhu, 2010). Special PT lanes, for example, can reduce bus travel time and improve schedule accuracy by eliminating interactions with passenger cars (Wu & Pojani, 2016). However, Abroshan et al. (2021) and Yang et al. (2020) have argued that dedicating lanes for buses could cause traffic bottlenecks and lead to longer travel times for passenger cars. Redman et al. (2013) suggested that increasing bus frequency and speed could reduce car usage from 34% to 22%.

Research has shown that an effective PT system can lead to increased passenger satisfaction, retention of existing passengers, and attraction of potential passengers (Ni et al., 2020; Hussain et al., 2018; Jen et al., 2011; Lai & Chen, 2011; Zhao et al., 2013; van Lierop and El-Geneidy, 2016; Vincente et al., 2020; Zefreh et al., 2020). Passenger satisfaction in PT services refers to the level of contentment or disappointment experienced by passengers when comparing their expectations with the actual services provided (Zhang et al., 2020; Pei et al., 2006). Several studies have identified key factors influencing passenger satisfaction. These include line reliability, bus stop locations, time precision, connection to other operators' lines, service frequency (Castillo & Benitez, 2013), travel speed, on-time performance, and service frequency (Mouwen, 2015), service frequency, departure time accuracy, stop safety, vehicle cleanliness, seating capacity, and pre-trip information provision (Zefreh et al., 2020), travel duration, on-board crowding, and accessibility (Mugion et al., 2018), reliability, frequency, speed, accessibility, price, information provision, transfer ease, vehicle condition, comfort, safety, convenience, and aesthetics (Redman et al., 2013), ride quality, fare level, bus frequency, and network coverage (Hasan et al., 2021), frequency, speed, and intermodality (De Oña et al., 2020). Furthermore, studies have shown that factors such as traffic congestion during peak hours, waiting time at stops, and trip time precision can decrease passenger comfort and satisfaction (Zhang et al., 2020). Recommendations to enhance passenger satisfaction include increasing bus frequency during peak hours and improving bus travel comfort. Additionally, frequent PT users prefer services with higher frequency (De Vos et al., 2020).

Istanbul, with a population of approximately 16 million, is facing significant transportation challenges (TUIK, 2024). According to the Urban Mobility Report from INRIX, Istanbul ranked sixth globally for the highest traffic delays, with an average delay of 96 hours per driver in the 2023 Global Traffic Scorecard This represents a 12% increase from 2022 and a 20% rise from pre-COVID levels (INRIX, 2024). Bagdat Street, one of Istanbul's busiest urban corridors, is particularly affected by congestion. A study conducted on Bagdat Street revealed that the main reasons for visits include walking, shopping, leisure, or commuting, with traffic conditions being a major complaint (Hacıoğlu et al., 2021). Another study found that shopping and general recreational purposes were the main reasons for visiting

Bagdat Street, with PT being the preferred mode of access due to road congestion (Sami and Dincer, 2020).

The main purpose of the study is to evaluate the performance of pedestrian and vehicle networks on Bagdat Street, which has one of the heaviest traffic volumes on the Anatolian side of Istanbul. The study examines the effects of IETT (Istanbul Electric Tramway and Tunnel Establishments) bus lines on traffic flow and passenger satisfaction and reveals the effects of changes in bus frequency, capacity, number of stops, and dedicated lanes on the network. The study utilized the PTV VISSIM 2023 traffic simulation, which provides detailed analysis of current performance, enables the evaluation of possible scenarios, and contributes to policy and strategy development processes. The study area was modelled with field studies, metropolitan municipality datasets, and traffic flow camera recordings. Basic components such as mapping the road network, creating traffic control programs, and modeling vehicle movements were integrated into the simulation (Ziemska-Osuchn and Osuchn, 2022). With the help of PTV VISSIM, the effects of bus lines on traffic on Bagdat Street were simulated, and vehicle and pedestrian delays were analyzed. In this context, recommendations such as adjusting bus frequencies and capacities aim to help local governments and transportation authorities make decisions to reduce traffic congestion by providing more efficient PT services.

This study is driven by the need to inform PT policies aimed at addressing traffic problems on Bagdat Street. The study thoroughly analyzes the impact of PT on both traffic performance and passenger satisfaction, specifically focusing on high-traffic areas like Bagdat Street. While there have been studies on traffic congestion in Bagdat Street, none have examined the combined effects of PT on traffic flow and passenger satisfaction, as undertaken by the authors. In addition to private vehicles and taxis, Bagdat Street accommodates yellow minibuses (shared minibuses) that travel the same route as İETT buses and the Marmaray line. Observations suggest that scenarios designed around an interconnected PT network, rather than parallel systems, significantly enhance efficiency. This is particularly relevant for decisions regarding İETT bus operations, such as route planning, capacity management, and stop distribution, as these can be swiftly implemented by the local government. Buses with similar capacities are used in PT lines without laybys in the study area. In this context, using buses with different capacities and changing the arranged frequencies can be used to improve both PT services on Bagdat Street and the connections of Bagdat Street with other modes of transport. Additionally, this study aims to provide insights for the forthcoming Üsküdar-Kadıköy-Maltepe Tramway line, scheduled for implementation in the coming years. Bagdat Street, long a preferred and iconic public space, continues to serve multiple purposes for a diverse population, including recreation, shopping, various celebrations, and social interactions.

METHOD

Study Area

The study area was selected to be Bagdat Street in the Kadıköy District, one of the busiest streets on the Asian side of Istanbul. Bagdat Street was chosen due to its role as a central axis, linking major public spaces and offering a diverse range of commercial, office, cultural, and entertainment venues. While Bagdat Street extends from Kadıköy to Maltepe, the section between Bostancı Marmaray station and Kızıltoprak is the most associated with the street (Hacıoğlu et al., 2021).

This study, conducted on a street with high public space use and pedestrian traffic, indicates that the primary reasons for visiting Bagdat Street include walking, shopping, leisure, or commuting. The most common complaint among visitors relates to traffic conditions. Similarly, another study found that shopping and general recreational purposes were the main reasons for visiting Bagdat Street, with PT being the preferred mode of access due to road congestion (Sami and Dincer, 2020). The street operates as a one-way thoroughfare. Çetin Emeç Boulevard, running parallel to the Marmara Sea, also operates as a one-way street in the opposite direction to Bagdat Street, extending from Kadikoy to Bostanci. Additionally, the Marmaray line runs parallel to the study area. The primary modes of transportation to Bagdat Street include private vehicles, taxis, yellow minibuses, the Marmaray line, sea transportation to Bostanci, bicycles, scooters, and pedestrians.

A 3,572-meter section of Bagdat Street, spanning from the Bostancı Marmaray line stop to Göztepe 60th Yıl Park, was selected as the study area (see Figure 1). The area consists of three lanes, each 3.5 meters in width. Additionally, within the study area, there are 22 one-way merging side streets, 22 one-way diverging side streets, and 5 two-way side streets connecting to Bagdat Street.

Figure 1. Study Area

Methodology of the Study

Within the scope of this study, camera recordings, field study data, and datasets containing PT information were used for modeling. Camera recordings from the Istanbul Metropolitan Municipality were obtained from the Bagdat Street-Ethem Efendi Street intersection, capturing traffic flow on Bagdat Street upstream on October 5, 2023. These records were examined in an office environment, and hourly traffic volume and vehicle composition data were extracted during non-peak (14:00-17:00) and peak hours (17:30-18:30) (Table1). In addition, as part of the field study, the proportions of the traffic flows merging and diverging Bagdat Street from side roads were determined. Data regarding traffic signals were also recorded, including the locations of signalized intersections, traffic signal cycles, and phase durations. The field study was completed on a typical weekday under weather conditions that did not affect traffic.

Table 1. Traffic Volume and Vehicle Composition of Bagdat Street

Time Period	Road	Traffic Volume Entering the Network (veh/h)	Passenger Cars (%)	Minibuses (%)	Buses (%)	Heavy Vehicles (%)
14:00-15:00	Bagdat Street	1739	92	5	1	2
	Side Roads (All)	3995	98	0	0	2
17:30-18:30	Bagdat Street	1782	92	7	1	1
	Side Roads (All)	4077	99	0	0	1

The dataset provided by IETT includes comprehensive information about PT services, such as service hours, bus schedules, bus capacities, stops, and the number of passengers boarding and alighting at each stop. The dataset, current as of May 2, 2024, serves as a valuable resource for those seeking insights into the operational aspects of PT in Istanbul (refer to Table 2 for bus codes and their corresponding general characteristics, and Annex 1 for details on stop-based boarding and alighting).

Table 2. General Information of Buses

Departure time	Bus code	Capacity	Bus Length (m)	Occupancy in the first stop	Departure time	Bus code	Capacity	Bus length (m)	Occupancy in the first stop
14:00	4	31	7.720	22	17:30	16D	77	10.5	7
14:10	16D	76	10.5	38		4	98	12.134	17
	ER2	31	7.720	11	17:40	16D	76	10.5	17
14:15	4	104	12.134	11	17:45	4	96	12.134	5
14:20	16D	76	10.5	32	17:50	16	98	12.134	33
14:30	16D	76	10.5	30		16D	76	10.5	6
	4	31	7.720	7	18:00	16D	76	10.5	8
14:38	16D	76	10.5	12	18:10	16D	76	10.5	5
14:45	4	27	7.305	16	18:15	4	31	7.720	13
14:46	16D	76	10.5	22	18:20	16D	76	10.5	8
14:54	16D	76	10.5	26					

Figure 2. Simulation Model

The study area was modeled in PTV VISSIM 2023 software. PTV VISSIM is software widely used in the field of traffic engineering, offering flexible and adaptable modeling capabilities. It provides detailed analysis and reporting and is capable of simulating vehicle and pedestrian behaviors at a microscopic level. First, the section of Bagdat Street between Bostancı bus stop and Göztepe 60. Yıl Park, along with the streets merging/diverging from this section, was drawn in accordance with the geometric features of the road (number of lanes, lane width) using the map feature of the software. Then, hourly traffic volumes were defined for both Bagdat Street and side roads. Vehicle volumes were specified according to Table 1 for Bagdat Street and Ethem Efendi Street where PT lines pass. For other streets, volumes were defined as 98% passenger cars and 2% heavy vehicles from 14:00-15:00, and 99% passenger cars and 1% heavy vehicles from 17:30-18:30. The legal speed limit on the street is 50 km/h, which was set as the desired speed for all vehicles. Subsequently, the cycle and phases of 20 traffic lights in the area were defined in the model. Finally, PT stops, passenger waiting areas, bus routes, and departure times were defined based on the number of passengers boarding and alighting at each stop (Table 2, Annex 1).

In the subsequent phases of the study, the model was configured with 42 random seeds, a 900-second warm-up time, and 3600 seconds of run time. Traffic simulations involve processes influenced by random factors such as driver behavior, vehicle arrival, and reaction times within a certain range of conditions. Consequently, atypical results may occur due to this randomness. To mitigate this variability, each scenario was run 10 times, and the average was computed to enhance the model's sensitivity and accuracy. Since the bus capacity and length differ from the default bus models in the simulation, the length, width, and height values were defined based on specific models, including the Karsan Jest for buses with a capacity of 16, the Isuzu Novo for buses with a capacity of 25-29, the Otokar Sultan Maxi for buses with a capacity of 30-40, the Otokar Sultan LF 50-59 passenger bus, and the Isuzu Novociti Life for buses with a capacity of 60-69. The following bus models were used in the simulations: MAN Lion's City 10E for 70-79 capacity buses, Mercedes-Benz Conecto for 104-capacity buses, Mercedes-Benz Conecto G for 130-capacity buses, and Mercedes-Benz CapaCity L for 200-passenger buses.

Calibration is the process of aligning the parameters of the model with observed data in the real world. In this process, key parameters such as the road network drawn according to geometric characteristics, traffic controls, right/left turn percentages, PT routes, stops, schedules, vehicle volumes, vehicle types, vehicle compositions, and desired speed profiles were utilized. Adjustments were made to minimize the discrepancies between observed data and simulation outputs, ensuring a more accurate representation of real-world conditions. Validation ensures that predicted and current road conditions are as similar as possible. During the validation process, a data collection point was added to the model at the intersection of Bagdat Street and Ethem Efendi Street, based on the location where the camera footage was recorded. This allowed for a comparison between the simulation results (Q_{model}) and the real-world data ($Q_{observation}$) collected from the same point. Various statistical methods, such as GEH statistics, Root Mean Square Error, and Mean Absolute Percentage Error, are used to evaluate the model's validity. These statistics are used to evaluate the reliability of the model by statistically comparing the real values with the data obtained from the model. This helps determine how closely the model aligns with real-world conditions. The GEH statistic, a non-linear metric commonly used in traffic engineering, compares the volume values of model data with observed data. A GEH value of less than 5 indicates a strong model, interpreted as a difference of 0.5 vehicles per hour.

$$GEH = \sqrt{\frac{2*(Q_{model} - Q_{observation})^2}{(Q_{model} + Q_{observation})}}$$

The acceptance criteria for the GEH value are as follows: 0-5 indicates a well-established model, 5-10 requires further research, and >10 is considered unacceptable.

The data collected from the field ($Q_{observation}$) and the outputs of the model (Q_{model}) were divided into 15-minute intervals and assessed using the GEH statistic. The comparison of model with real-world data, yielding GEH values below 2, demonstrates the model's alignment with observed data and reinforces its reliability (Table 3 and Figure 2). As the simulation focused on evaluating overall network performance, the GEH statistic was deemed sufficient, and additional statistical analyses were not considered necessary.

Table 3. GEH Statistics Results

Time Period	Observed	Model	GEH	Time Period	Observed	Model	GEH
14:00-14:15	412	447	1.68	17:30-17:45	419	457	1.81
14:15-14:30	444	428	0.76	17:45-18:00	449	448	0.04
14:30-14:45	432	454	1.04	18:00-18:15	465	453	0.56
14:45-15:00	449	429	0.95	18:15-18:30	451	422	1.38

Figure 3. GEH Statistic Results of Model Validation

As illustrated in Table 3 and Figure 3, the proposed model was met with considerable acceptance, prompting the initiation of scenario-based studies. Following the base scenario representing the current situation, five different scenarios representing various PT operating conditions were created. Due to insufficient data, only buses were evaluated in the scenarios, and only passengers using buses were defined as pedestrians in the network. The second scenario incorporates the designation of a dedicated lane for buses. The third scenario entails reducing the number of bus stops. The fourth scenario involves increasing bus capacity and reducing service frequency. The fifth scenario proposes increasing the number of bus lines while reducing their capacity. The last scenario represents the removal of bus lines from the transportation system. All scenario results were evaluated for both vehicle and pedestrian network performance. Although PTV VISSIM provides a variety of network performance results, the metrics listed below are the key parameters that best reflect the overall efficiency of the PT system, travel time, and traffic flow. Travel time and delay directly impact passenger satisfaction and system performance, while the number of stops and standstill time play a critical role in understanding traffic congestion. Average speed indicates the overall flow efficiency within the network. While other parameters are important for more specific analyses, these metrics are sufficient to evaluate network performance and operational efficiency in PT.

- Total travel time of vehicles: The total travel time of vehicles within or that have already left the network.
- Total delay of vehicles: The total delay of all vehicles within or that have already left the network.
- Total number of stops: The total number of stops, indicating interruptions in traffic flow, of all vehicles within or that have already left the network.

- Total standstill time of vehicles: The total standstill time of all vehicles within or that have already left the network.
- Average speed of vehicles: The average speed of vehicles (km/h).
- Average travel time of PT passengers: The average traveling time of pedestrians within the network or who have already left the network.
- Average standstill time of PT passengers: The average standstill time per pedestrian during the evaluation.

Scenarios

Scenario 1: Base Scenario

The first scenario, serving as the baseline for comparison, represents the current status of the PT system. This scenario considers the existing number of trips, bus stops, bus capacity, and the number of passengers. It acts as a reference point for evaluating the performance of the existing PT system and assessing the impact of alternative scenarios.

The number of pedestrians entering the system was determined, based on the number of PT passengers at each bus stop during the period between the bus trip they take and the previous trip, indicating that passengers are distributed stochastically during this time interval. The average walking distance for pedestrians between stops was found to be 338 meters. While the theoretical maximum walking distance to a stop is 169 meters, a more general walking distance of 100 meters was deemed appropriate, considering that not all pedestrians originate from the farthest point.

In the base scenario, despite comparisons between peak and non-peak hours, there is no notable discrepancy in the performance of the vehicle and PT passenger networks (Table 4). This consistency can be attributed to Bagdat Street's vibrant location, which is utilized by both vehicles and pedestrians throughout the day.

Table 4. Table Base Scenario Network Performance Results: Vehicles and PT Passengers

	Total Travel Time of Vehicles (hr)	Total Delay of Vehicles (hr)	Total Number of Stops of Vehicles	Total Standstill Time of Vehicles(hr)	Average Speed of Vehicles (km/hr)	Average Travel Time of PT Passengers (sec)	Average Standstill Time of PT Passengers (sec)
14:00-15:00	330.29	210.52	14677	167.35	35.4	600.92	70.86
17:30-18:30	327.27	214.45	15081	169.41	34.71	780.92	75.48

Scenario 2: Special Lanes for PT Buses

To mitigate the impact of PT systems on urban traffic density and to reduce the travel times of PT vehicles, dedicated PT lane was implemented in this scenario. Specifically, the right lane of Bagdat Street is dedicated PT, enabling buses to travel the route with greater efficiency and minimal disruption from general traffic. The remaining lanes, the left and middle ones, are allocated for the use of other vehicles.

When comparing the results of this scenario with the base scenario (Table 5), it is evident that the total travel time for vehicles increases by 107% between 14:00-15:00 and by 120% between 17:30-18:30. Total delays rise by 182% and 195%, respectively. The separation of the special PT lane leads to a significant increase in the total number of stops, by 422% and 433%, respectively. The total standstill time for vehicles also increases by 155% during both periods, while average speeds decrease by 42% and 40%, respectively.

The negative impact on traffic flow is primarily due to the reservation of the right lane for PT, which forces general traffic into the left and middle lanes. These lanes are unable to accommodate the traffic demand, resulting in increased congestion. On the other hand, allocating the right lane to PT reduces the average travel time for passengers by 12% between 14:00-15:00 and by 1% between 17:30-18:30. The average standstill time per pedestrian decreases by 5% and 6%, respectively. While the scenario benefits PT passengers, the increased congestion for general traffic could have wider negative impacts on the economic, social, and ecological structure of the city.

Table 5. Scenario 2 Network Performance Results: Vehicles and PT Passengers

	Total Travel Time of Vehicles (hr)	Total Delay of Vehicles (hr)	Total Number of Stops of Vehicles	Total Standstill Time of Vehicles(hr)	Average Speed of Vehicles (km/hr)	Average Travel Time of PT Passengers (sec)	Average Standstill Time of PT Passengers (sec)
14:00-15:00	682.06	594.65	76672.	426.98	20.56	526.22	67.5
17:30-18:30	720.00	633.31	80348.	460.59	20.75	773.16	70.66

Scenario 3: Reducing the Number of Bus Stops

In the third scenario, the number of bus stops was reduced to enhance the efficiency of the PT system. This involved identifying underutilized bus stops and eliminating five existing stops within the study area, which currently has 11 bus stops. The goal was to reduce bus travel times by removing stops with low passenger density from the PT routes.

Table 6 shows the total number of passengers boarding at each stop under current conditions, along with the distances between each stop and those removed. In Scenario 3, the walking distance for pedestrians to reach the stops has been doubled to 200 meters compared to the base scenario.

Table 6. Table Current and Planned Number of Passengers Boarding at Stops

	14:00-15:00		17:30-18:30	
Stop name	**Passenger Boarding**	**Passenger Boarding (Adjusted)**	**Passenger Boarding**	**Passenger Boarding (Adjusted)**
Bostancı	58	85	45	52
Tatlıçeşme	27		7	
Çatalçeşme	6	12	3	8
Tan sokak	6		5	
Suadiye	25	47	27	36
Mücahit Sokak	22		9	
Şaşkınbakkal	49	82	30	65
Kantarcı	33		35	
Erenköy	49	109	54	106
Caddebostan	60		52	
Yeniyol	69	69	33	33

Table 7. Distance (m) Between Stops: Current and Scenario 3

Stop name Bostancı	**Distance Between Stops (m) (Current)**	**Stop Name Bostancı**	**Distance Between Stops (m) (Scenario 3)**
Tatlıçeşme	450	Çatalçeşme	800
Çatalçeşme	350		
Tan sokak	230	Suadiye	580
Suadiye	350		
Mücahit Sokak	270	Şaşkınbakkal	670
Şaşkınbakkal	400		

continued on following page

Table 7. Continued

Stop name Bostancı	Distance Between Stops (m) (Current)	Stop Name Bostancı	Distance Between Stops (m) (Scenario 3)
Kantarcı	280	Erenköy	530
Erenköy	350		
Caddebostan	400	Yeniyol	700
Yeniyol	300		

The results of this scenario (Table 8) indicate that reducing the number of PT stops has a negligible impact on the overall performance of the traffic network, with changes of less than 2%. However, the reduction in stops results in a 31% increase in the average travel time of PT passengers between 14:00-15:00 and a 16% increase between 17:30-18:30. Additionally, the average standstill time for PT passengers increases by 57% and 17% during these respective time frames. While the reduction in stops may not significantly impact traffic flow, it does have a negative effect on PT passengers. Although reducing bus stops has minimal impact on traffic flow, it significantly increases travel and standstill times for PT passengers.

Table 8. Scenario 3 Network Performance Results: Vehicles and PT Passengers

	Total Travel Time of Vehicles (hr)	Total Delay of Vehicles (hr)	Total Number of Stops of Vehicles	Total Standstill Time of Vehicles(hr)	Average Speed of Vehicles (km/hr)	Average Travel Time of PT Passengers (sec)	Average Standstill Time of PT Passengers (sec)
14:00-15:00	330.08	210.36	14583	167.43	34.87	785.71	111.15
17:30-18:30	335.05	214.28	15074	169.37	34.69	904.5	88.45

Scenario 4: Reducing Frequencies & Increasing Capacities

In this scenario, the frequency of bus lines was reduced by half, while the capacity of buses was doubled. The objective was to enhance operational efficiency by reducing the number of trips. Table 9 presents the capacities, lengths, and occupancy at the initial stop of the newly implemented bus lines.

Table 9 (a). Scenario 4 Bus Capacities, Lengths and Occupancy at First Stop

Departure Time	Bus Code	Capacity	New Capacity	Bus Length (m)	Occupancy in First Stop
14:00	4	31	62	7.8	33
14:10	16D	76			
14:10	ER2	31	62	7.8	11
14:15	4	104			
14:20	16D	76	150	18.124	70
14:30	16D	76			
14:30	4	31	62	7.8	23
14:38	16D	76	150	18.124	42
14:45	4	27			
14:46	16D	76			
14:54	16D	76	150	18.124	48

Table 9 (b). Scenario 4 Bus Capacities, Lengths and Occupancy at First Stop

Departure Time	Bus Code	Capacity	New Capacity	Bus Length (m)	Occupancy in First Stop
17:30	16D	77	130	12.134	30
17:30	4	98			
17:40	16D	76	200	21	33
17:45	4	96			
17:50	16	98	98	12.134	16
17:50	16D	76			
18:00	16D	76			
18:10	16D	76	200	21	16
18:15	4	31	76	7.720	19
18:20	16D	76			
18:20					

Comparing the scenario results with the base scenario, the total number of vehicles stops decreased by 1%, and the average vehicle speed also decreased by 1% during the 14:00-15:00 period. In the 17:30-18:30 hour period, only the total travel time of vehicles increased by 3%. Other traffic network performance parameters were not affected by the scenario conditions. On the other hand, the reduction of PT frequency caused an increase of 21% in the average travel time of PT passengers between 14:00-15:00 hours, 14% in the 17:30-18:30 hour period, and an increase of 12% and 6% in the average standstill time of PT passengers, respectively. While the scenario had minimal impact on general traffic, it resulted in a notable increase in travel and standstill times for PT passengers.

Table 10. Scenario 4 Network Performance Results: Vehicles and PT Passengers

	Total Travel Time of Vehicles (hr)	Total Delay of Vehicles (hr)	Total Number of Stops of Vehicles	Total Standstill Time of Vehicles(hr)	Average Speed of Vehicles (km/hr)	Average Travel Time of PT Passengers (sec)	Average Standstill Time of PT Passengers (sec)
14:00-15:00	329.64	210.12	14503.	167.43	34.9	726.99.	79.13.
17:30-18:30	335.90	215.07	15142.	170.14	34.87	886.58.	80.31.

Scenario 5: Increasing Frequency & Reducing Capacity

This scenario examines the effects of doubling the frequency of PT lines while simultaneously reducing their capacity by half. The goal was to assess the impact of increased PT frequency on passenger waiting times and the use of smaller vehicles on overall traffic flow.

According to the scenario results (Table 11), reducing PT capacity and doubling its frequency between 14:00 and 15:00 resulted in a 3% decrease in total vehicle travel time, a 4% decrease in total vehicle delays, a 3% decrease in the total number of stops, and a 5% decrease in total standstill time. While these changes positively impacted traffic flow, average vehicle speeds decreased by 2%. For PT passengers, this scenario led to a 14% decrease in travel time and a 34% reduction in standstill time during this period.

Between 17:30 and 18:30, total vehicle travel time increased by 3%, and average speeds improved by 5%, with no significant effects on other traffic network performance parameters. Additionally, this scenario reduced the average travel time of PT passengers by 21% and the average standstill time by 55%.

Table 11. Scenario 5 Network Performance Results: Vehicles and PT Passengers

	Total Travel Time of Vehicles (hr)	Total Delay of Vehicles (hr)	Total Number of Stops of Vehicles	Total Standstill Time of Vehicles(hr)	Average Speed of Vehicles (km/hr)	Average Travel Time of PT Passengers (sec)	Average Standstill Time of PT Passengers (sec)
14:00-15:00	320.87	201.78	14255.	158.66	34.65	517.27	46.46
17:30-18:30	335.90	214.67	15136.	169.26	35.16	618.99	33.97

Scenario 6: Removal of Buses from PT

In this scenario, the existing bus lines were removed from the PT system, and the 404 passengers (from 14:00-15:00) and 300 passengers (from 17:30-18:30) who used these lines were redirected to other modes of transportation. Since buses were no longer available, PT passengers were directed to passenger cars and minibuses. An additional traffic volume, corresponding to the number of PT passengers, was added to the existing traffic, considering the vehicle composition provided in Table 1 (95% passenger cars & 5% minibuses for 14:00-15:00, and 93% passenger cars & 7% minibuses for 17:30-18:30).

Table 12. Scenario 6 Network Performance Results: Vehicles and PT Passengers

.	Total Travel Time of Vehicles (hr)	Total Delay of Vehicles (hr)	Total Number of Stops of Vehicles	Total Standstill Time of Vehicles(hr)	Average Speed of Vehicles (km/hr)
14:00-15:00	368.36	229.25	16229	176.97	31.54
17:30-18:30	367.90	231.30	16660	178.26	32.43

According to the scenario results (Table 12), the removal of bus lines from the PT system increases the total travel time of vehicles by 12% in both time periods compared to the base scenario. Total delays increase by 9% during the 14:00-15:00 period and by 8% during the 17:30-18:30 period. The total number of vehicles stops increases by 11% and 10%, respectively, across the two time periods. Additionally, the total standstill time of vehicles increases by 6% and 5%, respectively. The scenario also leads to a decrease in average speeds by 11% during the 14:00-15:00 period and by 7% during the 17:30-18:30 period.

Comparing Scenario Results

The base scenario presents the current status of the PT system, comparing traffic flows during peak and non-peak hours. Despite the vibrancy of Bagdat Street throughout the day, the results indicate that while traffic flow remains consistent, pedestrians experience longer waiting and travel times during peak hours.

Table 13 and Figure 4 illustrate the impact of the modeled scenarios on several key performance indicators, including total vehicle travel time, total vehicle delay, total vehicle stops, total vehicle standstill time, average vehicle speeds, average PT passenger travel time, and average PT passenger standstill time, disaggregated by peak hours.

Table 13. Scenario Results

		Scenario 1	Scenario 2	Scenario 3	Scenario 4	Scenario 5	Scenario 6
Total Travel Time of Vehicles (hr)	14:00-15:00	330.29	682.05	330.08	329.63	320.86	368.35
	17:30-18:30	327.27	719.99	335.05	335.90	335.90	367.89
Total Delay of Vehicles (hr)	14:00-15:00	210.52	594.65	210.36	210.12	201.78	229.25
	17:30-18:30	214.45	633.31	214.28	215.07	214.67	231.30
Total Number of Stops	14:00-15:00	14677	76673	14584	14504	14256	16229
	17:30-18:30	15081	80348	15075	15142	15137	16660
Total Standstill Time of Vehicles(hr)	14:00-15:00	167.35	426.98	167.43	167.43	158.66	176.97
	17:30-18:30	169.41	460.59	169.37	170.14	169.26	178.26
Average Speed of Vehicles (km/hr)	14:00-15:00	35.4	20.56	34.87	34.9	34.65	31.54
	17:30-18:30	34.71	20.75	34.69	34.87	35.16	32.43
Average Travel Time of PT Passengers (sec)	14:00-15:00	600.92	526.22	785.71	726.99	517.27	
	17:30-18:30	780.92	773.16	904.5	886.58	618.99	

continued on following page

Table 13. Continued

		Scenario 1	Scenario 2	Scenario 3	Scenario 4	Scenario 5	Scenario 6
Average Standstill Time of PT Passengers (sec)	14:00-15:00	70.86	67.5	111.15	79.13	46.46	
	17:30-18:30	75.48	70.66	88.45	80.31	33.97	

The implementation of a dedicated PT lane (Scenario 2), aimed at enhancing PT operational efficiency, was found to significantly negatively impact Bagdat Street's traffic network performance, with minimal benefits for PT passengers. The scenario results are consistent with the findings of Abroshan et al. (2021) and Yang et al. (2020). This scenario may only yield positive results if traffic demand on Bagdat Street decreases and PT passenger numbers increase.

Reducing the number of bus stops (Scenario 3) was anticipated to decrease stop-and-go movements and lane change maneuvers for PT vehicles, potentially improving overall traffic network performance. Contrary to the benefits proposed by El-Geneidy and Surprenant-Legault (2010) and Ercolano (2011), Scenario 3 results indicate that there is no significant improvement in traffic flow. However, a substantial increase in PT passengers' standstill and travel times has been observed. The reasons for this situation include the very low percentage of PT in the traffic flow on Bagdat Street and the high number of PT passengers.

Figure 4. Comparing Scenarios Results

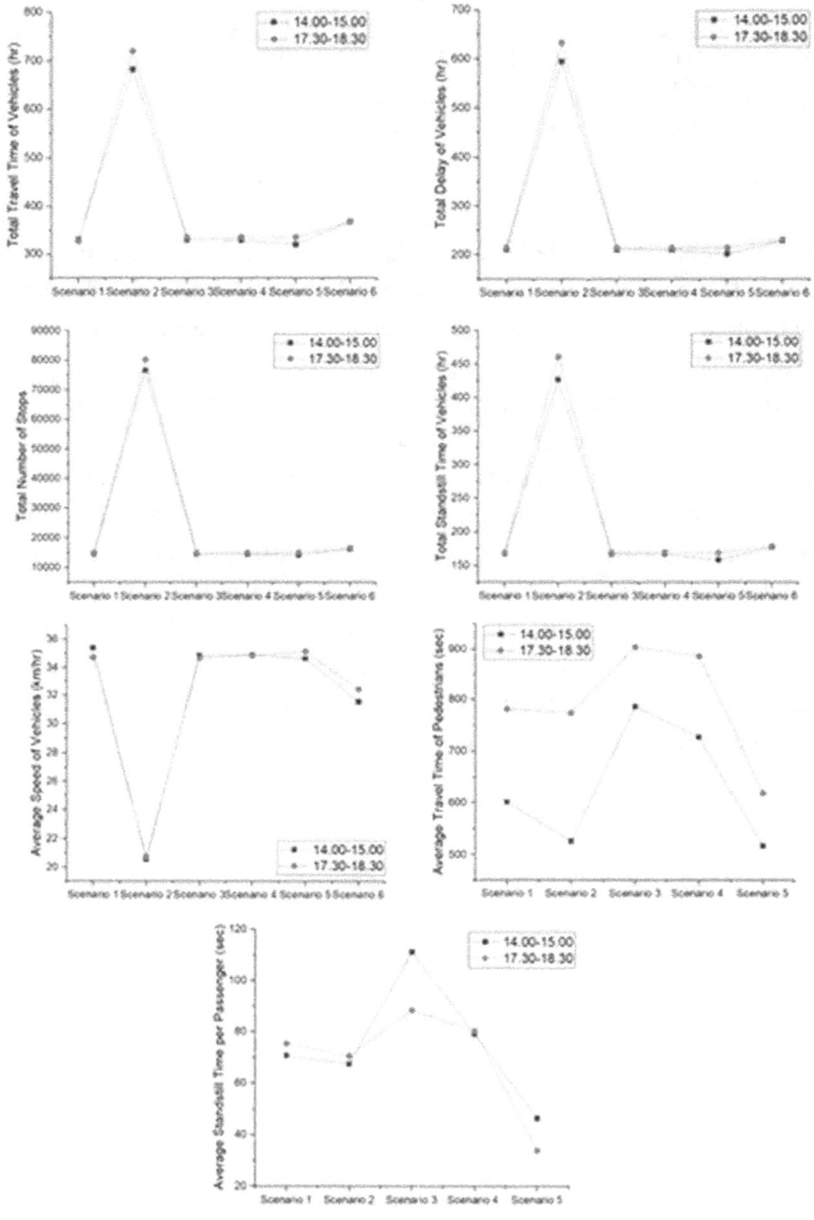

Additionally, to facilitate a clearer understanding of the scenario results, the speeds of road segments in each scenario have been mapped (Figure 5). This can be useful for observing the potential impacts of the policies implemented in these scenarios on traffic flow.

Figure 5. Scenario Results with Speed Maps

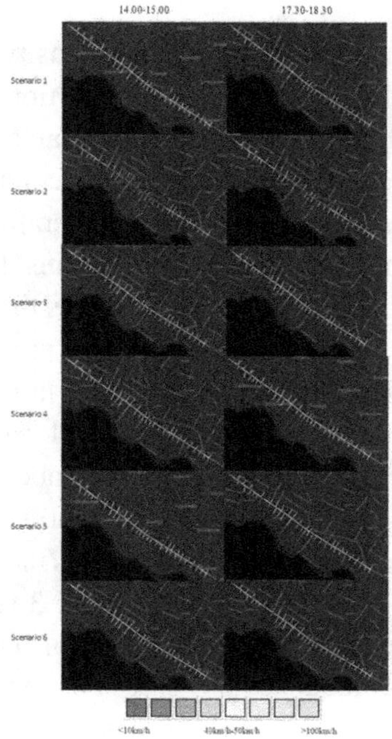

Modifications to PT frequency and capacity (Scenarios 4 and 5) have implications for both traffic network performance and PT passengers. Scenario 5, which increases PT frequency and reduces capacity, improved traffic network performance between 14:00 and 15:00. Although Scenario 5 slightly negatively influenced traffic network performance between 17:30 and 18:30, it significantly benefited PT passengers in both time periods. Given the limited impact of Scenario 5 on traffic network performance, its implementation is expected to enhance PT passenger satisfaction by improving travel and waiting times.

Scenario 6, which involves eliminating bus lines from the transportation system and redistributing PT passengers to passenger cars or minibuses, adversely impacted all traffic network parameters under consideration.

DISCUSSION

Bagdat Street exhibits a relatively high traffic density compared to other streets, regardless of the time of day. Currently, the proportion of PT buses in the traffic composition is only 1%. However, the size and maneuverability of buses in the current traffic can significantly impact traffic flow compared to passenger cars.

In this study, alternative scenarios for PT bus operations were explored, analyzing network performance for vehicles and pedestrians. Passenger satisfaction was assessed based on waiting times at bus stops and travel time. Additionally, a scenario was tested where bus lines were removed due to the presence of privately operated yellow minibuses and parallel rail systems. It's important to recognize that bus stops located near health and educational institutions with easy pedestrian access align with inclusive transportation policies. The total number of passengers boarding in this area is 400 during off-peak hours (14:00) and 300 during peak hours (17:30).

To summarize the scenarios: Scenario 1 serves as the base scenario. In Scenario 2, a dedicated lane is allocated for buses. Scenario 3 reduces the number of bus stops by half. Scenario 4 decreases the frequency of buses while increasing their capacity. In Scenario 5, the frequency of buses is increased, while their capacity is reduced. Finally, in Scenario 6, buses are completely removed, and passengers are redirected to other vehicles, excluding heavy vehicles.

As illustrated in Figure 4, Scenario 2 has the most adverse impact across all time intervals. During peak hours, total vehicle travel time increased by 2.2 times, total vehicle delay by 3 times, the total number of stops by 5.3 times, and total standstill time by 2.7 times, while average vehicle speed decreased by 0.6. Meanwhile, the average travel and standstill times for PT passengers remained relatively unchanged. The other scenarios do not notably affect vehicle-related measurements. However, Scenario 3, which reduces the number of stops by half, increases the average travel time of PT passengers by 1.3 times and their average standstill time by 1.6 times. Conversely, Scenario 5 leads to a decrease of 0.8 times in the average travel time of PT passengers and 0.5 times in the average standstill time of PT passengers. The results of Scenario 6 indicate that removing the bus line does not positively impact traffic congestion.

Given these findings, the most effective strategy for PT buses would be to increase the frequency of trips and use smaller vehicles. On a road where yellow minibuses do not have scheduled departure times and only depart when full, it can be said that an increase in the number of yellow minibuses within the traffic composition reflects the increase in demand. An analysis of the current traffic composition reveals that the 5% off-peak rate of yellow minibuses rises to 7% during peak hours. This increase is due to yellow minibuses reaching full capacity faster during peak times, thereby driving higher passenger demand. Therefore, optimizing the capacity of PT buses would not only reduce passengers' wait and travel times but also be more economical, as passengers can use transportation cards and transfer to other PT systems, enhancing overall passenger satisfaction.

FUTURE RESEARCH DIRECTIONS

The rapid expansion of urban areas brings inevitable traffic congestion and transportation challenges, especially in densely populated regions. To address these issues, many authorities advocate for PT as a solution. However, optimizing PT systems requires a comprehensive approach that includes not only physical improvements but also enhancements in passenger satisfaction to attract and retain users. In this study, traffic simulation was run to analyze the PT system on Bagdat Street under six different scenarios, identifying an increase in frequency and a reduction in capacity as the most effective solutions for minimizing total transportation time.

Future research could explore how these findings can be generalized to other urban areas with similar traffic dynamics, particularly those with more pronounced peak-hour congestion. Additionally, examining the long-term impacts of the proposed PT strategies on passenger satisfaction and overall transportation efficiency could provide further insights into optimizing PT systems in urban environments. Understanding the relationship between traffic flow conditions, PT operations, and travel demands during both peak and off-peak hours could also yield valuable data for urban planners and transportation engineers.

CONCLUSION

The findings of this study offer practical solutions for enhancing PT services on Bagdat Street, with potential applicability to other urban areas facing similar challenges. By increasing bus frequency and reducing vehicle capacity, PT systems can reduce total transportation time, contributing to a more efficient and sustainable urban transportation network. This research provides valuable data for urban planners and policymakers who aim to optimize PT systems and improve urban quality of life.

Ultimately, these strategic approaches will support the development of more livable, sustainable, and accessible cities. Continued research and thoughtful implementation of these strategies are essential for addressing the transportation needs of growing urban populations and achieving long-term urban sustainability goals.

ACKNOWLEDGMENT

This study was supported by Scientific and Technological Research Council of Turkey (TUBITAK) under the Grant Number 123M063, and the Istanbul Metropolitan Municipality. The authors thank to TUBITAK, Istanbul Metropolitan Municipality, and IETT for their supports.

REFERENCES

Abenoza, R. F., Cats, O., & Susilo, Y. O. (2017). Travel satisfaction with public transport: Determinants, user classes, regional disparities and their evolution. *Transportation Research Part A, Policy and Practice*, 95, 64–84. DOI: 10.1016/j.tra.2016.11.011

Abroshan, H., Luckana, R., & Wheat, D. (2021). Planning a dedicated bus lane to decrease traffic congestion considering policy implementation. In *39th International System Dynamics Conference*.

Bjertnæs, G. H. (2019). Efficient taxation of fuel and road use. *CESifo Working Paper* No. 8019

Conlon, M., Foote, P., O'Malley, K., & Stuart, D. (2001). Successful arterial street limited-stop express bus service in Chicago. *Transportation Research Record: Journal of the Transportation Research Board*, 1760(1), 74–80. DOI: 10.3141/1760-10

De Oña, J., Estévez, E., & De Oña, R. (2020). Perception of public transport quality of service among regular private vehicle users in Madrid, Spain. *Transportation Research Record: Journal of the Transportation Research Board*, 2674(2), 213–224. DOI: 10.1177/0361198120907095

De Vos, J., Waygood, E. O. D., & Letarte, L. (2020). Modeling the desire for using public transport. *Travel Behaviour & Society*, 19, 90–98. DOI: 10.1016/j.tbs.2019.12.005

Del Castillo, J. M., & Benitez, F. G. (2013). Determining a public transport satisfaction index from user surveys. *Transportmetrica A: Transport Science*, 9(8), 713–741. DOI: 10.1080/18128602.2011.654139

El-Geneidy, A. M., & Surprenant-Legault, J. (2010). Limited-stop bus service: An evaluation of an implementation strategy. *Public Transport (Berlin)*, 2(4), 291–306. DOI: 10.1007/s12469-010-0030-9

Ercolano, J. M. (2011). Limited-stop bus operations: An evaluation. *Transportation Research Record: Journal of the Transportation Research Board*, (994), 24–29.

European Commission. (2017). Clean transport, Urban transport. http://ec.europa.eu/transport/themes/urban/urban_mobility/urban_mobility_actions/public_transport_en. Accessed 27th January 2017.

Feldman, O. (2012). The GEH measure and quality of the highway assignment models. *Association for European Transport and Contributors*, 1-18.

Hacıoğlu, E., Eren, İ., & Polatoğlu, Ç. (2021). Enhancing socio-spatial activities in urban areas through urban furniture: Case of Bagdat Avenue-Istanbul. In *Contemporary Issues in Architecture and Urban Planning* (pp. 70–95). Architectural & Urban Forms.

Hasan, U., Whyte, A., & Jassmi, H. A. (2021). Public bus transport service satisfaction: Understanding its value to urban passengers towards improved uptake. *Transactions on Transport Sciences*, 12(1), 25–37. DOI: 10.5507/tots.2021.002

Héran, F., & Ravalet, E. P. J. (2008). *La consommation d'espace-temps des divers modes de déplacement en milieu urbain: Application au cas de l'île de France.* PREDIT.

Hussain, B., Zefreh, M. M., & Torok, A. (2018). Designing the appropriate data collection method for public transport passenger satisfaction analysis. *IJTTE. International Journal for Traffic and Transport Engineering*, 8(2), 177–183. DOI: 10.7708/ijtte.2018.8(2).03

INRIX. (2024). INRIX 2023 Traffic Scorecard Report. https://inrix.com/scorecard-city-2023/?city=Istanbul&index=5

Isa, N., Mohamed, A., & Yusoff, M. (2015). Implementation of dynamic traffic routing for traffic congestion: A review. In *Soft Computing in Data Science: First International Conference, SCDS 2015, Putrajaya, Malaysia, September 2-3, 2015* [Springer Singapore.]. *Proceedings*, 1, 174–186.

Jara-Díaz, S., Fielbaum, A., & Gschwender, A. (2017). Optimal fleet size, frequencies and vehicle capacities considering peak and off-peak periods in public transport. *Transportation Research Part A, Policy and Practice*, 106, 65–74. DOI: 10.1016/j.tra.2017.09.005

Jara-Díaz, S. R., & Gschwender, A. (2009). The effect of financial constraints on the optimal design of public transport services. *Transportation*, 36(1), 65–75. DOI: 10.1007/s11116-008-9182-8

Jen, W., Tu, R., & Lu, T. (2011). Managing passenger behavioral intention: An integrated framework for service quality, satisfaction, perceived value and switching barriers. *Transportation*, 38(3), 321–342. DOI: 10.1007/s11116-010-9306-9

Kaffash, S., Nguyen, A. T., & Zhu, J. (2021). Big data algorithms and applications in intelligent transportation system: A review and bibliometric analysis. *International Journal of Production Economics*, 231, 107868. DOI: 10.1016/j.ijpe.2020.107868

Lai, W. T., & Chen, C. F. (2011). Behavioral intentions of public transit passengers: The roles of service quality, perceived value, satisfaction and involvement. *Transport Policy*, 18(2), 318–325. DOI: 10.1016/j.tranpol.2010.09.003

Li, J. Q., Song, M. K., Li, M., & Zhang, W. B. (2009). Planning for bus rapid transit in single dedicated bus lane. *Transportation Research Record: Journal of the Transportation Research Board*, 2111(1), 76–82. DOI: 10.3141/2111-10

Mathew, S., Pulugurtha, S. S., & Mane, A. S. (2020). Effect of toll roads on travel time reliability within its vicinity: A case study from the state of North Carolina. *Transportation Letters*, 12(9), 604–612. DOI: 10.1080/19427867.2019.1671043

Mokonyama, M., & Venter, C. (2013). Incorporation of customer satisfaction in public transport contracts–A preliminary analysis. *Research in Transportation Economics*, 39(1), 58–66. DOI: 10.1016/j.retrec.2012.05.024

Mouwen, A. (2015). Drivers of customer satisfaction with public transport services. *Transportation Research Part A, Policy and Practice*, 78, 1–20. DOI: 10.1016/j.tra.2015.05.005

Mugion, R. G., Toni, M., Raharjo, H., Di Pietro, L., & Sebathu, S. P. (2018). Does the service quality of urban public transport enhance sustainable mobility? *Journal of Cleaner Production*, 174, 1566–1587. DOI: 10.1016/j.jclepro.2017.11.052

Ni, A., Zhang, C., Hu, Y., Lu, W., & Li, H. (2020). Influence mechanism of the corporate image on passenger satisfaction with public transport in China. *Transport Policy*, 94, 54–65. DOI: 10.1016/j.tranpol.2020.04.014

Paricio, A., & Lopez-Carmona, M. A. (2019). Urban traffic routing using weighted multi-map strategies. *IEEE Access : Practical Innovations, Open Solutions*, 7, 153086–153101. DOI: 10.1109/ACCESS.2019.2947699

Pei, F., Tang, W., & Cheng, K. (2006). Review on research and application of customer satisfaction. *World Standardization and Quality Management*, 10, 4–7.

Redman, L., Friman, M., Gärling, T., & Hartig, T. (2013). Quality attributes of public transport that attract car users: A research review. *Transport Policy*, 25, 119–127. DOI: 10.1016/j.tranpol.2012.11.005

Renne, J. L., Hamidi, S., & Ewing, R. (2016). Transit commuting, the network accessibility effect, and the built environment in station areas across the United States. *Research in Transportation Economics*, 60, 35–43. DOI: 10.1016/j.retrec.2017.02.003

Sami, Z., & Dincer, M. E. E. (2020). Study of physical, social and functional structures in Bagdat Street (Istanbul) as public space. *Civil Engineering and Architecture*, 8(3), 224–236. DOI: 10.13189/cea.2020.080305

Stefanello, F., Buriol, L. S., Hirsch, M. J., Pardalos, P. M., Querido, T., Resende, M. G., & Ritt, M. (2017). On the minimization of traffic congestion in road networks with tolls. *Annals of Operations Research*, 249(1-2), 119–139. DOI: 10.1007/s10479-015-1800-1

Storchmann, K. H. (2001). The impact of fuel taxes on public transport—An empirical assessment for Germany. *Transport Policy*, 8(1), 19–28. DOI: 10.1016/S0967-070X(00)00033-0

TUIK. (2024). Adrese Dayalı Nüfus Kayıt Sistemi Sonuçları, 2023. https://data.tuik.gov.tr/Bulten/Index?p=Adrese-Dayali-Nufus-Kayit-Sistemi-Sonuclari-2023-49684

Van Lierop, D., & El-Geneidy, A. (2016). Enjoying loyalty: The relationship between service quality, customer satisfaction, and behavioral intentions in public transit. *Research in Transportation Economics*, 59, 50–59. DOI: 10.1016/j.retrec.2016.04.001

Vicente, P., Sampaio, A., & Reis, E. (2020). Factors influencing passenger loyalty towards public transport services: Does public transport providers' commitment to environmental sustainability matter? *Case Studies on Transport Policy*, 8(2), 627–638. DOI: 10.1016/j.cstp.2020.02.004

Waygood, E. O. D., Sun, Y., & Susilo, Y. (2014). Transportation carbon dioxide emissions by built environment and family lifecycle: Case study of the Osaka metropolitan area. *Transportation Research Part D, Transport and Environment*, 31, 176–188. DOI: 10.1016/j.trd.2014.06.001

Wu, I., & Pojani, D. (2016). Obstacles to the creation of successful bus rapid transit systems: The case of Bangkok. *Research in Transportation Economics*, 60, 44–53. DOI: 10.1016/j.retrec.2016.05.001

Yang, G., Wang, D., & Mao, X. (2020). Modelling the modal shift effects of converting a general traffic lane into a dedicated bus lane. *Promet (Zagreb)*, 32(5), 625–637. DOI: 10.7307/ptt.v32i5.3375

Zefreh, M. M., Hussain, B., & Sipos, T. (2020). In-depth analysis and model development of passenger satisfaction with public transportation. *KSCE Journal of Civil Engineering*, 24(10), 3064–3073. DOI: 10.1007/s12205-020-1871-7

Zhang, M., Li, Z., Si, H., Cheng, L., Zhou, X., & Wang, B. (2023). Urban travel time and residential location choice: The impacts of traffic congestion. *Sustainable Cities and Society*, 99, 104975. DOI: 10.1016/j.scs.2023.104975

Zhang, X., Liu, H., Xu, M., Mao, C., Shi, J., Meng, G., & Wu, J. (2020). Evaluation of passenger satisfaction of urban multi-mode public transport. *PLoS One*, 15(10), e0241004. DOI: 10.1371/journal.pone.0241004 PMID: 33079972

APPENDIX

Table 14 (a).

	4- 14:00		16D- 14:10:00	
Stop's name	**Passenger boarding**	**Passengers Alighting (%)**	**Passenger boarding**	**Passengers Alighting (%)**
Bostancı	11	23%	2	5%
Tatlıçeşme	1	0%		
Çatalçeşme	3	0%		
Tan sokak	1	0%		
Suadiye	2	10%	1	15%
mücahit sokak	3	18%	5	3%
Şaşkınbakkal	7	17%	8	5%
Kantarcı			3	0%
Erenköy	4	6%	4	21%
Caddebostan	2	10%	5	2%
Yeniyol	1	7%	17	0%
	16D- 17:30		**4- 17:30**	
Stop's name	Passenger boarding	Passengers Alighting (%)	Passenger boarding	Passengers Alighting (%)
Bostancı	1	14%	6	18%
Tatlıçeşme			1	0%
Çatalçeşme	1	0%		
Tan sokak			2	0%
Suadiye	5	8%	2	0%
mücahit sokak			1	0%
Şaşkınbakkal	3	15%	4	0%
Kantarcı	4	6%	5	0%
Erenköy	8	14%	9	2%
Caddebostan	6	8%	10	11%
Yeniyol	3	0%	6	0%

Table 14 (b).

ER2-14:10		**4-14:15**		**16D- 14:20**	
Passenger boarding	**Passengers Alighting (%)**	**Passenger boarding**	**Passengers Alighting (%)**	**Passenger boarding**	**Passengers Alighting (%)**
		2	0%	6	19%
		3	0%	1	14%
		1	29%		
		3	0%	3	45%
		2	0%	1	0%
				5	0%
		3	0%	8	0%
2	18%	8	0%	4	48%
3	8%	6	0%	12	0%
5	6%	6	0%	13	4%
16D- 17:40		**4- 17:45**		**16- 17:50**	
Passenger boarding	Passengers Alighting (%)	Passenger boarding	Passengers Alighting (%)	Passenger boarding	Passengers Alighting (%)
8	47%	2	0%	12	30%
2	12%	1	0%		
		2	0%		
2	19%	5	0%	4	0%
2	0%			2	0%
6	9%	2	0%	2	0%
4	8%	6	0%	1	0%
6	43%	9	0%	3	0%
6	0%	14	0%	1	0%
		6	0%	1	0%

Table 14 (c).

16D-14.30		4-14:30		D16- 14:38:00	
Passenger boarding	**Passengers Alighting (%)**	**Passenger boarding**	**Passengers Alighting (%)**	**Passenger boarding**	**Passengers Alighting (%)**
7	23%	5	71%	2	17%
1	24%	5	0%		
1	0%			1	8%
		2	0%		
5	3%	2	0%	1	44%
		4	5%	3	0%
4	3%	4	0%	2	17%
2	9%		7	0%	
6	31%	4	6%	2	0%
3	0%	6	9%	2	0%
3	3%	10	2%		

16D- 17:50		16D- 18:00		16D- 18:10	
Passenger boarding	Passengers Alighting (%)	Passenger boarding	Passengers Alighting (%)	Passenger boarding	Passengers Alighting (%)
3	50%	1	13%		
				2	40%
				2	0%
				1	0%
1	17%	2	0%	2	0%
3	0%	2	0%	1	0%
7	23%	3	0%		
				1	33%
1	0%	1	7%	1	0%
4	6%	8	0%		

Table 14 (d).

4-14:45		16D- 14.46		16D-14:54	
Passenger boarding	**Passengers Alighting (%)**	**Passenger boarding**	**Passengers Alighting (%)**	**Passenger boarding**	**Passengers Alighting (%)**
14	81%	5.00	23%	4	15%
5	0%	7.00	0%	4	7%
1	0%				
2	0%				
1	0%	5.00	3%	2	15%
4	60%				
10	9%	8.00	3%	1	13%
	5	0%	2.00	5%	3
12	0%	3.00	54%		
10	0%	4.00	0%	7	13%
8	0%	3.00	0%	3	0%
4- 18.15		**16D- 18:20**			
Passenger boarding	Passengers Alighting (%)	Passenger boarding	Passengers Alighting (%)		
10	69%	2	25%		
1	100%				
1	0%	3	10%		
4	0%				
3	0%	4	0%		
5	0%				
18	0%				
12	0%				
4	0%	1	15%		

Chapter 9
Road Aesthetics and Its Effect on Road Users

Anuj Jaiswal
Maulala Azad National Institute of Technology, Bhopal, India

Arshi Parashar
https://orcid.org/0000-0002-1483-5223
Maulala Azad National Institute of Technology, Bhopal, India

Vicky Lalramsangi
https://orcid.org/0000-0002-3792-4544
Maulana Azad National Institute of Technology, Bhopal, India

ABSTRACT

Aesthetics is synonymous to beauty. Anything which adds beauty to the road is a component of road aesthetics. Like safety, utility and comfort; road aesthetics is of equal importance for our road system. Road aesthetics is having a great value to the public and to the visitors who use it every day and in-turn becomes a field of great interest to planners, engineers, architects and designers. It influences road user's perception and behavior and becomes important as effective time spent on roads has increased manifold with increasing globalization and better transport and communication facilities. This paper investigates into the need and the technicalities of an aesthetically positive road environment taking into consideration human characteristics. A possible relationship between aesthetics and behavior is described through interdisciplinary research within arts & aesthetics, environmental psychology and traffic safety.

DOI: 10.4018/979-8-3693-6695-0.ch009

INTRODUCTION

Aesthetics is one of the main factors for visual comfort. Apart from visual, it gives a quotient of psychological and mental well-being. An aesthetically pleasing environment aids in keeping our minds calm and balanced. One such field where aesthetics play an important role is road design. Roads carry traffic and often remain to become just the functional lifeline of the city without having its soul. Thus, the need for road aesthetics has emerged in the field of transportation planning and design as a viable topic for research. Various components contribute to Road Design, and interestingly, each can be seen with a sense of beauty. The Inclusion of landscape elements, vibrant streetscape, apt lighting, colourful wall art or interactive art installations along the street all contribute to road aesthetics. Ecosystem services (ES), effectively address urban, social, and environmental concerns and benefit from such interventions (Bolund and Hunhammer, 1999; Gomez et al., 2013).

Generally, municipalities are in favour of introducing a variety of nature-based solutions (NBS) into their streets and are even willing to give up a sizable amount of space in exchange for implementing NBS. The characteristics of street greening preferences were revealed by a survey found to be significantly correlated with age and car ownership. Moreover, the result of focus group interviews (FGI) indicated that living on a very green street was preferred and the implementation of NBS was linked to several concerns such as changes to mobility, safety, noise and potential conflicts with other users (Philips et al., 2023).

Dzhambov (2018) made recommendations for enhancing street-scrape's perceived safety and visual appeal both during the day and night. These can help with better streetscape design in cities. These findings should be taken with caution as the in-situ situation might not be accurately reflected using photograph-based analysis. Measuring the shifts in people's presence has been crucial for reconstructing affective environments (Abusaada and Elshater, 2021). According to a study, walking preferences were found to be strongly correlated to the variety and perceived pleasantness of the environment which includes features like facades, colour, aspect ratios, maintenance and greenery (Balasubramanian et al., 2022). Lately, researchers have demonstrated the application of each particular criterion as well as how the idea of sustainable entire streets can improve safety, liveability, environment and urban aesthetics (Montella et al., 2022). The degree to which pedestrians are satisfied with their constructed surroundings, thermal perception, perceived air quality and loudness can all be used to measure pedestrian comfort and the combined effect of thermal perception. Individual recreational walking comfort was found to be mediated by thermal feeling, perceived air quality and loudness in that order concerning objectively measured factors (Ma et al.,2016).

In this study first, the components of Road Design that contribute to aesthetics and user behaviour are discussed. Then, the experience of road users is illustrated with live case examples followed by elements of Road Aesthetics. Finally with the help of some case studies the effect of road aesthetics on road users is discussed. In the end, the study concluded with good and bad road environment for the road users.

COMPONENTS OF ROAD DESIGN CONTRIBUTING TO AESTHETICS AND USER BEHAVIOUR

Road design is a function of both aesthetics and the surrounding environment. Creating a pleasant road environment consists of seeing and understanding these qualities and incorporating them into our aesthetic experience of the road. It consists of the following components:

- ***Design***

  Design is composed of aesthetic, functional, technical & economic deliberations and physical conditions of a site. The definition of design is frequently given as the balancing and coordinating of technological, utilitarian, and aesthetic factors. Even more closely linked to a specific location and set of concrete conditions is road design. This places requirements on illumination, safety, visibility, and technical design. There aren't many options for customization because these premises have to take aesthetics into account.
- ***Planning***

  The planning of roads is crucial to the design of roadways. Planning controls land allocation and urban development by defining and placing roadways long before their actual design is considered. Planning of road layout and their placement decides what all road touches in the immediate vicinity as built, unbuilt and natural features. It decides upon which major activities are connected through it, which in turn these connections surroundings and activity connections dictate the mood, aspiration and expected behaviour on the road. For example, a road connecting a major housing hub with a movie theatre, which passes through a lake on one edge, will have different user behaviour and aesthetic needs than a road connecting a religious space with a housing hub, which passes through a concrete jungle.
- ***History***

Every place has a history, and so does every road. The place, its environs, climate, and culture all constitute its history. When selecting the ideal architectural design for the road environment, consideration can be given to both the history of the road itself and its surroundings. If a place has historical qualities or any episodic characteristics, design elements that remind of the rich past can be installed on the road to remind its users of the road's history.

- ***Safety***

  Safety becomes very important in road design, as it contributes majorly to both functionality and aesthetic pleasure. Experiencing safety on the road allows users to indulge in aesthetic pleasure and successfully complete the purpose of commuting without fear. Safety for road users is the most prior quality in road design. A road, no matter how beautiful, cannot be beneficial if it is not safe. To aid in orientation for all users, roads and streets should be planned with clarity and simplicity in mind. Getting a general idea of the road, walkways, bike paths, and side-road intersections should be simple for drivers. A road should be equally safe for its all users whether motorists, car users, cyclists or pedestrians. A possible relationship between aesthetics and safety may be explored through interdisciplinary research within arts and aesthetics, environmental psychology and traffic safety. This may reveal less tangible areas of road design in which both architects and engineers can find potential support when planning for an aesthetically positive road environment taking into consideration human characteristics (Drottenborg, 2001).

- ***Utility***

  Road design, without utility, would be a failed design attempt. So, the utility of commuting (for all categories) must not fail, but it could bring with it the inherited quality of aesthetics to make a utility plus aesthetic road design. This would always create a positive impact on road users. For example, roadside light fixtures become necessary for night-time commuting, so an aesthetically soothing gesture of the same can be made. Figure 1 illustrates similar gestures of street-scaping.

Figure 1. Utility and road aesthetic

- ***Comfort***

  Comfort while commuting becomes an important component of road design. It includes the comfort of vision, in making decisions, in waiting, in moving, tactile comfort, and other similar aspects. Aesthetic elements as part of the design may contribute to sensory comfort to road users. Retaining or planting trees along the edges of roads may bring the comfort of shade and vision for slow-speed commuters and bring visual comfort to fast commuters. Figure 2 illustrates comfort as a component of road design.

Figure 2. Road section of a Trivandrum street, trying to retain maximum trees to increase visual comfort and shade for commuters

- ***Ecology***

  In addition to being a major cause of air and noise pollution, road transportation has other drawbacks such as decreased scenic value, insecurity, accidents, and unhealthy environments. An overall analysis of the interaction between the environment and the traffic system should be done at an early design stage since some of these negative consequences can be mitigated by careful planning. This will allow ecology to be included in road aesthetics, which will benefit to the users.

- ***Road Legislation***

  Although, good architecture, beautiful roads, and beauty are not the function of laws or regulations, but it has the potential to indirectly increase the prerequisites and opportunities for integrating aesthetic elements into road design. It could also govern the future maintenance, usage patterns and guidelines for road and road users.

- ***Aesthetics***

  Road aesthetics deal with all those features that add beauty, pleasantness, and attraction to the road indirectly charming the users including citizens and tourists. With stringent planning and civil rules, efforts to beautify the roads have decreased. However, in the recent past, many studies have been done to find the relationship between road aesthetics and its effect on users.

  All these components contribute fairly to user experience and road aesthetics. To understand the effect of road aesthetics on the users, the prerequisite is to find out the actual experience of the traffic and road design by the users and their reactions to it. These experiences are explained in the following section.

Experience of road users

The effect of traffic and road elements on users is often perceived by the users as an experience. From the perspectives of those inside motor vehicles as drivers or passengers and those outside of them as bicycles or walkers, this section provides a phenomenological analysis and account of the aesthetic experience of the road and its traffic. The experience includes dynamic traffic and static elements of road design. Dynamic traffic impacts user experience through sound, visual appearance, speed, and its composition. With these factors taken into account, two components of the aesthetic experience are described: the sensory experience of the traffic and road elements, and the cognitive or logical way that the mind experiences the road and its traffic.

Figure 3. Range of senses by Malnar and Vodvarka

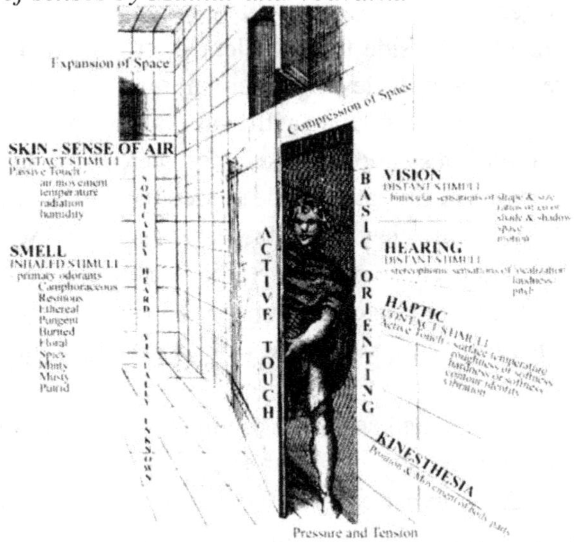

- ***Sensory Experience***

  The experiential phenomena deal with the process of sensation, perception and conception. It's not just a vision but all other senses also like smell, touch, sound, and sense of obstruction to your body together participate in the whole process of experience (Fig. 3).

A sensory experience of a road involves all senses, with the factors of a user being in motion or the surroundings in motion. A sensory experience of a road even includes perception at a subconscious level. These various layers of experience have been classified into three categories and elaborated.

- ***Multisensory***

  It is the experience of finding a road aesthetically pleasing because of the combined effect of visual, auditory, tactile and olfactory sensations from all the senses.

- ***Cyclists/Pedestrians:***

  For non-motorized road users, the experience of moving on a road with all their senses working is more unpleasant. Since traffic today is synonymous with more noise, smoke, dust, and cacophony, in such an environment the overall message conveyed by eyes, nose and ears is full of irritation and dissatisfaction. In the case of roads with limited traffic, green belts alongside and a pleasant view from it, the scope to enjoy and take pleasure from the road increases in the case of pedestrians or cyclists.

- ***Drivers/Passengers:***

  Experience from inside the vehicles is completely different. It can be compared to a movie viewing where all things are experienced without sensing it. Considering a road with heavy traffic volume, for people inside a vehicle there is less smoke intake, a complete noise barrier is created and there is less tension created by the cluttered environment. For a calm and organized road, the view offered is less, there is less scope to capture the beauty around because of the metallic body around. Comparatively, for drivers with higher speed, their vision is focused on the way forward and they are forced to curb their senses unless the road is quiet and non-risky.

- **Kinaesthetic**

  The terms kinetic and aesthetic combine to form the word kinaesthetic. The sensation in motion is connected to this one. As a matter of fact, the kinaesthetic experience of moving around the city is essentially a sensory amalgam of sight, sound, and touch. Because of the motion's consistency, economical use of time, money, and effort, and satisfaction of a goal fulfilled, the experience of motion is pleasant and gratifying.

- *Cyclists/Pedestrians:*

  Cyclists or pedestrians spend a lot of energy on walking or paddling for their movement; thus, this satisfaction of constant motion is not experienced by them unless they are on a slope, or at ease. Moreover, their experience of motion is always felt in relation to other vehicles which are mostly motorized. Consequently, this feeling of slow motion can be unsatisfying. In the case of dense coagulated traffic, when one can see huge cars or metal boxes planted at their positions for hours, pedestrians or cyclists find an easier way to escape without breaking their motion. This feeling of being moving in comparison to those stagnated in congestion is extremely happy and gratifying. This kinaesthetic happiness is often interrupted when cyclists or pedestrians stop for bigger vehicles to pass at times of usual traffic flow.

- *Drivers/Passengers:*

  As discussed earlier, motion without physical effort is happy and rather exciting at great speed for some. But at times of traffic jams or congestion, it is irritating and even frustrating when extended to hours. Since the expectation of reaching a destination is high in the case of motorized vehicles, it is even more irritating. Moreover, the feeling of being trapped in a box while ingrained in a jam is disturbing. It is this reason of high expectations and frustration that follows that one finds more accidents of motorized vehicles.

- **Vicarious or Indirect**

  This is the sensation of the road and its environment which we perceive subconsciously or rather effortlessly. All those sensations which are felt but are not given enough attention by the users, fall in this category. The presence of a clean or light surface road is pleasant to travel on or a road with an exciting view is always a pleasure to commute on. On the contrary, a road with ill views, bad smells, or uninviting surface colour give dull and unpleasant experience. The difficulty of communicating on a busy road or the mind being too busy to make oneself safe on the road are other examples of this sensation.

- **Cognitive experience**

  Cognitive experience is what we judge by our mind and reasoning and not what our senses make us feel. The reason we hold handles in buses or even cars is primarily "cognitive," as it stems from our awareness of the potential consequences of not getting support, rather than just feeling the presence of moving vehicles. The sense of experiencing the city through signage is also cognitive in nature. The users relate to moving motor vehicles as if they are moving *objects*, not moving people. Anything which comes in their way is felt as an objection whether it is an object or people or vehicles with humans inside them. *Also,* on the road everyone whether drivers, cyclists, or pedestrians is found alert to save themselves from getting harmed. Here, all sensations are busy in being alert with little or no scope to appreciate the beauty of the trip. At times of high fast-moving traffic, road users are extremely active or alert since a moment of negligence can spark accidents. The usual road users, especially in the metro cities are so used to the fast pace that they get bored when they are without work e.g. waiting at the bus stop, etc. On the contrary, if the traffic is slow moving, it makes the senses numb and dull. This stifling of sensory and cognitive experience is made worse for drivers and passengers in crowded streets by the fact that when they are seated and stationary, their tactile, auditory, and olfactory sensations are put "on hold."

From the above experience of the road user, a few points have been analysed. Firstly, road aesthetics should make the experience both pleasant and safe, so no road aesthetic element should be too loud or distracting that it results in road chaos. Subtle aesthetic elements with finer details, so that it never distracts a fast commuter but a person who is going slow and wants to stay and give things a closer look could enjoy the detail (Figure.4).

Figure 4. Subtle aesthetic elements of road

Secondly, the aesthetic experience of the road users is quick, alert, and full of risks and emotions. In this situation, the need is to provide a road environment which is pleasant and soothing on one hand and safe and orienting on the other. It is the imprint that a road environment leaves on the user's mind. Further, our intention is to study few techniques which help improve the road environment and aesthetics, eventually enhancing the road users' experience.

Elements Of Road Aesthetics

Street Furniture

As the name suggests it is the furniture of the road. It is the name given to all those accessories that enhance the road and add to its beauty and function. It includes bus stops, benches, lamp posts, dustbins, planters, cycle stands, signage, public sculpture, post boxes, etc. They are installed on the road and can be designed with different colour schemes, themes, or concepts (Figure 5). They add an aesthetic pinch to the street and create interest for onlookers and travelers. These are the elementary components of the street without which a street lacks composition and utility.

Figure 5. Innovative roadside furniture, incorporating both functionality and aesthetics

Lighting

Lighting is an important architectural function. It is a feature that is required to add beauty to everything around. Street lighting is even more important since it is a fundamental requirement for the safety and beauty of streets at night hours. The characteristics of the light fixtures like type, colour, height, dimensions, and spacing determine the experience of the users. Dimensions must be harmonized with the surrounding scale and the type of fixture must illuminate what it is intended to. The light should be of good hue with little illumination of the surroundings, and attention should be paid to the night sky to avoid glare. When a road's lighting fixtures are consistent and harmoniously emphasize its magnitude, the surroundings become more appealing *(Beauty Full Road handbook)*.

Road Surface

Road paving is one of the elements which forms a major part of the road appearance. Asphalt, granite, concrete, gravel, and pavers are all options for paving the city's floors. Patterns and materials can serve to define the street area and make it more comprehensible, appealing, and intriguing. By varying the paving type, colour or level, you can also improve traffic flow and separation.

Plantation

Green cover in the form of plantations, grass, or trees is a major factor that inserts beauty in the road environment. The visual expression of the road is primarily influenced by the roadside vegetation. They can also significantly alter their settings, drawing attention to or hiding features and structures from their surroundings. All types of vegetation found beside highways are referred to as road plantings, including avenues, wayside trees, forests, groves, hedgerows, hedges, shrubs, grass, and flowers. The general layout of the road and the surrounding landscape or metropolis should inform the usage of vegetation in the road environment. Plantings should be considered in relation to other areas of road planning, such as safety, in both urban and rural development.

Public Art/ Facades

Art done in public areas is called public art. Recently public art has gained momentum in defining city aspirations and supporting city branding. Such artistic facades or surfaces add inventive flavour to the road and make dull places exciting. Public art on long walls or on any surface that defines a road boundary contributes to the aesthetics of the road. These can be huge walls along the road on an edge (Lynch K.) which are designed with vibrant colours, shapes, forms or textures, or a combination of them. They give an aesthetic appeal to the road and cut the monotony of the otherwise repetitious road trip. Studies have also shown that such art makes the journey seem shorter than it is.

Effect on Road Users

Considering the reality that road plays a significant role in our landscape, there is barely any discussion on the concept of road aesthetics. (Blumentrath & Tveit, 2013). According to Matijosaitiene and Stankevice (2014), one of the major problems of tourism and sustainable land management is continuing to create an attractive road infrastructure that, in turn, satisfies & pleases road users. Aesthetics is a measurement of how well an environment appears to the eye. The movement of people and things is a social and identity-forming process as much as a physical one. These days, a brand-new idea called landscape services has even emerged because of this understanding. According to their definition, they "contribute to human well-being through landscapes and landscape features" (Bastian et al. 2014).

Effect of Trees and Plantation

The iinclusion of trees and other streetscape characteristics into the roadside environment helps prevent crashes and injuries on urban highways. Naderi (2003) investigated the safety implications of aesthetic streetscape upgrades installed along the roadside and medians of five arterial roads in downtown Toronto. In his study, it was discovered that the inclusion of roadside features such as trees and concrete planters resulted in statistically significant reductions in the number of mid-block crashes along all five roadways, with the number of crashes decreasing to 5% from 20% as a result of the streetscape improvements. (Naderi, 2003).

According to the study of Lovell & Sullivan (2006), tree planting (for example, between a roadway& a commercial strip can improve the attractiveness of the road's landscape for road users as well as assist in better integrating a road into the landscape (Blumentrath & Tveit,2013).

Dense Green cover promotes the microclimate for pedestrians' thermal comfort & visual satisfaction in urban environments (Louafi et al. 2017). In their empirical investigation, Akbar et al. (2003) found that instead of selecting one type of plantation, most respondents preferred a variety of vegetation types besides highways. According to Crankshaw N. (2009), a different variety of plants would produce a range and diversity of texture patterns, which highlights one of the environmental aspects linked to psychological serenity and user appeal (Macdonald, 2005). Additionally, research demonstrates how vegetation shapes the environment and functions of space on a human scale. A sense of enclosure is also produced by the row of trees placed along the streets (Ja'afar et al., 2012).

Urban street trees form vertical barriers that define an edge, allowing motorists to guide their movements and judge their speed, resulting in overall speed reductions. When comparing roadway tree sections to identical treeless streets, run-off-the-road crashes and total crash severity are reduced. Visual barriers made from planted trees provide prominent boundaries to walkways, allowing cars to distinguish between pedestrian and vehicular zones. This forbids automobile owners from interfering with pedestrian traffic, hence avoiding accidents.

Shaded areas from trees create pleasant walking environments, bringing about increased talking, association, and natural surveillance of neighbourhoods and other civic spaces. In cities with good tree coverage, there is less need for chemical sun-blocking agents. Temperature differentials of 5-15 degrees are felt when walking under tree-canopied streets. These temperature increases significantly impact energy costs for homeowners and consumers. A properly shaded neighbourhood, mostly from urban street trees, can reduce energy bills for a household by 15-35%. Other research and observations confirm that motorists' perception of the time it takes to get through treed versus non-treed environments has a significant differential. A road stretch without trees or landscaping is perceived to be longer than one that has trees. Trees are also known to reduce anxiety or calm the drivers leading to fewer road fatalities. Also, trees and other landscaping supports biodiversity and other ecosystem. Trees increase the value of property and enhance quality of life.

Effect of Street Furniture

Street furniture includes benches, trash cans, signs, bollards, lights, and other things that give people a sense of comfort. When properly incorporated into a public space's design, street furniture gives that area a unique identity and fosters a feeling of place (Telford, 2000). Street furnishings also create environments for resting, engaging and eating and promote social interactions. Elderly people, people with limited mobility, and adults with young kids may find these arrangements to be of great importance. However, serving a practical purpose, urban types of furniture like seats and tables in squares & parks can serve a social purpose (Deakin et al., 2007), bringing people together by creating a welcoming atmosphere. Outdoor spaces in an urban area should be equipped to meet the requirements of the young, old & disabled. They should also be connected to facilitate triangulation, which encourages people to chat with strangers and promotes interactions between individuals. In open areas, the arrangement and selection of street furniture can generate stimuli; for instance, if seating, garbage bins, and telecommunication systems are distanced from one another, they may foster social separation, whereas a close arrangement with amenities such as a tea cart can effectively promote social interaction (Yücel, 2013). Remaining stationary within an urban environment provides an opportunity for reflection and fosters a physical or personal connection to the place that differs from the experience of standing or walking through it. (Dee, 2001).

Signage systems serve multiple critical functions: they offer directions and information to assist users in navigating a location, thereby enhancing the location's image and coherence, while also promoting learning. (Yücel, 2013) There are four fundamental types of signs: educational, regulatory, warning, and informational (Flink et.al, 2001) People feel safer when they are aware of their location and can navigate to where they need to go.

One of the least obtrusive ways to stop vehicles from invading pedestrian areas is with bollards, which are vertical barriers. The application of bollards should be restricted to situations where vehicles seeking to park are damaging sidewalk infrastructure, vegetation, private property and furniture, particularly along narrow roadways. Aesthetically pleasing bollards can be used to designate certain areas, such as shared public pathways or pedestrian-only streets. For pedestrians in median refuges, illuminated bollards can add to the illumination (Yücel, 2013).

The purpose of street furniture is to combine and coordinate scale, form, placement, and materials to create aesthetic impact, safety, and accessibility, by acknowledging the user's requirements (Ji, 2012). The quality, distribution, and arrangement of road furniture reflect the quality of an urban environment and can establish benchmarks for prospective development (Carmona et. al., 2010). Street furniture mostly is taken by road users (inside or outside the vehicles) in a positive manner. Earthy colours, clean surfaces, and harmless features are appreciated by road users mostly pedestrians who are most likely to use them. A theme maintained by the urban designer for the street furniture is always better since it makes up the mind of the road user to identify the road. Deep or dull colours, too many items, a heterogeneous variety of elements, etc can create confusion and be disliked by most road users.

Effect of Artwork and Facades

Public artwork refers to sculptures and other pieces of art that are placed in outdoor public areas and are physically and freely accessible to the public (Coutts & Jokela, 2008). Public artwork, which can range from sporadic pieces to a complete streetscape, enhances and complements the surroundings & brings public areas to life (Cartier & Willis, 2008). While human-scale works might offer areas of visual appeal for walking people, large-scale works can unify a district's theme or mark a neighbourhood's entrance. Hence, public art that blends in with its surroundings can also establish a context for removing obstacles, fostering various types of social connections, and encouraging creative expression (Worth). Public artworks can act as a focal point on roads & in public spaces, particularly at important intersections and places where people gather or move around a lot. However, they might not be suitable for busy walking areas like pedestrian thoroughfares, except if they serve as an item of road furniture. In addition, they can provide a particular feeling of belonging in less frequented locations (Yücel, 2013).

Topographic features like elevation, waveforms, flat forms, and climax forms are crucial in determining how sequential experiences are created on roads & highways (Blumentrath & Tveit, 2014). Any art form needs time to be observed, appreciated, and then absorbed or felt. This is contrary to vehicular motion which is based on speed, motion and concentration (on the route). Thus, any art form on the road that requires serious attention is an absolute failure for motorized vehicles. Whereas if the art is such that it creates vicarious happiness or the sense of it being there makes the driver relaxed and happy, it is useful. This can be when the art is a part of the driver's vision. Moreover, light, fresh and earthy colours distract less and add to the beauty of the road. Cyclists and pedestrians derive the greatest enjoyment from this experience. They perceive the art form from a significantly closer proximity, resulting in a more profound impact on them. In addition, cyclists can appreciate the diverse public artworks with minimal risk owing to their reduced speed. Art forms with too striking or too dull colours can be a negative point for road users since too dull will leave the art unnoticed and too striking would create flashes and attraction for vehicle drivers, increasing the risks of accidents.

Effect of Lighting

Bright, clear, and candescent lighting is best suited for road users. White light is the best of all. Incandescent lights invite insects and warmth and create a depressing environment. Roads with better visual character are found to be safer for road users. This character is imparted by white light. It is suited for all climates and all zones. Site lighting's main function is to offer safety and security after dark, while it can also give colours or vibrancy to a space at night. It increases the user's sense of security. There is a systematic order of significance in lighting, just similar to that of signage. At the top priority level, primary sidewalks and activity areas should be illuminated to serve as the hub of pedestrian activity after dark. Bollards, street-lights, façade lights, shop fronts, and other urban elements can serve as sources of illumination. Site illumination helps to promote security and deter vandalism, improve pedestrian and vehicular crossing safety, and serve as a warning system for potential dangers (Simonds,1998).

The correlation between a light source's brightness and an observer's distance influences light quality; as the light becomes more diffused with increased distance, the quality improves. Consequently, the light source should be positioned within the specified height range for optimal brightness. More lights will be required if luminaries heights are adjusted and lowered to a pedestrian scale, therefore the positioning of the poles and luminaries will have a greater impact on the streetscape (Project for Public Spaces).

Effect of Pavement

The primary infrastructure for facilitating pedestrian movement is pavement (Shaaban, 2019). Any street will become more walkable when pedestrians have access to secure, comfortable, and convenient infrastructure (Litman, 2016). Altering the walking surface's texture at strategic locations can also make it easier for people with vision impairments to locate public facilities (Yücel, 2013).

- A tranquil surface serves as an effective backdrop for the city's frequently intricate buildings.

- An excessive variety of paving materials and transitions between them can draw attention and render the street space appear puzzling or disorderly.
- Excess of level differences is a problematic issue for cyclists and motorists and hence should be avoided.
- The city square or plazas should have different paving to differentiate between pedestrians and vehicular ways.
- Paving colour, texture, expanse, etc are integral parts of road aesthetics. Too dark grey or a continuous expanse of a similar colour is considered negative for road users. Moreover, the same paving for long distances creates monotony which numbs the senses leading to many accidents. Thus, patterns in paving with proportionate markings for pedestrians or cyclists or any other public use and constant variations are more fit for vehicle users. Bright-coloured roads, clean surfaces, smooth textures, and earthy colors are best for all road users.

CASE STUDIES

For a better assessment of the above studies, some case studies were discussed to show the impact of road aesthetics on road users.

Lake View Road, Bhopal

The Lake View Road (VIP Road) in Bhopal is the sole attractive connection between the airport and the city of Bhopal. All guests are transported to the city via this route. It features a lake on one side and hills on the opposite side. The incorporation of bright white light, sky blue railings designed in wave patterns to symbolize water, multi-coloured welcoming flags, and an exquisite hand-painted mural representing Madhya Pradesh Tourism culminates in an exemplary road design. The presence of multi-coloured bougainvillea flowers along the edges, along with various vibrant bushes and hedges on the street divider, enhances its aesthetic appeal (Figure 6).

Figure 6. Views of lake view road, Bhopal

The unsettling aspect of this road is a recently erected colossal statue of Raja Bhoj situated on the remnants of the ancient Fatehgarh fort on the edge of the lake. This statue, owing to its size and popularity, draws numerous residents and tourists daily, causing them to congregate on the standard-sized footpath. This generates a crowd that encroaches onto the roadway, resulting in traffic issues. Furthermore, the statue's ability to attract the attention of drivers leads to crowding and congestion, occasionally resulting in minor collisions.

The installation of road furniture and the integration of natural elements in the beautification of VIP Road have rendered it a favoured route for commuters, despite its greater length compared to the old city road (Fig 6). The visually appealing environment enhances the travel experience, resulting in decreased congestion on the shorter, yet less attractive, old city road. Therefore, it can be interpreted that enhancing the visual appeal of the road increases the likelihood that individuals will select it over alternative routes, leading to a smoother and less congested commute.

Link Road, Bhopal

The Link Road is one of the oldest and well-maintained roads in Bhopal. It is a part of the new city area and connects the central city to the newer developments. The most significant feature of this road is its old trees (Fig 7). These trees are quite old, having multiple branches that offer shade and provide a temperature drop of a few degrees for road users. During summer, this road is an asset to its users. It has a rich belt of hedged dividers and wide roads. The roads are flanked by wide footpaths which have been adorned by benches, lampposts, dustbins and garden lights, all following a cohesive design theme. The initiatives undertaken in urban design can be regarded as a success, as numerous individuals are observed relaxing on benches illuminated by soft garden lights beneath the canopy of mature, healthy trees. Consequently, it enhances the comfort and relaxation of road users while also ensuring safety. A significant reason for this is the regulated traffic on this road, attributed to the presence of similar parallel link roads.

Figure 7. Views of link-road, Bhopal

CSMT Junction, Mumbai, India

Almost 5000 square meters of pedestrian areas at the junction of a significant North-South route in front of Chhatrapati Shivaji Maharaj Terminal (a UNESCO World Heritage Site) in Mumbai was transformed under the direction of the Global Designing Cities Initiative, through travel lane alignments and space-making (GDCI 2014). The project aimed to transform the spatial arrangement of Mumbai's streets to more effectively meet the needs of pedestrians and public transport users, garnering a 93% approval rating from both the public and the media. However, barriers were repeatedly moved to improve traffic flow (Fig 8). This difficulty prevented the implementation from being finished and caused the trial time to be shortened. The five Design Strategies that are considered for the project (GDIC 2014) are travel lane alignment, new shorter and direct pedestrian crossings, adjustment of signal cycle time for safe crossings, curb extensions to delineate traffic movement and creation of public spaces in alignment with local street vendors and food joints (Figure 8).

Figure 8. Before and after transformation

Pune, India

An 800 m long road stretch of Kharadi South Main Road was selected along which three schools are located. Along the trial stretch, a safe pedestrian and cycle track route was created. Separate parking arrangements for school buses, school vans and parent drop-off were created to avoid chaos at school gates. As a part of traffic calming measures and minimizing vehicle speed, Zensar Chowk junction was squeezed, and rumble strips and zebra crossing were marked. The implementation of the safe pedestrian and cycle track route and separate parking has greatly reduced the chaos that was previously experienced during school hours. The traffic calming measures have also contributed significantly to this positive change. The safe pedestrian and cycle track route encourages active transportation and diminishes vehicular traffic, thereby fostering a cleaner and healthier environment (Figure 9).

Figure 9. Street transformation in Pune

Copenhagen City

Beginning in the 1960s, as traffic volume increased, the interior portions of the narrow historic streets, the growing shopping districts surrounding central Copenhagen, and the sidewalks all saw an increase in the number of pedestrians bumping into one another and obstructing traffic. In 1962, Copenhagen's City Council resolved to establish a car-free pedestrian zone extending from the western Town Hall Square to the eastern Kongens Nytorv, commonly referred to as 'Stroget'. This area encompasses historical squares and an intricate web of narrow streets that collectively radiate from the Stroget, as well as the medieval part of Copenhagen. With almost 3.2 km of total length, this section is the world's longest and oldest pedestrian street system (Figure 10).

Figure 10. Before and after changing of pavement (GDCI)

All the central Copenhagen's squares and streets were heavily used for vehicular traffic and parking until 1962, and the fleet of private vehicles was expanding quickly. A lot of public discussions preceded, and the 1.15 km long main street was then converted to a pedestrian route because it was viewed as a pioneering and remarkable effort (Figure 11). There was a belief in Scandinavia that pedestrian streets would never be successful. Owners of local businesses said that no cars will lead to no customers, and no customers will result in no business. Businesses soon realized that traffic-free environments generate higher financial revenues, making Stroget a huge success (GDCI, 2014).

Figure 11. Pedestrian street in central Copenhagen (great extent of outdoor entertainment)

Key elements of the project were the complete closure of the street to traffic, removing pavements and curbs, installing new paving, and integrating street furniture to facilitate pedestrian movement. The objectives of the above project were to enhance city center connectivity, create a beautiful and quality environment, create an area that encourages businesses, encourage a variety of people to reside in and visit the city core, and transform unused alleyways into active laneways to revitalize the city. The following evaluations were observed:

- After the conversion, the number of pedestrians rose by 35% in the first year only.
- Stopping and staying actions increased by 400% between 1968 and 1996.
- 20% more people will be walking each day, or 15 minutes, across the entire city.
- Pedestrian space increased by 600%, from 15,800 sq.m to 99,700 sq.m respectively.
- From 2970 seats in 1986 to 7020 seats in 2006, outdoor café seating increased by 81%.

In 53 years as a pedestrian street, Stroget went through multiple repairs and renovations, using materials whose quality has constantly improved, repurposing plazas and public areas to enhance pedestrian comfort, and adding outdoor uses. Local artist Bjørn Nørgård renovated Amager Square in 1993. It is currently the second most populated urban area of the city (GDCI, 2014)

Portland, Oregon

In Portland city of Oregon, concrete cuts are done to remove concrete paving and add trees to the otherwise dry sidewalks. They have 317 concrete cuts in the year, increasing the number of cuts to 1000 since 2008 when the program started. That entails planting thousands of trees where nothing existed previously (Figure 12).

Figure 12. SE Grant street's sidewalk, Portland, Oregon, tree planting - before and after

Making room for urban trees has many advantages, including the management of thousands of gallons of stormwater and the significant cooling effect they have on nearby neighbourhoods during the summer. This is crucial as the number of hot summer days increases due to climate change. The cooling impact of trees is depicted in the thermal image to the right from Portland State University's research on urban heat islands and public health (Figure 13).

Figure 13. Thermal images of Portland's park blocks reveal heated rooftops and cool trees

Koogaan-de-Zaan, Amsterdam

'Koogaan de Zaan', a town in the metropolitan area of Amsterdam is located 10 kilometres N-W direction of the city. A new motorway was brutally built through the urban fabric of the town in the 1970s. The area had been abandoned for many years, mostly being used for surface parking. The project's objectives included revitalizing the area beneath the elevated freeway and re-establishing the connection between the two sides of the city. A proposal to redesign the area beneath the building to incorporate a fish shop, a flower shop, recreational amenities and a supermarket, such as a park, skate park or water-sport marina that would link the city's two sides and connect them to the adjacent river, emerged from a collaborative process involving local government, business owners and residents (Figure 14).

Figure 14. Before and after adding water element (GDCI, 2014)

Key components are the conversion of a surface parking lot into a lively, mixed-use area beneath an elevated structure. new paving materials and pavement enlargement. Container stores and flexible recreational areas are shown in Figure 15. The objectives were to re-establish connectivity between the two banks of the Koogaan de Zaan River, to revitalize the area beneath the elevated structure, and to provide recreational spaces and community amenities for local residents. Insights gained from the collaboration between the community and local government, driven by political will, facilitated the revitalization of a space that had remained neglected for over 30 years. Business owners' and residents' participation in the process of design provided the municipality and designers with information about the needs of the community and aided in the creation of a successful civic space.

Figure 15. Plan and Sections of the transformed street (GDCI)

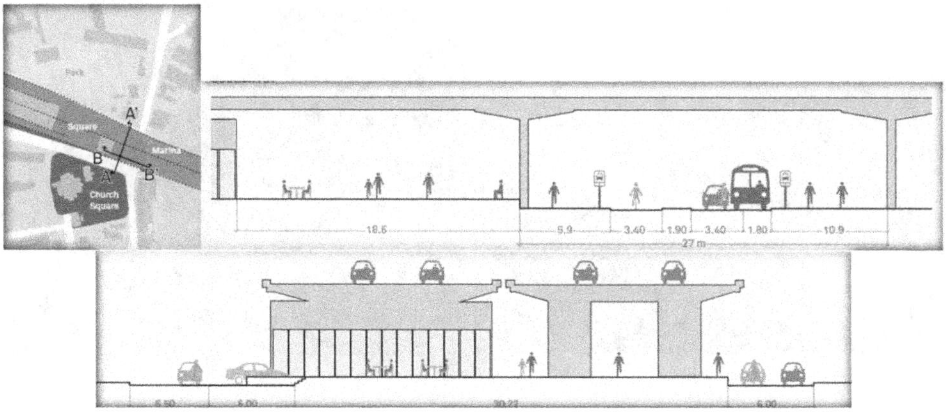

Seoul, South Korea

The 10-lane road and the four-lane elevated road along the Cheonggyecheon stream, which daily carried over 1,70,000 vehicles, were demolished by the Seoul Metropolitan Government. The modified street prioritizes public transport over personal vehicles, fostering a more pedestrian-friendly and environmentally-friendly space. Between 2003 and 2008, the project helped to increase bus ridership by 3.3% and 15.1%, respectively. Today, 64,000 people visit the revitalized street every day (Fig. 16 and 17).

The following are the observed key components:

- Demolition of an elevated concrete highway structure.
- Uncovering a previously hidden urban stream.
- Construction of a sizable new open space along the stream that receives natural light.
- Construction of pedestrian-friendly amenities and leisure areas (two plazas, eight themed locations).
- Building 21 new bridges to connect the city's infrastructure.

Figure 16. Before and after adding landscape (GDCI)

The objectives of the project are to enhance the quality of life, including the air and water, and reconnect two sections of the city that were previously separated by a road system.

Evaluation

- 10.3% less air pollution and 76% more pedestrian activity;
- Bus ridership rose by 45% while subway ridership fell by 3.3%
- Reduced vehicle volume by -45% and the effect of urban heat island by -4.5%

Figure 17. Section of the Street after transformation

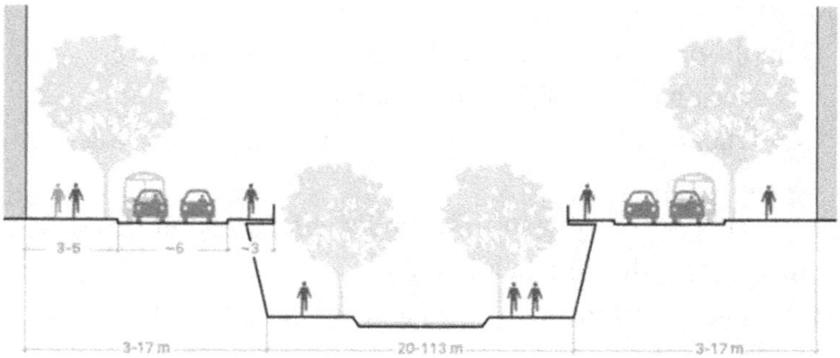

New York, USA

Following an increase in vehicular accidents near Times Square Street, New York, during the summer of 2009, the city's Department of Transportation imposed a temporary restriction on Broadway, permitting access exclusively to pedestrians (Figure 18). The city resolved to institutionalize this modification due to its success. In the subsequent year, they engaged Snhetta to design a new permanent plaza, the first segment of which was revealed in 2014 (Figure 19).

Figure 18. Before and after the transformation

Figure 19. Before and after the addition of pedestrian space and seating

Fortaleza, Brazil

The Cidade-do-Gente (City of People) program aims to create safer and livelier communities for individuals in locations where there is a lot of conflict between motorized vehicles and vulnerable road users. City leaders adopted dramatic street designs and proved their efficacy by using inexpensive and quick-to-build materials like planters and paint. The goal of this project was to put pedestrian safety first while lowering the traffic speeds in Drago do Mar's crowded cultural quarter. The street design, which includes wide driving lanes, remained constant over time despite changes in land use. With a focus on pedestrians, the renovation intended to modernize the spatial arrangement to accommodate current uses (Figure 20).

Figure 20. Before and after introducing urban art (GDCI, 2022)

Pittsburgh, USA

Strawberry Way, an alley traversing the northern part of Downtown from Liberty to Grant streets, has historically served as a shortcut for certain pedestrians in Pittsburgh, although not always a preferable one. Trash was usually strewn across its asphalt and the blank, high walls of skyscrapers hurried walkers through as fast as their feet could carry them. Colourful designs were painted on the blacktop by local artist Deanna Mance, roadblocks were put up to stop cars from driving through, and picnic tables and planter boxes were placed on the street to encourage people to enjoy their new Downtown Oasis (Figure 21).

Figure 21. Before and after painting the street surface

In addition to being more aesthetically pleasing than before, the project has resulted in a 43% increase in pedestrian traffic through the alley, according to a study completed by Envision Downtown. Also, there has been an outstanding 462% increase in people spending time in Strawberry Way.

CONCLUSION

From all the above exploration, we conclude that a lot of factors combine to leave an impact on the road user. The road aesthetics can positively and negatively affect its users and thus each road has a customized quality list that makes it suitable for its user. But for the sake of generalization, the good and bad road environment can be concluded as below:

Good Road Environment

A good road (beautiful, safe and healthy) environment is

- Clearly defined.
- Consists of soft landscape features such as flowers, trees etc. in appropriate proportions.
- If vegetation is absent, the shape and colour of the artificial environment establish aesthetic satisfaction.
- The colours are often bright and yet exhibit natural, earthy, and distinct qualities.

Bad Road Environment

On the contrary, the bad road (ugly, unsafe and unhealthy) environment is

- Not clearly defined.
- Devoid of soft vegetative landscape. If vegetation is present, it is disregarded due to its considerable distance.
- Predominant spaces delineated for road traffic.
- If huge road spaces are absent, the shape and colour of the built environment establish aesthetic unsatisfactory experiences.
- The buildings appear grey, likely due to considerable distances.
- The environment features extensive hard landscape elements of concrete or asphalt.

Generalized Conclusions

- Thermal comfort is vital for a road to function properly.
- Any material or pattern repeatedly used in large areas is not valued.
- Soft landscape elements such as vegetation, water, etc are appreciated by all road users.
- Environment containing numerous concrete elements.

- Too many elements of concrete in the environment are not valued.
- The perception of "beauty" may be interpreted as "ugliness" when the subjective experience of safety is elevated.
- Road utility cannot be compromised in road aesthetics.

FURTHER RESEARCH

While the subject of road aesthetics is longstanding, the study of its relation with users' psychology, preferences, and behaviour changes, etc is recent. Numerous domains exist where similar research can be conducted. Other topics like the study of a road before and after the addition of road aesthetic elements and the behavioural changes of its users, or the study of accidents before and after the implementation of road aesthetic features, may serve as viable avenues for further research in related areas.

REFERENCES

Abusaada, H., & Elshater, A. (2021). 'Effect of people on placemaking and affective atmospheres in city streets'. *Ain Shams Engineering Journal*, 12(3), 3389–3403. DOI: 10.1016/j.asej.2021.04.019

Akbar, K. F., Hale, W., & Headley, A. D. (2003). Assessment of scenic beauty of the roadside vegetation in northern England. *Landscape and Urban Planning*, 63(3), 139–144. DOI: 10.1016/S0169-2046(02)00185-8

Antonson, H., Mårdh, S., Wiklund, M., & Blomqvist, G. (2009). Effect of surrounding landscape on driving behaviour: A driving simulator study. *Journal of Environmental Psychology*, 29(4), 493–502. DOI: 10.1016/j.jenvp.2009.03.005

Balasubramanian, S., Irulappan, C., & Kitchley, J. L. (2022). 'Aesthetics of urban commercial streets from the perspective of cognitive memory and user behavior in urban environments'. *Frontiers of Architectural Research*, 11(5), 949–962. DOI: 10.1016/j.foar.2022.03.003

Bastian, O., Grunewald, K., Syrbe, R.-U., Walz, U., & Wende, W. (2014). Landscape Services: The Concept and Its Practical Relevance. *Landscape Ecology*, 29(9), 1463–1479. DOI: 10.1007/s10980-014-0064-5

Blumentrath, C., & Tveit, M. S. (2014). "Visual characteristics of roads: A literature review of people's perception and Norwegian design practice," Transportation Research Part A. *Policy and Practice, Elsevier*, 59(C), 58–71.

Bolund, P., & Hunhammar, S. (1999). 'Ecosystem services in urban areas'. *Ecological Economics*, 29(2), 293–301. DOI: 10.1016/S0921-8009(99)00013-0

Burden, D. (2006). *22 Benefits of Urban Street Trees*. Glatting Jackson and Walkable Communities, Inc.

Carmona, M., Heath, T., Oc, T., & Tiesdell, S. (2010). *Public Places Urban Spaces:The Dimensions of Urban Design*. Published Elseiver Ltd.

Cartiere, C., & Willis, S. (2008). *The Practice of Public Art*. Routledge. DOI: 10.4324/9780203926673

Coutts, G., & Jokela, T. (2008). *Art, Community and Environment: EducationalPerspectives*. Intellect Books. DOI: 10.2307/j.ctv36xw54v

Crankshaw, N. (2009). *Creating Vibrant Public Spaces; Streetscape Design in Commercial and Historic Districts*. Islandpress.

Deakin, M.; Mitchell, G.; Nijkamp, P.; Vreeker, R. (2007) Sustainable UrbanDevelopment: Volume 2; The Environmental Assessment Methods, Routledge Press; p365

Dee, C. (2001). *Form and Fabric in Landscape Architecture*. Taylor & Francis.

Drottenborg, H., (1999). Aesthetics and Safety in Traffic Environments.

Dzhambov, A., Hartig, T., Markevych, I., Tilov, B., & Dimitrova, D. (2018). 'Urban residential greenspace and mental health in youth: Different approaches to testing multiple pathways yield different conclusions'. *Environmental Research*, 160, 47–59. DOI: 10.1016/j.envres.2017.09.015 PMID: 28961469

Egebjerg, U. (2002). *Beautiful Roads - A Handbook of Road Architecture*. Danish Road Directorate.

Flink, A. C.; Olka K.; Searns, R. M. (2001)Trails for the Twenty-First Century:Planning, Design, and Management Manual for Multi-Use Trails, Island Press;p88,89.

Global Street Design Guide. (2014). National Association of City Transportation Officials.

Ja'afar, N. H., Sulaiman, A. B., & Shamsuddin, S. (2012) The Contribution of Landscape Features on Traditional Streets in Malaysia,Procedia - Social and Behavioral Sciences,Volume 50,Pages 643-656,ISSN 1877-0428.

Ji, Y. G. (2012). *Advances in Affective and Pleasurable Design (Advances in HumanFactors and Ergonomics Series)*. CRC Press.

Litman, T. A. (2016). *Evaluating Transportation Land Use Impacts: Considering the Impacts, Benefits and Costs of Different Land Use Development Patterns* (Vol. 1). Victoria Transport Policy Institute.

Louafi, S., Abdou, S., & Reiter, S. (2017). Effect of vegetation cover on thermal and visual comfort of pedestrians in urban spaces in hot and dry climate. *Nature & Technology*, (17), 30B.

Ma, X., Chau, C. K., & Lai, J. H. K. (2021). 'Critical factors influencing the comfort evaluation for recreational walking in urban street environments'. *Cities (London, England)*, 116, 103286. DOI: 10.1016/j.cities.2021.103286

Macdonald, E. (2005). Street-facing Dwelling Units and Livability: The Impacts of Emerging Building Types in Vancouver ' s NewHigh-density Residential Neighbourhoods. *Journal of Urban Design*, 10(1), 13–38. DOI: 10.1080/13574800500062320

Matijosaitiene, I., Stankevičė, I., & Ribeiro, J. (2014). Road Landscape as a Product: Does it Satisfy Consumers' Aesthetic Needs? *The Baltic Journal of Road and Bridge Engineering*, 9(4), 297–305. DOI: 10.3846/bjrbe.2014.36

Montella, A., Chiaradonna, S., Mihiel, A. C. D. S., Lovegrove, G., Nunziante, P., & Rella Riccardi, M. (2022). 'Sustainable complete streets design criteria and case study in Naples, Italy'. *Sustainability (Basel)*, 14(20), 13142. DOI: 10.3390/su142013142

Naderi, J.,R. (2003), 'Landscape Design in Clear Zone: Effect of Landscape Variables on Pedestrian Health and Driver Safety', *Transportation Research Record Journal of the Transportation Research Board 1851(1851):119-130*

Phillips, A., Da Schio, N., Canters, F., & Khan, A. Z. (2023). 'A living street and not just green": Exploring public preferences and concerns regarding nature-based solution implementation in urban streetscapes'. *Urban Forestry & Urban Greening*, 86, 128034. DOI: 10.1016/j.ufug.2023.128034

Qiu, W., Li, W., Liu, X., Zhang, Z., Li, X., & Huang, X. (2023). 'Subjective and objective measures of streetscape perceptions: Relationships with property value in Shanghai'. *Cities (London, England)*, 132, 104037. DOI: 10.1016/j.cities.2022.104037

Shaaban, K. (2019). Assessing Sidewalks and Corridor Walkability in Developing Countries. *Sustainability (Basel)*, 11(14), 3865. DOI: 10.3390/su11143865

Simonds, J.O. (1998) Landscape Architecture: A Manual of Land Planning andDesign, McGraw-Hill Professional; p155,158

Sullivan, W. C., & Lovell, S. T. (2006). Improving the visual quality of commercial development at the rural-urban fringe. *Landscape and Urban Planning*, 77(1-2), 152–166. DOI: 10.1016/j.landurbplan.2005.01.008

Taylor, N. (2003). The Aesthetic Experience of Traffic in the modern city [Carfax Publishing.]. *Urban Studies (Edinburgh, Scotland)*, 40(8), 1609–1625. DOI: 10.1080/0042098032000094450

Transport & Regional Affairs Department of the Environment. (2000). *By Design, UrbanDesign in the Planning System: Towards Better Practice*. Thomas Telford Ltd.

Van Renterghem, T., & Botteldooren, D. (2016). 'View on outdoor vegetation reduces noise annoyance for dwellers near busy roads'. *Landscape and Urban Planning*, 148, 203–215. DOI: 10.1016/j.landurbplan.2015.12.018

Wang, R., Lu, T., Wan, C., Sun, X., & Jiang, W. (2023). 'Measuring the Effects of Streetscape Characteristics on Perceived Safety and Aesthetic Appreciation of Pedestrians'. *Journal of Urban Planning and Development*, 149(3), 05023020. DOI: 10.1061/JUPDDM.UPENG-4314

Worth, M. Creating significance through public places art an inclusive+interdisciplinary practice, Public Art and Urban Design. Interdisciplinary and SocialPerspectives Waterfronts of Art III, web version on the w@terfront nr, 4.http://www.ub.edu/escult/epolis/WaterIII.pdf. p51.

Yücel, G. F. (2013). Street Furniture and Amenities: Designing the User-Oriented Urban Landscape. In Özyavuz, M. (Ed.), *Advances in Landscape Architecture*. IntechOpen., DOI: 10.5772/55770

Chapter 10
Have Rebranding Strategies Improved the Economy and Environment in an Old District of a Metropolis? A Pilot Study in Wanchai, Hong Kong

Giovanna Chiu
German Swiss International School, Hong Kong

Chloe Kwun Yee Lee
https://orcid.org/0009-0003-3390-5009
Hong Kong Baptist University, Hong Kong

Dickson K. W. Chiu
https://orcid.org/0000-0002-7926-9568
The University of Hong Kong, Hong Kong

Chloe Tsz Yu Tse
https://orcid.org/0009-0000-5561-004X
German Swiss International School, Hong Kong

ABSTRACT

Wanchai was the first old commercial district in Hong Kong to undergo redevelopment with much economic activity and environmental changes. This study con-

DOI: 10.4018/979-8-3693-6695-0.ch010

ducted fieldwork, including urban land use surveys, environmental quality surveys, questionnaires, and cross-examined secondary in newspapers, government reports, and online information to find out if these rebranding strategies through building reconstruction improved the economy and environment of Wanchai. Results indicated that though the economy has grown significantly with such development strategies, the quality of the environment has not been improved much. Some suggestions are given accordingly.

INTRODUCTION

Urban rebranding refers to the process of altering the perception of an urban area through strategic initiatives aimed at reshaping its image (Ashworth & Kavaratzis, 2007). The goals include enriching a city's reputation (or branding), boosting economic growth, and improving the living environment for citizens. Such rebranding strategies are particularly beneficial for underdeveloped regions or areas associated with negative connotations, as they can be revitalized to address urban challenges such as decay, poverty, and pollution. Consequently, economic development is stimulated, enhancing the quality of life for residents and fostering sustainable living practices (Herstein & Berger, 2013).

Wanchai, the first historic commercial district in Hong Kong to undergo redevelopment, has undergone a series of regeneration policies, making it a pertinent case study for this research. This study investigates the effectiveness of these policies in improving Wanchai and identifies which aspects of development could be optimized to benefit other districts in Hong Kong requiring rebranding.

Previous studies of urban rebranding have mainly focused on the West but have rarely focused on Asia, especially metropolises in this area. Therefore, this study explores the possible outcomes and influences of implementing city rebranding strategies in Hong Kong. The first part of this chapter reviews the importance of urban rebranding and the influences that urban rebranding contributes to. Next, the method of conducting this study is detailed, followed by the analysis and results. Finally, this chapter discusses the practical implications before concluding with suggested research directions.

LITERATURE REVIEW

What is Urban Rebranding?

Urban rebranding refers to redesigning an area with a competitive advantage *(forming an impressive brand)* that captivates more investment, tourists, consumers, and other business activities (Herstein & Berger, 2013). Urban rebranding is also vital for distinguishing and refining the positioning of a city (Ashworth & Kavaratzis, 2007). To achieve these goals, setting rebranding strategies is the determinant of reconstructing and enhancing the value of old districts. Suitable and well-planned rebranding strategies can improve the district's characteristics and maintain its historical culture (Daldanise, 2020).

Rebranding strategies such as "City marketing" are among the most effective and predominant strategies adopted. "City Marketing" focuses on constructing, communicating, and managing a city's image (Kavaratzis, 2004). Also, urban rebranding through the tourism industry is common for obtaining revenue through consumers and flourishing external investment in the city.

Additionally, "Green city branding" is now another direction and theme of rebranding project in Hong Kong (Chan & Marafa, 2014). Thus, there are various strategies for urban rebranding, which is essential in deciding suitable strategies for effective rebranding.

Indication of Effectiveness of Urban Rebranding

To examine the effectiveness of urban rebranding, constructing indicators to assess the perception and willingness of major stakeholders such as residents, travelers, and investors is significant. The current emphasis mainly comprises economy and environment, as detailed in the following subsections. Only after these two fundamental dimensions are fulfilled can reputation and branding be built.

Dastgerdi and De Luca (2019) gathered a classification of factors commonly used as indications in urban rebranding research. Economic performance includes enhancements that cultural activities bring to a city, policy-making, media and advertising, and urban planning. Under the "urban planning" category, "strategic plan" digs into the relationship behind urban rebranding strategies. These indicators emphasize the recognition of stakeholder interests, the identification of the city's characteristics, the balance between global and local needs, etc.

Furthermore, Florek et al. (2021) introduced the City Brand Effectiveness Measurement System (CBEMS), which incorporates diverse indicators to assess urban branding comprehensively. This model considers city-specific goals and plans, internal and external effects, subjective perceptions, and objective factors, addressing short-term and long-term measures.

By gaining valuable analysis results by implementing these indicators, the potential issues and results of urban rebranding strategies could be addressed more precisely, giving a more desirable direction for possible improvement strategies.

Economic Indicators and Tools for Urban Rebranding Strategies

The enabling dimension of urban rebranding is economic indicators and tools in designing necessary strategies because of the enormous costs (Zuo et al., 2013). Regarding economic influence, urban rebranding aims to reshape public perceptions of cities and boost economic competitiveness by promoting investments, businesses, technology, and tourists (Herstein & Berger, 2013).

One of the key economic benefits of urban rebranding is the increase in property values and housing prices. An example of Elmwood Village in New York is that the rebranding is concentrated in and around the central business district (Willer, 2021). Housing prices in Elmwood experienced a vast surge where the average rental went up from US$700 to 1,200 - 1,500 (Willer, 2021), attributed to the improved image and desirability of the rebranded area, attracting residents and driving up demand. Thus, an increase in housing prices due to urban rebranding can benefit property developers and owners who can sell or rent their real estate at higher prices (Priemus, 2004).

A successful rebranding campaign can elevate a city's global profile and status as a destination (Gerosa & Tartari, 2021). Insch and Bowden (2016) examined the rebranding city Brisbane in Australia, reconstructed and repositioned into a city with lower living costs compared to Sydney and Melbourne. Brisbane still enjoys a decent lifestyle with improved education infrastructure. To external commercial investors, Brisbane is an attractive alternative to relocating business activities and households for lower costs.

Besides, Umhlanga in South Africa effectively rebranded the city through technological advancements, such as rebuilding the previous network infrastructure and placing more development to transform Umhlanga into a "Smart City" (Erwee et al., 2021). As a result, it gains a favorable outcome in enticing companies to have continuous capital investment and attracting educated workers to the city (Erwee et al., 2021). These cases illustrate that a location or city can be considered a favorable investment destination and achieve economic growth if the rebranding strategies capitalize on its unique strengths and advantages.

While suitable urban rebranding can effectively contribute to investment activities, enhancement of visitors and traveling is another area that urban rebranding creates. By crafting a new narrative highlighting unique cultural and heritage assets, rebranding seeks to attract more leisure and business travelers (Zenker et al., 2017; Kavaratzis & Hatch, 2013). Several case studies have demonstrated the positive impacts of successful rebrands on tourism numbers. For example, Malaga, in Spain, has undergone a rebranding process, transforming into a city filled with contemporary and conventional multicultural elements rather than solely focusing on promoting its sun and beach as tourism characteristics (Erwee et al., 2021). This shift caters to various types of visitors, including those seeking the sea and a pleasant climate or those looking to immerse themselves in cultural environments. Malaga has become a distinctive tourist destination, providing desirable profitability and continuous economic development for the residents and society. Thus, favorable urban rebranding strategies can create continuous capital investment and revenues for the local government and its economic structure.

Further, several studies have also pointed out that the economic impacts may vary depending on the location's image, competitiveness, and rebranding policy (Dastgerdi & De Luca, 2019). Some researchers also show that the failure of urban rebranding is created, in turn, by economic loss or low economic value (Miller et al., 2013). Therefore, it is crucial to set up a proper rebranding strategy by observing and complying with the actual characteristics and culture of that city or location (Cleave & Arku, 2020).

Environmental Indicators and Tools for Urban Rebranding Strategies

Another important dimension of urban rebranding is environmental indicators and tools in designing strategies because of the recent emphasis on development sustainability (Zuo et al., 2023). Besides the economic benefits, urban rebranding can also refine the environment, such as the hygiene and living conditions of the old districts. Hygiene and green facilitation policies should be included during urban rebranding to improve the city's image (Braiterman, 2011). Beautification efforts also enhance urban sustainability. Investing in parks, trails, and other recreational areas encourages active lifestyles and walkability, reducing residents' environmental footprint (Dupre, 2019). Port Said in Egypt conducted community-driven projects as part of the rebranding, such as rebuilding and expanding the waterfront, to help improve air and water quality in cities (Rehan, 2014). Therefore, expediting the promotion of green products, such as trees, water, and flowers, and improving water facilities is often emphasized in urban rebranding projects. Such approaches can enhance the city's environment and image.

Rebranding often involves the development of green spaces and parks, which play a crucial role in improving urban living conditions and environmental quality. For example, Wolch et al. (2014) highlighted that cities like Los Angeles, which undertook rebranding initiatives, substantially increased urban green spaces, improving air quality and public health benefits. The introduction of green spaces not only enhances the aesthetic appeal of a city but also contributes to better air quality and provides recreational areas for residents.

Apart from that, "Green" is often used as a theme for city rebranding, which increases the proportion of implementing policies or activities related to the environment. According to Andersson (2016), the Växjö government in Sweden would like to brand its city as the "Leading Green City." Consequently, the Växjö government has launched various policies and programs to reduce carbon emissions and improve the current environmental status. For instance, cleaning two lakes has enhanced water quality and eliminated odors.

Enhancing perceptions of the urban environment through initiatives such as improving open or green spaces, increasing cleanliness, and redeveloping run-down areas benefits both the environment and citizens significantly.

METHOD

Wanchai has now evolved into a modern district featuring international exhibition centers, diverse bars and restaurants, and an art center in its north while preserving its historical architecture and cultural heritage in the south. Renowned as one of the most bustling commercial hubs in Hong Kong, it accommodates the offices of numerous small and medium-sized enterprises, seamlessly blending its vibrant cultural past with a dynamic business environment. Yet, the government and the public consider the southern portion underdeveloped and have been promoting recent development and branding, which is the focus of this study. Further details on Wanchai are available at https://en.wikipedia.org/wiki/Wan_Chai.

This study employed a mixed-method approach to investigate the extent of development and rebranding in Wanchai. Four distinct indicators and tools were utilized to collect primary data at various locations within the district. (1) Urban land use surveys were conducted to assess the distribution and utilization of land within Wanchai, providing insights into the changes in land use patterns over time. (2) Environmental quality surveys evaluated various environmental factors, such as air and water quality, noise levels, and green spaces, to gauge the impact of redevelopment on the local environment. (3) Structured questionnaires were administered to residents, business owners, and visitors to gather their perceptions and experiences regarding the changes brought about by the redevelopment efforts. (4) Photographs were taken at different points in Wanchai to visually document the physical and aesthetic transformations resulting from the redevelopment initiatives

In addition to the primary data collection, secondary data was also gathered to supplement and contextualize the findings. This secondary data included newspaper reports, which were reviewed to capture public discourse and media perspectives on the redevelopment of Wanchai. Governmental reports provided official documents and statistics on the redevelopment efforts, offering an authoritative perspective on the extent of development and the success of various measures. Online data from reputable sources enriched the study with contemporary and diverse viewpoints.

Finally, old photographs from the Internet were compared to the historical and current states of the district, highlighting the physical changes over time. This comprehensive approach ensured a robust analysis of the redevelopment of Wanchai, enabling a thorough evaluation of its success and areas for improvement.

Sampling

A systematic approach was used for most techniques, and random sampling was used for photos, both of which worked well. An improvement that could be implemented would be using a stratified systematic approach. Stratified sampling is where the population is separated into different groups called strata in terms of categories such as age and gender. This could allow for extrapolated trends between strata and help better understand how development has directly impacted citizens of different demographics.

Our sampling methods were systematic and random. First, we divided Wanchai into four transects. Along those lines, we conducted EQS every 10 meters on one side of the street. We interviewed random people using the structured questionnaire every 20 meters on one side of the street. The urban land use was identified for every ground-floor building along the street. Photographs were taken at notable points along the transects.

A map transect was used to find locations suitable for sampling. Four different transects were made on the four most frequented roads: Ship Street, Lee Tung Avenue, Wanchai Street, and Burrows Street (roads highlighted in blue). Data was then sampled every 10 meters (systematic sampling) in each transect and processed for comparison.

Urban Land Use Surveys

These were qualitative assessments done to determine the types of infrastructure in different transects. There were five categories - residential, commercial, industrial, recreational, and transport. A collection point was made every 10 meters, and they were then color-coded into their respective categories on the maps.

These surveys were necessary to get a consensus on the type of economic and recreational activities in particular areas in Wanchai. The level of economic activity could be determined by looking at the type of infrastructure present in the transects, and applications of the government's development strategies could be judged.

Environmental Quality Surveys (EQS)

At every collection point, an environmental quality survey was done to evaluate the general atmosphere of every transect, using a bipolar grading system - 1 as worst, 5 as good. Five categories were rated: buildings, environment, culture, traffic, and commerce and business. The points of each collection point were added and counted out of a total score - 25 for buildings, 30 for the environment, 20 for traffic, 25 for commerce and business, and 15 for culture. Points were given according to visual indication. EQS was necessary as details about the general environment of Wanchai could be determined and built on from the urban land use surveys.

Questionnaires

Pedestrians were asked verbally at each transect about Wanchai's current state and situation, and answers were recorded on a Google Form, which was then processed. Unlike the EQS and the urban land use survey, questions were not asked at specific collection points but at the transects. The questionnaire provided various qualitative and quantitative data, which can broaden the range of information collected. A mixture of yes/no and open-ended questions was included. Insight about current opinions and trends can also be extracted from data and compared with secondary data to determine their validity.

Photos

Photos were taken at every transect and, like the questionnaires, were not taken at specific collection points. A random sampling system was used since they were also taken outside transects. These were needed to visually assess environmental and economic conditions (i.e., number of closed/open shops, amount of pollution) and to compare them among the different transects. The level of city development could then be judged to reveal the differences among various areas of Wanchai.

ANALYSIS AND RESULTS

Urban Land Use Surveys

Each point was categorized on maps (see above in methods and justifications) to determine which type of activity was the most prominent in Wanchai. Table 1 summarizes the overall results. Results indicate that residential usage at Ship Street is slightly more prominent, while Wanchai Street and Burrows Street have more commercial usage. Lee Tung Avenue has some contributions to transportation, with more commercial buildings than residential ones. Burrows Street also has some industrial usage.

Table 1. Demographics of respondents (n=50)

	Commercial	Residential	Transportation	Industrial
Ship Street	44%	56%	0%	0%
Lee Tung Avenue	50%	33%	17%	0%
Wanchai Street	69%	31%	0%	0%
Burrows Street	57%	29%	0%	14%
Overall	57%	37%	3%	3%

Various services are offered in Wanchai, mainly dedicated to commercial and slightly less for residential usage. As over half of the usage is commercial, this indicates good economic activity in Wanchai. A high density of residential facilities also shows a large population in Wanchai to support economic growth. The diminishing industrial facilities also demonstrate that Wanchai has moved on from the secondary sector and into the tertiary sector, with evidence of development strategies implemented by the government in place as its emphasis is on commercial usage.

Transport and recreational facilities were lacking from the four transects, indicating a lack of planning from earlier development. The lack of recreational facilities may be a manifestation of the confined space available in Wanchai.

Environmental Quality Surveys (EQS)

Scores from each EQS were scaled and put into bar graphs based on their data points. Data points 1-5 are in Ship Street, 6-10 in Lee Tung Avenue, 11-17 in Wanchai Street, and 18-22 in Burrows Street. The results are summarized in Figure 1, covering five aspects plus the overall scores: Business, Traffic, Culture, Environment, and Buildings.

Figure 1. Environmental quality survey (EQS) results

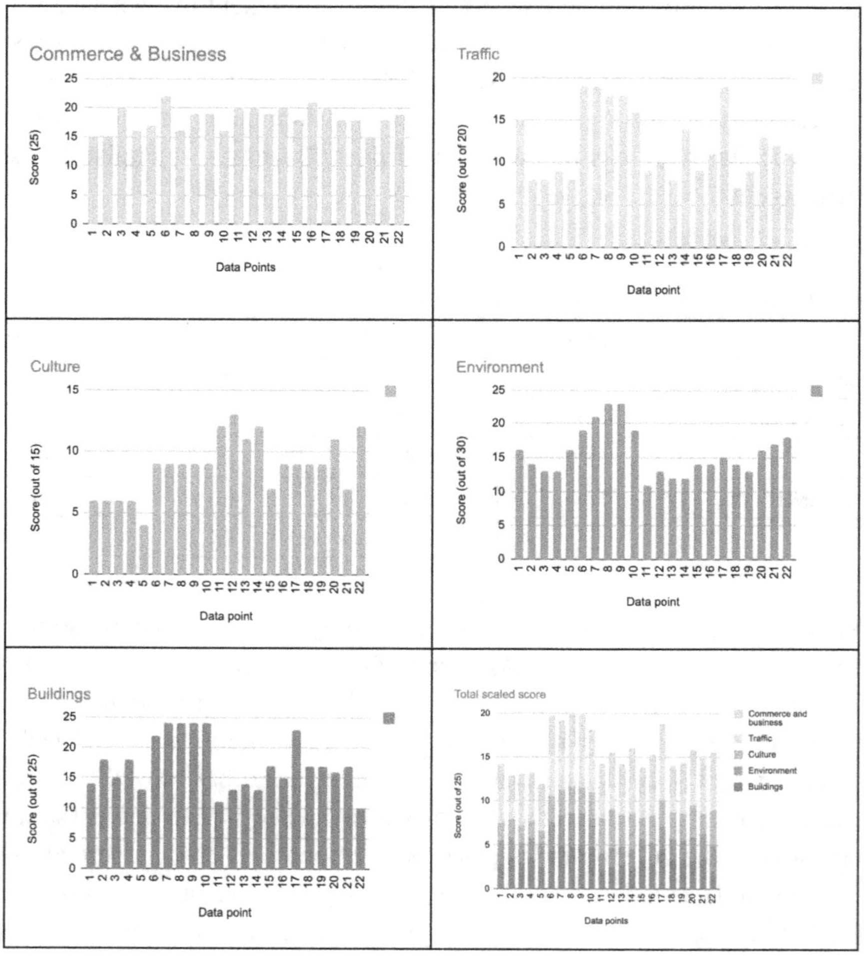

For business, all data points scored high in this section and relatively similar. No transect had apparent leverage over the others, which means a good level of tertiary facilities distributed among Wanchai, which are all of good quality. The mean scaled score was 3.6/5 (the highest of all five categories), indicating Wanchai's development and economic activity.

For traffic, the distribution of scores was uneven. Data points at Lee Tung Avenue scored much higher than other transects, showing how improvement in traffic was only congregated in one area and overlooked in others. The mean score was 3.07/5, meaning the traffic quality is average overall in Wanchai. Such a condition may impact citizens' overall quality of life, though not significantly.

For culture, most data points did not score well in this category. Points at Lee Tung Avenue scored higher than others, with anomalies at points 20 and 22 in Burrows Street. Results indicate that most of Wanchai has become completely modernized except for some old buildings repurposed into new structures. With a mean scaled score of 2.3/5, the lowest of all five categories, this serves as evidence for cultural erosion.

For the environment, most data points had low scores, except for those on Lee Tung Avenue. This result demonstrates the lack of care in environmental settings in Wanchai, rendering issues such as pollution and smog prevalent in most areas. Again, this reflects an uneven distribution of development in Wanchai. Only Lee Tung Avenue has a good environment, while other areas have not been as regulated and are neglected. A mean scaled score of 2.6/5 shows that the overall environment is inadequate, and much improvement is necessary to maintain an agreeable environment.

For buildings, data points scored average except for Lee Tung Avenue. Issues such as vandalism, lack of maintenance, and security were common problems identified in most buildings, indicating an uneven amount of attention the government gave to the rest of Wanchai. A mean scaled score of 3.4/5 indicated that most buildings were only average.

Overall, data points 6-10 (Lee Tung Avenue) and 17 (Wanchai Street) scored highest among the 22 points. Lee Tung Avenue had the highest gross score, indicating a good overall environment. However, the difference in total accumulated points is evident compared to places like Ship Street and Burrows Street. Results imply that governmental strategies were only effective in some parts instead of the entirety of Wanchai.

There is one anomaly, point 17, which was on Wanchai Road. The point was on a junction between a narrow avenue and the main section of the road, which explains the contrast between total scores.

The mean total scaled score was 15.8/25, indicating that the overall environment of Wanchai is average. More development and change should be brought more evenly to all parts of Wanchai.

Questionnaires

Figure 2. Questionnaire results

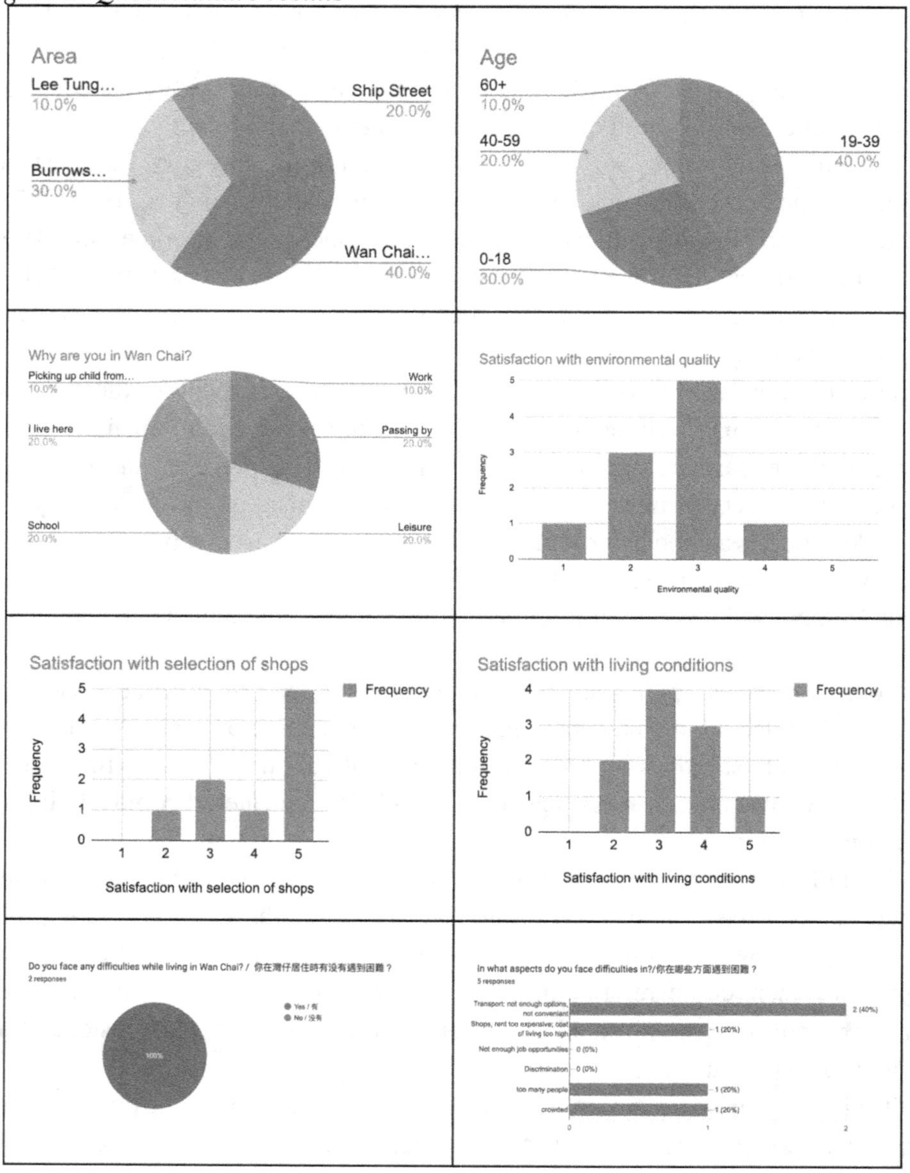

Most of the people involved in the questionnaire were on Wanchai Road, and most people were aged 19-39. The reasons for the people presented in the studied area were split: leisure, school, residents, and passers-by, demonstrating that Wanchai accommodates many attractions that non-locals may desire. Wanchai is fully developed in that aspect. Most people were moderately satisfied with the environmental quality of WanChai, meaning that efforts are still necessary to improve general conditions. Compared to the EQS, these results were fitting and confirmed that the environmental quality is not particularly satisfactory.

Questionnaire results indicate that most people were delighted with the selection of shops in Wanchai, which showed diversified services and prosperous economic activities. Most respondents were moderately satisfied with their living conditions, but this was closely followed by those satisfied with their living conditions. While this may show a disparity between people, this reflects that Wanchai generally provides reasonable living conditions and that respondents can generally afford livable environments.

Results indicate all respondents faced problems while living in Wanchai. The main problems included the lack of transport options and overpopulation in the area. Only one respondent reported having financial difficulties, which may show that people in Wanchai generally can afford the cost of living. Most respondents liked the diversified options for shopping while facing pollution and congestion problems. Results show that although the economy in Wanchai is well-developed, the environment remains poor, and much improvement is still necessary.

Selected Photos

Figure 3 shows that Wanchai Street is highly frequented and has many stalls selling groceries and foodstuffs. Though the photo above shows only a small portion of the street, this section presents issues such as pollution and congestion.

Figure 3. Photo at Wanchai street (taken by authors)

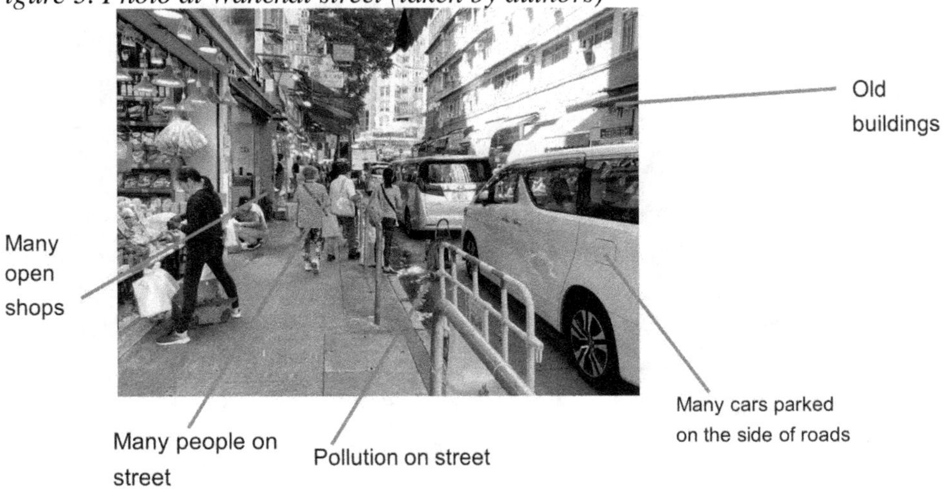

Figure 4. Photo at Burrows street (taken by authors)

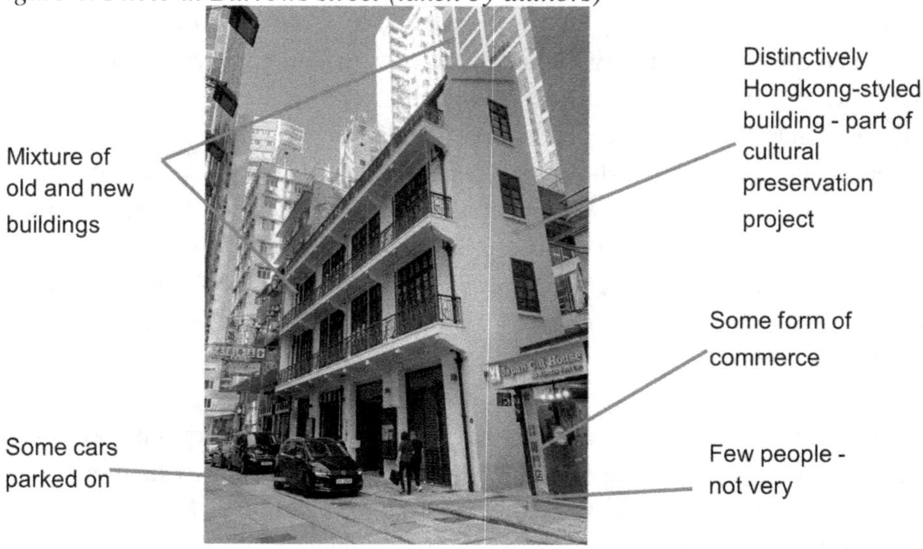

Figure 4 shows that Burrows Street is short and not heavily frequented by locals. It houses a few commercial facilities like shops and has an old building repurposed to be a mall. However, the lack of people indicates that it is not very popular as a spot, so the mall generally goes unnoticed.

Figure 5. Photos at Lee Tung avenue (taken by authors)

Figure 5 shows that Lee Tung Avenue is the most developed (and new) out of all four transects, having many high-rise modern residential buildings and many restaurants dedicated to tourists and relaxation. As a tourist attraction, the general environment is good as there is little pollution or vandalization.

Figure 6. Photos at Ship street (taken by authors)

Figure 6 shows development neglect in Ship Street, with vandalism and old infrastructures. This shows a lack of consideration for the environment in the area, which is still underdeveloped. In other areas, such forms of neglect were also seen (see Figure 7). There is a stark contrast between the general environment in Lee Tung Avenue and the other transects. The government has focused on developing new tourist attractions but has not renewed other areas that need attention in Wanchai.

Figure 7. Photos at Cross lane, directly opposite a trash collection venue (taken by authors)

Secondary Data

Government rebranding strategies in Wanchai started in 2006, with the official name "Wanchai Development Phase II," focused on extracting land to build more transport. According to data from the Hong Kong Government, the median household income in Wanchai has a general increase from HK$25,000 to $40,300 (1US$=7.8HK$) from 2006 to 2017, meaning there has been an increase in living standards for citizens post-development Moreover, the population of Wanchai has significantly increased since 2007, indicating that more people have moved to Wanchai since development due to improved quality of life. According to a survey conducted in 2011, Wanchai ranked the happiest out of all Hong Kong districts, scoring 8.55. This achievement could be attributed to pleasant environments or good economic situations.

However, the environment has not been appealing to its citizens. A letter from a citizen to the SCMP in 2018 showed that the noise pollution was unbearable due to heavy construction. A poll by the HK government about the number of pollution-related complaints in 2017 showed that Wanchai logged 1000 complaints - the second highest in the HK island sector, most of which were air pollution. This shows that traffic and congestion were also significant problems in 2017. The environment has not seen much change from 2006 to now. The district still has many of the same old buildings and stores from the 1950s. Many cars are parked to the side, and many shops are open (see Figure 4). The streets have more people, and the same shops are still in business, indicating that shop owners were able to get a stable income from readily available customers. There is, however, still a lot of traffic on the sides.

DISCUSSION AND SUGGESTIONS

Recent developments in Wanchai have elicited significant concern among residents regarding the district's rapid transformation. The government's initiatives to modernize areas such as Star Street have sparked fears of eradicating Wanchai's unique cultural heritage (Development Bureau, n.d.). This modernization threatens long-standing shops and restaurants, often operated by elderly proprietors who have spent their lives serving the community. The potential replacement of these establishments with high-end dining and nightlife venues targeted at foreign expatriates not only jeopardizes the livelihoods of these small business owners but also undermines the social fabric that defines Wanchai (Development Bureau, 2007).

Addressing these issues necessitates concerted efforts from the government, businesses, and residents. Key concerns include congestion, preserving Wanchai's cultural identity, and pollution. The government should implement comprehensive policies to address environmental issues, including pollution from vehicles and noise from traffic and renovation activities. A more integrated approach to developing neglected areas of Wanchai is essential. It should involve consulting residents to ascertain their preferences regarding street renovations, such as those proposed for Star Street. It is crucial to consider the adverse effects of renovation, such as noise pollution, which residents find particularly distressing. Furthermore, preserving Wanchai's cultural heritage is imperative to prevent globalization from diluting Hong Kong's unique characteristics. Rebranding efforts that displace local vendors can be mitigated by offering support and incentives to these vendors.

Traffic congestion is one of the foremost issues exacerbated by these developments (Transport Advisory Committee, 2014). Wanchai's narrow streets and high population density already strain the district's infrastructure, and the influx of new businesses and increased vehicular traffic only intensifies this problem (Lee, 2019). Congestion leads to longer travel times, reduced air quality due to increased vehicle emissions, and heightened noise pollution, all of which diminish the quality of life for residents.

To avoid air pollution and ensure the hygiene of the streets for people and the neighborhood (The Government of the Hong Kong Special Administrative Region, 2019), it is recommended that active engagement in raising awareness of their concerns and the preservation of a diverse range of goods and traditional products be pursued by businesses, particularly for small enterprises. Restaurants must also address air pollution by reducing the emission of greasy smoke from their vents (Environmental Protection Department, 2018).

Additionally, improving air quality by encouraging electric or hybrid vehicles through incentives would help reduce tailpipe emissions. Transport companies can contribute by providing incentives for public transport usage, increasing the frequency of subway trains, and transitioning buses, minibusses, and taxi fleets to electric vehicles to reduce emissions.

Residents play a crucial role in public discourse (Transport Advisory Committee, 2014), utilizing public transportation to alleviate congestion, avoiding illegal parking, which exacerbates traffic issues, and adhering to responsible litter disposal practices.

Building developers should prioritize the renovation of existing structures over the construction of new luxury shops and bars. Supporting small local businesses and placing less emphasis on attracting affluent clientele is critical. Developers should also consider incorporating more recreational spaces, such as parks, and providing affordable housing options. Retaining Hong Kong's cultural elements in new developments is essential to preserve the city's unique identity.

A collaborative approach involving the government, businesses, residents, transport companies, and building developers is vital to ensure Wanchai retains its cultural heritage while addressing contemporary challenges. This multifaceted strategy will help preserve the authentic experiences of Hong Kong residents and foster a more socially meaningful environment for all.

LIMITATIONS AND RESEARCH OUTLOOK

Some techniques employed in this pilot study needed improvement in execution and general method. The urban land use survey was executed well because samples were collected every 10 meters. However, some discrepancies were not recorded in the data. Many buildings were a hybrid of both commerce and residency but were recorded as "commercial" due to only the ground floor being commercial. This should have been a separate category or colored as both commercial and residential.

The EQS was detailed and had a wide range of questions regarding different aspects of the general atmosphere in Wanchai. Though that may be true, many errors resulted in a reduction of data validity and reliability. The criteria for a "good" score and a "bad" score were subjective. We judged them based on observations, and no rules were set on how they would be judged with non-biased opinions. More objective methods are necessary to some categories to avoid this, such as using a decibel counter to measure noise pollution and counting pieces of garbage on the street to judge pollution levels. Besides, some categories were not a good indication of development/improvement. The "culture" section was a reflection of this.

The photos were reflective and could visually indicate economic and environmental conditions in Wanchai. To improve the photo sampling, they could have been taken systematically and at every data point, along with the urban land use surveys and the EQS. To increase the range of data, more methods of collection could be used.

Not recording anomalies in the urban land use survey would reduce data validity as they are a mixture of both categories. The highly subjective nature of the EQS may have also decreased this as everyone in the group may have had different opinions on what is "good" or "bad". The vague questionnaire would have also reduced the validity as the people interviewed may have had different perspectives about issues in Wanchai. The fact that only 20% of interviewees were locals would also reduce the validity as they would be the ones most familiar with the area.

A problem encountered in this fieldwork was the small sample size. Only one small area of Wanchai was investigated, and only four transects with 22 data points were made, which is by no means representative of the actual conditions in the whole of Wanchai (though the authors are Hongkongers familiar with the area). This problem could be solved with more transects, data points, and people to answer the questionnaire.

Reliability was a big issue as the fieldwork was only conducted from 11:00-15:00 on a weekday. This does not reflect the general population, as most people are still at work or school during those times. To increase the reliability, the fieldwork could be repeated at different times.

CONCLUSION

In conclusion, though the economy of Wanchai has grown significantly since the start of development strategies, the environment has not seen much change, as issues such as pollution and vandalism are still prevalent in many areas.

In all sampling methods, Wanchai showed good economic growth after rebranding. The land use surveys demonstrated that commerce was the most prevalent form of infrastructure, and the EQS also provided positive results for commerce and business as the quality was consistent overall. Photos around the district also showed that many shops were densely populated, proving that economic activity is robust in Wanchai.

However, the environment has not been developed much. Pollution is still prominent and is worsening as construction increases. Congestion was also a frequent problem in the EQS, the questionnaires, and photos from secondary data.

Regular and quality maintenance of a rebranded area is crucial to sustaining environmentally friendly conditions for residents, commercial, visitors, and other stakeholders. Sufficient efforts and investment are necessary to maintain sustainability, such as waste management, reducing pollution, cleaning the outside building walls, repairing and updating traffic and public facilities to meet user requirements, and balancing the environment and public needs. Upholding a quality environment is crucial to maintaining the brand and reputation of a place to attract new visitors, residents, and investors and retain existing ones, which is essential for economic sustainability. Once achieved, such development achievement should be promoted to the public along with the attractions and events. Our recent publication has discussed such promotion strategies (Zuo et al., .2023), but differences in urban and rural branding are worth further investigation.

REFERENCES

Andersson, I. (2016). 'Green cities' going greener? Local environmental policy-making and place branding in the 'Greenest City in Europe.'. *European Planning Studies*, 24(6), 1197–1215. DOI: 10.1080/09654313.2016.1152233

Ashworth, G., & Kavaratzis, M. (2007). Beyond the logo: Brand management for cities. *Journal of Brand Management*, 16(8), 520–531. DOI: 10.1057/palgrave.bm.2550133

Braiterman, J. (2011). City Branding through New Green Spaces. In *Palgrave Macmillan UK eBooks* (pp. 70–81). https://doi.org/DOI: 10.1057/9780230294790_9

Chan, C., & Marafa, L. M. (2014b). Rebranding Hong Kong "Green": The potential for connecting city branding with green resources. *World Leisure Journal*, 56(1), 62–80. DOI: 10.1080/04419057.2013.876587

Cleave, E., & Arku, G. (2020). Immigrant attraction through place branding? Evidence of city-level effectiveness from Canada's London. *Cities (London, England)*, 97, 102502. DOI: 10.1016/j.cities.2019.102502

Daldanise, G. (2020). From Place-Branding to Community-Branding: A Collaborative Decision-Making Process for cultural heritage Enhancement. *Sustainability (Basel)*, 12(24), 10399. DOI: 10.3390/su122410399

Dastgerdi, A. S., & De Luca, G. (2019). Strengthening the city's reputation in the age of cities: An insight in the city branding theory. *City. Territory and Architecture*, 6(1), 2. Advance online publication. DOI: 10.1186/s40410-019-0101-4

Development Bureau. (2007, December 20). Press Release: Secretary for Development talks about Wan Chai old area revitalization project [:]https://www.devb.gov.hk/tc/publications_and_press_releases/press/index_id_3836.html

Development Bureau - Urban Renewal Strategy Review. (n.d.). Wanchai DAS final report. Retrieved from https://www.ursreview.gov.hk/eng/doc/Wanchai%20DAS%20final%20report_080410.pdf

Dupre, K. (2019). Trends and gaps in place-making in the context of urban development and tourism. *Journal of Place Management and Development*, 12(1), 102–120. DOI: 10.1108/JPMD-07-2017-0072

Environmental Protection Department. (2018, November 2). *Solutions for Pollution Problems*https://www.epd.gov.hk/epd/english/greenrestaurant/solutions/solutions_water.html

Erwee, C., Chipungu, L., & Magidimisha-Chipungu, H. (2021). Rebranding Umhlanga as an intelligent city. In *Springer eBooks* (pp. 391–437). https://doi.org/DOI: 10.1007/978-3-030-69698-6_77

Florek, M., Hereźniak, M., & Augustyn, A. (2021b). Measuring the effectiveness of city brand strategy. In search for a universal evaluative framework. *Cities (London, England)*, 110, 103079. DOI: 10.1016/j.cities.2020.103079

Gerosa, A., & Tartari, M. (2021). The bottom-up place branding of a neighborhood: Analyzing a case of selective empowerment. *Space and Culture*, 120633122110323. Advance online publication. DOI: 10.1177/12063312211032355

Herstein, R., & Berger, R. (2013). Much more than sports: sports events as stimuli for city rebranding. *Journal of Business Strategy/˜the œJournal of Business Strategy, 34*(2), 38–44. https://doi.org/DOI: 10.1108/02756661311310440

Insch, A., & Bowden, B. (2016). Possibilities and limits of brand repositioning for a second-ranked city: The case of Brisbane, Australia's "New World City", 1979–2013. *Cities (London, England)*, 56, 47–54. DOI: 10.1016/j.cities.2016.03.003

Kavaratzis, M. (2004). From city marketing to city branding: Towards a theoretical framework for developing city brands. *Place Branding*, 1(1), 58–73. DOI: 10.1057/palgrave.pb.5990005

Kavaratzis, M., & Hatch, M. J. (2013). The dynamics of place brands. *Marketing Theory*, 13(1), 69–86. DOI: 10.1177/1470593112467268

Lee, W. L. (2019, February 13). Illegal driving in Wan Chai causes congestion. District councilors advocate strengthening video law enforcement [ ]. *Hk01.com*https://www.hk01.com/18 /293958/ - - #

Miller, D., Merrilees, B., & Yakimova, R. (2013). Corporate Rebranding: An integrative review of major enablers and barriers to the rebranding process. *International Journal of Management Reviews*, 16(3), 265–289. DOI: 10.1111/ijmr.12020

Priemus, H. (2004). Changing urban housing markets in advanced economies. *Housing, Theory and Society/Housing. Theory and Society*, 21(1), 2–16. DOI: 10.1080/14036090410032872

Rehan, R. M. (2014). Urban branding as an effective sustainability tool in urban development. *HBRC Journal/HBRC Journal, 10*(2), 222–230. https://doi.org/DOI: 10.1016/j.hbrcj.2013.11.007

Ruiz, E. C., De La Cruz, E. R. R., & Vázquez, F. J. C. (2019). Sustainable Tourism and Residents' Perception towards the Brand: The Case of Malaga (Spain). *Sustainability (Basel)*, 11(1), 292. DOI: 10.3390/su11010292

The Government of the Hong Kong Special Administrative Region. (2019, April 15). *Three restaurants in Western District and Wanchai fined for illegally emitting excessive cooking fumes*. The Government of the Hong Kong Special Administrative Region - Press Releases. https://www.info.gov.hk/gia/general/201904/15/P2019041500362.htm?fontSize=1

Transport Advisory Committee. (2014). Report on Study of Road Traffic Congestion in Hong Kong. In *Transport Advisory Committee*. https://www.tlb.gov.hk/eng/boards/transport/land/Full_Eng_C_cover.pdf

Willer, C. J. (2021). Rebranding place "to build community": Neighborhood branding in Buffalo, NY. *Urban Geography*, 43(9), 1350–1371. DOI: 10.1080/02723638.2021.1927323

Wolch, J. R., Byrne, J., & Newell, J. P. (2014). Urban green space, public health, and environmental justice: The challenge of making cities 'just green enough.'. *Landscape and Urban Planning*, 125, 234–244. DOI: 10.1016/j.landurbplan.2014.01.017

Zenker, S., Braun, E., & Petersen, S. (2017). Branding the destination versus the place: The effects of brand complexity and identification for residents and visitors. *Tourism Management*, 58, 15–27. DOI: 10.1016/j.tourman.2016.10.008

Chapter 11
The Circularity in Transportation Completed

Eminegül Tekerci
https://orcid.org/0009-0002-9538-167X
Izmir Institute of Technology, Turkey

Yavuz Duvarcı
https://orcid.org/0000-0002-5496-2358
Izmir Institute of Technology, Turkey

Volkan Emre Uz
https://orcid.org/0000-0002-9328-4756
Izmir Institute of Technology, Turkey

Nicel Saygın
https://orcid.org/0000-0001-7773-1563
Izmir Institute of Technology, Turkey

ABSTRACT

This study explores the need for radical change in the transport sector, focusing on circularity and its implementation through innovative technologies. The unsustainable nature of current transport systems is discussed, highlighting the urgent need for change due to environmental impacts such as the climate crisis exacerbated by fossil fuels. The research presents a framework for 'Transport Circularity' that integrates solar energy, autonomous vehicles and other intelligent transport systems to create more sustainable, efficient and socially inclusive mobility solutions. Beforehand, drawing on philosophical perspectives, particularly Spinoza's views and Marxist critiques on conventional transportation, the study argues that a systemic overhaul

DOI: 10.4018/979-8-3693-6695-0.ch011

is inevitable. This transition's social and economic implications are also discussed, including its potential to dismantle existing power structures that benefit from the status quo. The study concludes with a vision of a resilient, sustainable and equitable future for urban and regional transport systems, giving a concrete example of Izmir.

INTRODUCTION

As is known, nowadays, production/consumption processes carried out in unnatural ways are being questioned. There is sufficient evidence suggesting that awareness has been developed within the public sector concerning this issue. In contrast, circular economy models have begun to be formulated as a new approach, and their projection onto the transportation sector will be discussed through possible models.

It is evident that the existing systems are neither natural; therefore, they cannot fit into any known philosophical and logical paradigm. It is imperative for transportation to undergo a transformation through "natural" way (which will be called Spinozan), ensuring that its effects are at an acceptable level. Thus, this study focuses rather what ought to be with a new strategy and with new technologies from this point on.

Although energy efficiency from unnatural ways has seemed the most convenient one compared to others, fossil fuel cannot be viewed positively at all due to their harmful and cumulative external effects on the nature because humanity is aware after all that irreversible points have already been exceeded as a result of the climate crises it faces (White Paper 2011). All those adverse effects mentioned above have been questioned for a long time, since the irreversible effects have continued to increase. In the perspective of future "green" transportation (or mobility), no "green light" should be given to fossil fuels anymore; on the contrary, now it will be given to, for example, electric vehicles and autonomous vehicles in various forms (White Paper, 2011). Nevertheless, the societal acceptance of systemic issues, the loss of life resulting from accidents, and the recurrence of such incidents impede the resolution of these problems and exacerbate the situation. This cultural perspective fosters the perpetuation of problems and hinders the development of effective solutions. Therefore, there is a need for a radical (revolutionary) change in the existing systems, and if possible, it should not be postponed (Buchanan, 2005). Due to the insufficiency of time, the transformation will need inevitably be radical and revolutionary one.

The concept of change has been a fundamental aspect of human thought since antiquity. The idea that "the only thing unchanged is the change" reflects the understanding that all established structures, including those in transportation, will inevitably evolve. As horse-drawn carriages, which were the dominant mode of transportation a century ago, gave way to railways and then to motorised vehicles, it seems reasonable to posit that the current transportation infrastructure will undergo significant transformations in the future. Therefore, it must be accepted that all these transportation forms currently used will inevitably undergo significant changes in the near future. Especially in the very short term, where applications like artificial intelligence are rapidly evolving, this is now quite likely when the clues emerging today are considered.

Therefore, in this study, the clues of these changes will first be examined, along with what the change (and technological transformation) can bring together, how the future can be channeled in a good direction (with concrete examples), and which necessary steps need to be taken further. Since a more "joyful" perspective of Spinoza will be adopted, such an approach will naturally have a positive/hopeful (normative/what should be) outlook on the future. From a philosophical perspective, such a "returning to nature", the necessity of seeking solutions to problems in a strict natural way, can initially be based on a Spinozan perspective (solution for sorrow through joy) (Spinoza, 2012). In this sense, it should be emphasized why a simple approach to environmental problems arising from transportation, especially why it is necessary to center Baruch Spinoza's philosophy of "joy" (natural solutions) and "sorrows" (unnatural developments) as an introduction. The Marxist criticism accompanies such a solution-oriented transformation in a radiacal way for historical progress (Schmitt, 1988).

In addition to the environmental problems, it becomes inescapable to bring more resilient, and more human-centered transportation structure to the agenda when it is considered that current systems create permanent systemic problems, in addition to such Marxist class issues, their vulnerability to pandemics, etc; neither private car-based nor conventional public transportation systems whose competitive power are weakened every year, have reached a point where they can not respond to accumulated problems and new mobility needs. In summary, the traditional transportation paradigm as it stands has now begun to collapse.

At the final point reached, many smart transportation applications, autonomous vehicles, and developments such as systems that obtain energy from the road are coming to the agenda as the special ways to overcome some of these problems. In line with these developments and the mentioned perspective, in this book chapter, sort of Spinozan natural analysis will be put forward as a macro solution model that complements with the new agenda of "circularity", and all developments in the world will inevitably have to evolve to this circular schema, whether they want to, or not.

As known, the term circularity refers to creating a sustainable and self-renewing system in transport that minimises waste and maximises resource efficiency while producing re-usable ouputs to others (sectors, etc.). This concept aims to reduce dependence on fossil fuels through the integration of renewable energy sources such as solar energy.

In practice, within the context of this study, circularity involves transferring the energy of solar panels strategically placed in urban areas to electric and autonomous vehicles. This model also supports sustainable practices such as energy harvesting and rainwater management. The aim is to create a transport network that supports environmental sustainability, where resources are constantly replenished and used efficiently. The inevitable (or, natural) "Transport Circularity" referred to should operate within the following cycle:

Figure 1. A triangular positive-directional circular scheme supporting each other

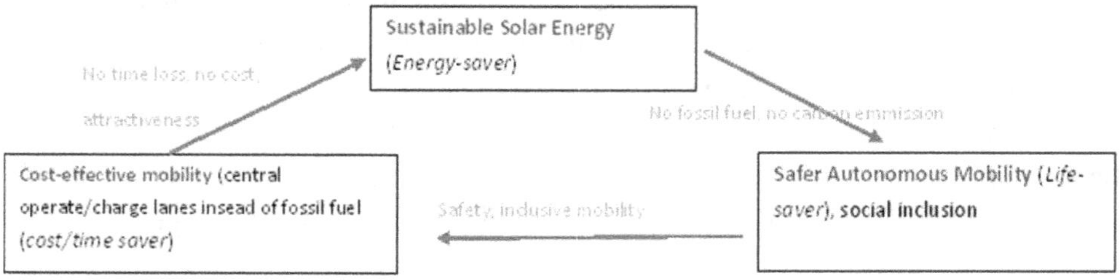

The issue is to seek a guide of solutions to the transport problem of the future and to ensure that the non-renewable energy currently used is replaced with renewable energy, which aligns with the sustainability principles of Europe. According to the European Commission's Sustainable and Smart Mobility Strategy document, the targets for 2030 include having 30 million zero emission vehicles on the roads, 100% carbon-neutral cities, making journeys under 500 km carbon neutral and scaling autonomous mobility. For 2050, zero-emission vehicles, functional and multimodal Trans European Networks (TEN-T) and almost zero deaths in all modes of transport are targeted (European Commission, 2020). As part of the European Green Deal, these strategies aim to contribute to achieving climate neutrality in the transport sector.

This chapter focuses on the need for a fundamental transformation of current transport systems due to their unsustainability and environmental impacts. First, the need for radical change is justified by arguing that current systems are unnatural and therefore do not fit into a philosophical and logical paradigm. Then, how solar energy and autonomous systems can be integrated is examined, considering innovative technologies and the concept of 'circularity' to enable this transformation. As an example of this integration, this triple structure is discussed in practice within the scope of a specific project. Finally, the social and economic impacts of this transformation are assessed, and how existing power structures can resist this change is discussed. All these issues are addressed to support the argument that it is inevitable to create a sustainable, efficient and socially inclusive transport system.

As the scope of the study, Spinoza's basis for the necessity of complying with the nature provides hope for the future of transport, and the modern Critical Theory provides basis for the criticism of the today's wrong-going. However, as a limitation, it cannot be delved deeper into the theory and philosophical details within the limited scope of the study; only necessary connections will be made at the relevant points.

The following section presents a concise overview of the theoretical and philosophical foundations of the critique of the prevailing transportation paradigm, identifying the core philosophical tenets. Subsequently, the following sections will address the practical and physical implementation of innovative technologies that can facilitate the transition to a circular transport system. This will involve an examination of three key technologies that can be combined to achieve this goal, accompanied by an analysis of global case studies and a review of relevant literature. Ultimately, this section examines whether a conceptualised circular transport system can yield positive outcomes in a real-world urban context, specifically in the Karşıyaka district of Izmir. In conclusion, the findings are synthesised to derive the final statements.

CRITICISM OF TRANSPORTATION THROUGH SPINOZAN AND MARXIST LENSES

The modern transportation system, shaped by capitalist structures, has long perpetuated a series of societal and environmental problems, including accidents, air pollution, noise, and social inequities. Critical Theory, particularly the Frankfurt School, critiques this conventional system, arguing that it is grounded in positivist sciences that treat the world in a detached and particularistic manner. This approach neglects the deeper economic and societal contradictions, as well as the environmental degradation caused by these systems (Marcuse, 1964). Critical Theory, thus, calls for a reevaluation of technology and transportation systems, urging a move away from systems that primarily serve economic interests and elites, to those that benefit both humanity and nature.

Marcuse (1964) points out how technology, rather than being a tool for human liberation, has increasingly detached from its ethical and social purpose. Technological advancements, such as megaprojects like the hyperloop, often appear disconnected from cohesive urban and societal needs, further highlighting this disconnect (Yavuz & Öztürk, 2021). These projects, driven by capitalist imperatives, are poorly integrated into existing urban systems, and they lack a broader strategy to address the systemic problems that the traditional transportation model exacerbates. Critical Theory, therefore, challenges this trajectory, advocating for transportation systems that align with social and ecological well-being.

Philosophically, both Spinoza and Marxist critiques offer a critical lens to evaluate the mobility crisis. Spinoza's rational ethical framework calls out for solutions that are pragmatic yet aligned with nature's order, challenging the existing elite-dominated systems that prioritize private interests over the public good (Zelyut, 2021). Spinoza's philosophy, in this sense, critiques the inherent inequality of current transportation systems, where mobility predominantly benefits the wealthy at the expense of broader societal needs. This ethical approach suggests that rational and sustainable solutions—such as the use of renewable energy—should be prioritized in transforming urban transport into a more equitable and sustainable model.

From a Marxist perspective, Gramsci's concept of hegemony further reveals how transportation systems reinforce the dominance of the elite class. These systems, by design, cater to the privileged, while imposing negative consequences, such as pollution and limited access, on lower-income communities. The resistance to transformative, sustainable models of mobility reflects the struggle between the ruling class and the working class (Gramsci, 1997). As innovative and more equitable transportation models emerge, the existing power structures that benefit from the status quo face increasing challenges.

The intersection of Spinoza's ethics with Marxist critique, along with insights from thinkers like Habermas, illustrates the inherently political nature of transportation systems. Habermas emphasizes the role of rational-critical debate in challenging dominant power structures, which applies to how transport systems and public spaces are designed. Both he and Gramsci underscore the importance of public participation and democratic engagement in reshaping these systems to ensure equitable access and sustainability (Benson, 2009). Current transport systems often reflect the interests of those in power, limiting the potential for innovative solutions that could address the environmental and social crises of our time.

In conclusion, both Critical Theory and these philosophical reflections emphasise the imperative for a re-evaluation of the manner in which we conceptualise and approach the issues of transportation and mobility. The growing demand for sustainable solutions, driven by environmental concerns, is likely to intensify political tensions as elite resistance to change persists. However, if rational, ethical solutions—drawing from the ideas of Spinoza, Marx, Habermas, and Gramsci—can be implemented, a more equitable, sustainable transport system can be realized. The integration of renewable energy technologies, such as solar panels in transport systems, represents one such solution that is both in harmony with nature and socially just. Moreover, dismantling the entrenched power structures that perpetuate inequality is essential to achieving a future of mobility that benefits all.

CONCEPT OF "CIRCULARITY" IN TRANSPORTATION

Once the philosophical and political context of the preceding section has been grasped, which elucidates the rationale behind the imposition of an unnatural mode of transportation upon society, it becomes possible to proceed to an examination of the concept of circularity in urban transport. This concept promises a future characterised by healthy and natural modes of transportation, in a manner that aligns with the ideas of Spinoza and Gramsci.

The concept of 'circularity' in transportation revolves around creating a sustainable and regenerative system that minimises waste and maximises resource efficiency. This is in line with wider environmental principles, particularly the concept of using renewable energy sources such as solar power to power transport systems (Jaiswal et al., 2022). The integration of solar energy into urban transport infrastructure can significantly reduce reliance on non-renewable energy sources, in line with the concept of circularity.

In this approach to this new Transport Circularity, the focus is on harnessing natural renewable energy, particularly direct solar energy, and integrating it into the urban landscape. Solar panels can be installed in urban areas using streets, open spaces and other public areas. The energy collected by these panels can be directly transferred to autonomous transport systems, especially electric and driverless vehicles (Liu et al., 2023). Here, the specific meaning of circularity in transport will be conveyed as a proposal in line with the most natural way as it can be within the limited context of this research, based on some concrete application examples from real life.

The strategic placement of solar panels in public spaces, such as streets and open areas, has the potential to generate a substantial amount of renewable energy for cities. This approach not only optimises land use, but moreover provides a direct source of energy for electric vehicles, thereby reducing energy transmission losses and infrastructure costs (Fagnant & Kockelman, 2015). To illustrate, the solar energy harvested from panels situated in public spaces can be conveyed directly to electric vehicles, thus reducing transmission losses and operational costs.

To further explore the feasibility of this concept, it is necessary to calculate the amount of energy required to meet urban travel demand. This calculation involves determining the area required for solar panel coverage, measured in either kCal or kW. For example, a city could assess its daily travel demand and calculate the energy requirements of its autonomous electric vehicle fleet. From this, the area required for solar panels can be determined, ensuring that the energy supply matches the transport demand (Doğan, 2022).

This research also includes the development of a new generation of surface materials. These materials are designed to optimise land use as well as to contribute to energy production. In particular, the use of areas such as impermeable pavements, road surfaces and medians for rainwater harvesting (RWH) and sustainable urban drainage systems (SUDS) is one of the potential application areas of these materials. In line with the principle of circularity, such design ensures efficient management of water by directing rainwater into specific channels, thus supporting urban green spaces and contributing to pollution prevention. At the same time, the ability of the surfaces to collect solar energy enables the integration of renewable energy sources, reducing dependence on fossil fuels and thus maximising resource efficiency. In line with the work of Eames et al. (2013), this design contributes to sustainable water management and energy practices that reinforce circularity in urban design. All those will lead to new research areas for scientists, and practicioners as well.

Consequently, the idea behind this concept of circular transport emphasizes the seamless integration of renewable energy sources, particularly solar energy, into urban infrastructure. By designing spaces that serve multiple functions - such as energy generation and water management - cities can create more efficient, cost-effective and sustainable transport systems. Ultimately, this approach contributes to a broader vision of circularity (in terms of both the physically circular process and the efficient use of outputs between related sectors), where urban resources are continuously regenerated and used to their full potential. It is also defined as reduction of waste, and futility use (European Parliament, 2024).

This idea draws on principles of sustainability and strategic urban planning, as discussed by Dixon et al., (2018), who highlight the importance of foresight techniques and strategic placement of infrastructure in urban environments.

INTEGRATED TECHNOLOGICAL AND SOCIAL IMPACTS OF FUTURE TRANSPORTATION SYSTEMS

The transformation of the transportation sector is being driven by cutting-edge technologies with the potential to significantly enhance both environmental sustainability and social equity. Among these advancements, the widespread implementation of solar panel systems, electric autonomous vehicles, and wireless energy transfer represents a crucial step toward reducing carbon emissions, improving efficiency, and creating more inclusive transportation systems.

Solar energy, harnessed through photovoltaic panels integrated into urban environments and transport infrastructure, provides a renewable energy source capable of fueling public transportation while generating surplus energy for other urban uses. This shift from fossil fuels to clean energy is pivotal for addressing climate change, and its integration into transportation will play a major role in reducing cities' carbon footprints (Lobaccaro et al., 2019; White Paper, 2017). Similarly, electric autonomous vehicles offer a sustainable alternative to gasoline-powered cars. These vehicles optimize routes, reduce traffic congestion, and minimize accidents caused by human error, contributing to both safety and sustainability. Furthermore, the lower operational and fuel costs of electric vehicles will likely drive their widespread adoption, especially when paired with wireless energy transfer systems that allow for continuous, on-the-move charging (Duvarcı & Akpınar, 2023; Bi et al., 2016).

However, while these technologies hold great promise, their success relies on the integration of these innovations into a cohesive system. Solar power, electric autonomous vehicles, and wireless energy transfer must work together to create a comprehensive ecosystem that maximizes resource efficiency and minimizes downtime. Real-world testing and collaboration between governments, private companies, and the public are essential to ensure the smooth adoption and scalability of these technologies (Gotzler et al., 2022).

Social outcomes, including equity and affordability, are equally critical in shaping the future of transportation. New transport systems must prioritize accessibility for vulnerable groups, such as the elderly, disabled, and low-income populations. Electric vehicles, with their lower operating costs and reduced maintenance needs, offer a more affordable option compared to gasoline-powered cars, making mobility more accessible to a broader segment of the population (International Energy Agency, 2020; KPMG, 2021). Shared mobility platforms and autonomous EV services further lower personal transport costs, especially in cities that have piloted Mobility as a Service (MaaS) systems, benefiting low-income individuals by eliminating the need for personal vehicle ownership (Goodall et al., 2017).

In addition to cost savings, the ease of use and time efficiency provided by autonomous vehicles contribute to a better quality of life. These vehicles reduce traffic congestion by optimizing travel routes, shortening commuting times, and improving overall productivity. They also increase independence for individuals unable to drive, such as the elderly and disabled, by providing greater access to transport options and services (Hasan & Van Hentenryck, 2021).

The potential social benefits, however, hinge on ensuring that advanced transport technologies are accessible to all, not just the wealthy. Government policies that promote affordable access to electric and autonomous vehicles through subsidies and inclusive pricing models are essential for preventing the deepening of social inequality. Moreover, these new systems can create job opportunities in the operation and maintenance of advanced vehicles, contributing to more inclusive economic growth (Duvarcı, 2024; Hasan & Van Hentenryck, 2021).

In conclusion, the future of transportation lies in the integration of technological advancements such as solar energy, electric autonomous vehicles, and wireless energy transfer. These innovations have the potential to reshape the sector by reducing environmental impact, lowering costs, and promoting social equity. For these systems to be successful, a collaborative effort from all stakeholders—governments, private companies, and the public—is required to ensure the benefits of these technologies are widely shared and contribute to a more sustainable and inclusive future.

FEASIBILITY OF THE NEW SYSTEM AND THE SUBSIDIZATION ISSUE

The new circularity system presents a significant economic opportunity, offering the potential for profitability across both the public and private sectors. The transition to sustainable transportation is not only an environmental imperative but also a profitable venture. One of the key economic advantages of the circularity system is its ability to reduce costs through energy efficiency and reduced reliance on fossil fuels. By lowering operating expenses, the system can pass savings on to consumers, making transportation more affordable and increasing demand for these new technologies. This, in turn, creates a positive feedback loop where increased adoption leads to greater profitability.

The figures in the Table 1 only shows the cost breakdown for APM (Automated People Movers, as monorail) systems similar to PRT, leaving aside other associated costs of the components of the integrated system.

Table 1. Percentage breakdown of APM (similar to PRT) costs

Component	Aver. Cost %
Guideway	26.5
Stations	11.4
Maint. & Support	5
Power & Utility	7
Vehicles	19.3
Control, Communic & Engin.	30.8

(Kerr et al, 2005)

However, in a PV-covered walkway integrated PRT system described in our own model, the guideway and station cost components will more likely be removed with various installation flexibilities. In a simulation study, PRT is found to be worthy alternative to BRT (Bus Rapid Transit) and LRT (Light Rail Transit) systems in service quality and environmental impacts, while the operator and consumers would save considerable amounts of money from low construction and operation costs (Juster & Schonfeld, 2013). When the infrastructure weights of a PRT, APM are to be compared to a LRT system, LRT seems ten times heavier system than, though having almost the passenger capacity it shown at Figure 2.

Figure 2. Comparison of superstructure weights (Kerr et al., 2005)

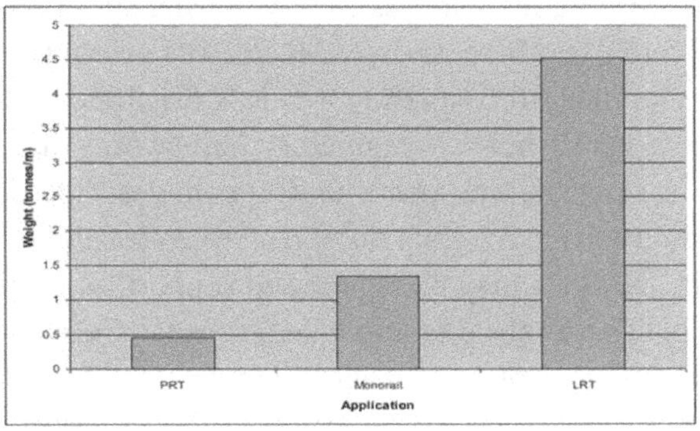

PRT construction cost is found to be around 4,5 million US$ per mile, while this is around 18 million for the APM systems (similar to PRT) (Kerr et al., 2005). Today, most sustainable solutions (as bicycle and pedestrianization programs, MaaS, etc.) have minor impacts and do not have the potential to alter the mainstream current system. When compared to other sustainable solutions, which are also piecemeal and detached, the new integrated circular solution would pose a radical alternative to the traditional one, yielding more comprehensive, cohesive, and substantial results.

Investments in the circularity system are expected to garner considerable interest from both public and private entities. Technologies such as electric vehicles (EVs) and wireless energy transfer systems are scalable and have broad applications, making them appealing to investors (Aggarwal & Singh, 2021; Li & Mi, 2014). Early investment in these technologies can provide companies with a competitive edge in the expanding market for sustainable transportation. Additionally, the circularity system introduces new revenue streams, such as the sale of surplus solar energy produced by integrated solar panels (Calzolari et al., 2022). Additionally, for investing in the production sectors (solar panels, electric road components, etc.) of each component of the integrated system can take place in the stockmarkets, as in the case of Railbus example. These supplementary revenue sources contribute to the system's financial sustainability, diminishing the need for continual government support. Also, government policies such as tax incentives to ease the deployment of such novel technologies, since they are environmentally friendly solutions, would support the widespread adoption of the circular transport. The governments may also support the related start-up projects, and pioneer the pilot projects, with the financing mechanisms from global investors.

Feasibility is essential for the successful deployment of the circularity system. Although the technology itself is viable, the success of the system hinges on the development of necessary infrastructure. For instance, widespread adoption of EVs necessitates a comprehensive network of charging stations (Aggarwal & Singh, 2021). Similarly, incorporating wireless energy transfer systems into roadways demands substantial initial investments (Li & Mi, 2014). Governments must prioritize infrastructure development to ensure the system's feasibility, which includes not only physical infrastructure but also regulatory frameworks that foster innovation and investment in sustainable transportation technologies.

Subsidization is crucial in bridging the gap between feasibility and profitability. Initial government subsidies may be required to support the circularity system as it gains momentum. These subsidies can manifest as tax incentives for companies investing in sustainable technologies, grants for research and development, or direct funding for infrastructure projects (Calzolari et al., 2022). Governments should consider these subsidies as investments in the future, given that the long-term benefits of the system—such as reduced environmental impact, improved public health, and economic growth—far outweigh the initial costs. By providing financial support during the early stages, governments can facilitate the success of the circularity system and stimulate further private sector investment.

The broader economic impact of the circularity system extends beyond profitability and feasibility. Reducing reliance on fossil fuels can stabilize energy prices, which are often volatile due to geopolitical factors. This stability enhances the system's attractiveness to investors and consumers, contributing to its long-term profitability. Moreover, the circularity system can generate specialized employment opportunities in areas such as renewable energy production, EV manufacturing, and smart grid management. These jobs benefit communities economically and help reduce economic inequality.

Furthermore, the circularity system supports the broader transition to a sustainable economy by promoting the use of renewable energy sources and reducing carbon emissions. This aligns with global efforts to combat climate change.

When the PRT system's ability to meet existing travel demands is considered (approximately one-third of vehicles with more effective use of the fleet), it can be seen to replace the majority of private cars on urban streets. This is particularly in comparison to other forms of transport, such as flexible buses, light rail transit (LRT), or bus rapid transit (BRT), which are conventional systems that are costly. It can be reasonably deduced that the lightweight pods and lighter infrastructure requirement (one tenth of LRT) will result in reduced construction costs, greater applicability, and enhanced safety and energy efficiency. The elimination of all corresponding costs, zero-accident rate, and the use of solar panels to meet at least one-quarter of existing energy needs (including the elimination of transporting oil fuel) are likely outcomes. It can be reasonably deduced that this transport model will be the most cost-effective, sustainable and plausible candidate for implementation if the aforementioned circular system, along with the necessary components (PV-paved surfaces and electric roads) are rolled out across the city.

One significant benefit will be the enhanced air quality, which is challenging to quantify in financial terms. Furthermore, as a fundamental attribute, the notion of inclusivity and accessibility for the most disadvantaged groups, namely the elderly, vulnerable, and children, would represent another significant cost-saving measure. Other improvements, such as the elimination of traffic congestion, the reduction of parking areas that occupy a significant amount of urban space, noise pollution and the associated health problems, and so on, would not be as straightforward to quantify (requiring more detailed and sophisticated analyses). Nevertheless, the advent of autonomous vehicles will precipitate a reduction in the number of drivers required, which will in turn give rise to an unemployment problem at the prevailing price level.

DISMANTLING THE EXISTING HEGEMONY STRUCTURE FOR THE “BRAVE” SPINOZAN TRANSPORT

The traditional transport system, shaped for many years by fossil fuels and conventional vehicles, is on the verge of a major transformation. This change is not only about the adoption of new technologies, but also about the process of dismantling (or, displacement) the status quo that has profited from the traditional system for decades, which makes it Marxist transformation. The shift to a more sustainable and efficient transport model is a revolutionary change that challenges entrenched interests and requires a rethinking of how societies move people and goods.

The revolution needed for a “brave” new transport is to dismantle this traditional system and replace it with one that prioritizes sustainability, inclusion and efficiency. This revolution requires not only technological, but also social, political and economic changes. It requires a fundamental change in the way societies approach transport, shifting from a model that prioritises profit and consumption to one that values the well-being of people and the planet.

One of the key drivers of this revolution is the adoption of renewable energy sources, such as solar and wind energy, which can replace fossil fuels in the transport sector. The transition to electric vehicles powered by renewable energy allows societies to significantly reduce their carbon footprint and mitigate the environmental impacts of transport (Hossain et al., 2022). Furthermore, this transition has the potential to decentralize energy production, reducing the power of large corporations and allowing communities to produce and control their own energy resources. But, the adoption of all novel "green" technologies, sustainable approaches, autonomous vehicles, etc. require an integration process and good orchestration.

Alongside technological advances, the transport revolution also requires changes in policy and governance. Governments must play an active role in breaking the status quo by implementing regulations that promote sustainability and penalise environmentally damaging practices. This can include steps such as introducing carbon taxes, providing subsidies for electric vehicles and renewable energy infrastructure, and investing in public transport systems that are accessible and affordable for all (Deakin, 2001). By creating a policy environment that favours sustainable transport, governments can accelerate the process of moving away from the conventional system.

A significant element of this revolutionary process is the necessity for social and economic equality. It is imperative that the alteration of the status quo does not result in the emergence of new forms of inequality. As a fundamental aspect of the traditional status quo, the integrated circularity system is inherently opposed to the formation of inequalities. Instead, it emphasises inclusivity. The anticipated benefits of the new transport system are expected to be distributed equitably across society. This encompasses the provision of affordable transport options for low-income communities and the accessibility of jobs created by the new system to all (Clairand et al., 2019).

It is similarly crucial to consider the requirements of vulnerable demographics, such as the elderly and individuals with disabilities, who may encounter distinctive obstacles in accessing transportation. By prioritising equity in the design and implementation of the new system, societies can facilitate the creation of a more equitable and inclusive transport model. The necessity to address climate change and the increasing demand for more sustainable and equitable transport options demonstrates that this revolution is inevitable. As an increasing number of cities and countries pledge to reduce carbon emissions and transition to renewable energy sources, the impetus for change will continue to intensify.

In conclusion, the dissolution of the status quo in the transport sector represents a pivotal moment in the broader movement for sustainability and social justice. The dismantling of the traditional system and the completion of the revolution needed for a "brave" new transport model will provide societies with the opportunity to create a transport model that is not only environmentally sustainable, but also socially and economically just. This revolution will require bold action from governments, businesses and individuals, but the rewards – clean air, healthy communities and a fairer society – will be well worth the effort.

SOME PROMISING INNOVATIONS IN MOBILITY

PRT System and Wireless Energy Charching

Personal Rapid Transit (PRT) systems consist of small electric vehicles with a capacity of two to six people. These vehicles operate on demand and are centrally organised on fully segregated infrastructure. PRT vehicles offer a high speed (up to 60 km/h) and a capacity of up to 10,000 passengers per hour, completely outside other traffic. The highest level of comfort and privacy is ensured by allocating the vehicles to a single user or group of users on request. Travel times are short because there are no intermediate stops between the origin and destination of a journey and no interaction with other traffic (Poojari & Uday, 2014).

These vehicles usually move along tracks or elevated lines on a defined network (but, this is not a necessary condition). PRT systems allow passengers to travel from one point to another in a direct and non-stop manner. The advantages of PRT systems include reducing traffic congestion, shortening journey times and reducing carbon footprint (Muller, 2011; Choromanski et al., 2013). The vehicles use electricity and automatically guided by advanced navigation systems. By providing on-demand service, these systems ensure that vehicles are allocated when passengers need them, thus increasing the efficiency of the system (Mogge, 2009). Yet, though PRTs are said to be more efficient and environmentally friendly, their comparative performance analyses have not been analysed firmly in a holistic manner (Mascia et al., 2016; Diez-Jimenez et al., 2020).

Wireless energy charging technology enables the wireless transfer of electrical energy from a source to a receiver through electromagnetic fields. This technology uses inductive and magnetic resonance methods for contactless energy transfer. Inductive charging works by utilising the magnetic field between two coils, while magnetic resonance charging provides more efficient energy transfer over greater distances (Doğan, 2022). The main advantages of wireless charging include ease of use, contactless charging and the ability for vehicles to receive energy on the go. However, there is still room for improvement in terms of cost and energy efficiency (McDonald, 2011).

In summary, PRT systems and wireless energy charging technologies are important innovative technologies for the future of urban transport. These technologies can both support environmental sustainability and improve the user experience. However, infrastructure investments and technological developments are required for the widespread implementation of these systems. In the future, the integration of these two technologies is expected to form the basis of a greener and more efficient transport system in cities (Mogge, 2009; Poojari & Uday, 2014).

Walkable Solar Panels

Solar panels, particularly in their innovative forms like "walkable solar panels," play a critical role in the shift towards renewable energy by directly converting solar energy into electricity, reducing reliance on fossil fuels, and promoting sustainability. Walkable solar panels, as developed by companies like Onyx Solar, are designed to be durable enough to withstand foot traffic while still efficiently generating energy. These panels, installed in various locations such as rooftops and pathways, offer a unique way to optimize urban spaces for energy production, thereby contributing to a city's renewable energy infrastructure (SEIA, n.d.)(Solar Panels Network, n.d.).

Such advancements not only maximize the usable space for solar energy but also allow for creative solutions in urban environments. For example, these panels have been used to create solar-powered walkways, demonstrating their ability to balance functionality and energy efficiency (SEIA, n.d.). Walkable solar panels present an innovative solution aimed at enhancing energy production, particularly in urban environments. Unlike conventional panels, these are constructed from durable and robust materials capable of supporting pedestrian use. They are designed for high-traffic areas such as parks, pavements, and bridges. These panels are engineered to maximize solar energy capture while ensuring effective energy production in confined spaces (Ma et al., 2019).

An important aspect of walkable solar panels is their integration with urban functionality and aesthetics. This feature is crucial for city centers and tourist areas, where maintaining visual appeal while generating energy is important. The efficiency of these panels may vary based on material quality and panel size, but they offer a dual benefit of energy generation and aesthetic preservation (Hu et al., 2022).

Although the energy output of walkable solar panels is generally lower compared to traditional panels, due to thicker protective layers and structural modifications for durability, their potential is significant. By being deployable over large areas and adaptable to various environments, these panels can contribute meaningfully to overall energy production. Particularly in dense urban centers and commercial districts, widespread adoption of walkable solar panels could have a substantial impact on energy supply (Vijayan et al., 2023).

Solar panels hold a crucial place among renewable energy sources, and walkable solar panels represent a promising new application of this technology. Designed to boost energy production in urban settings while preserving aesthetic values, they have the potential for broader adoption in the future. Continued research and development will enable cities to meet their energy needs more sustainably and efficiently.

CASE STUDIES FROM ALL OVER THE WORLD

PRT System Examples

Three notable examples of Personal Rapid Transit (PRT) systems have been selected for examination: the Masdar City PRT in Abu Dhabi, UAE; the Heathrow Airport Pod system in London, UK; and the Suncheon SkyCube PRT Ride in South Korea. These systems were chosen for their diverse applications and unique contributions to the field of urban transportation. The Masdar City PRT is recognized as a pioneering model of eco-friendly transport, reflecting the sustainable vision of Masdar City and its commitment to reducing carbon emissions. The Heathrow Airport Pods are highlighted for their practical integration within an airport environment, showcasing efficiency in passenger transport and seamless connectivity. The Suncheon SkyCube is noted for its innovative design and its role in enhancing urban mobility in a city renowned for its environmental focus. Collectively, these examples provide valuable insights into the adaptation of PRT systems to various contexts and objectives, offering a comprehensive view of their effectiveness and potential for broader implementation.

Masdar City PRT, Abu Dhabi, United Arab Emirates

Masdar City, located near Abu Dhabi, is an innovative project aiming to be a model eco-city of the future. Designed to accommodate around 40,000 residents and 1,500 businesses, the city is powered largely by renewable energy, featuring one of the largest solar farms in the Middle East. Its design also includes water recycling and waste management systems to minimize environmental impact. The system can use the existing road infrastructure as mixed with the other traffic. Transportation within the city includes a small-scale PRT system using driverless, solar-powered electric vehicles. However, due to financial challenges, some aspects of the project have been scaled back or delayed (Robles, 2006). This example points to the use of existing road infrastructures, requiring no additional infrastructure for the PRT system, which will eliminate the guideway costs. Since the city is already fed by solar farms, the PRT can utilize this renewable source.

Figure 3. Masdar city PRT system

Figure 4. Masdar city PRT system, part 2

Heathrow Airport Pod, London, United Kingdom

The only public Ultra pod system at Heathrow Airport, which entered service in May 2011, is known as the Heathrow pod system. It consists of 21 vehicles operating on a 3.9 kilometre (2.4 mile) route connecting Terminal 5 to the business passenger car park immediately north of the airport.

Ultra largely utilises commercially available technologies to reduce construction costs. These technologies include systems that allow tyres to move on an open guideway. This approach has resulted in a system that Ultra claims is economical (University of Bristol,n.d). The company reports that the total cost, including vehicles, infrastructure and control systems, is between £3 million and £5 million per guideway. However, the implementation at Heathrow is reported to have cost £30 million for 3.8 kilometres (2.4 miles) of guideway (Ultra, 2009).

This cost difference emphasises the cost advantage and flexibility of implementation afforded by Ultra's system, while the reasons for the higher cost at Heathrow may include factors such as greater structural complexity and additional safety measures. By being energy efficient and using zero-emission vehicles, this system minimises the environmental impact of traditional fossil fuel transport options. It also operates according to passenger demand, reducing unnecessary energy consumption and providing a sustainable transport solution. When designed to integrate with renewable energy sources, the Heathrow Pod system can fully contribute to the energy and resource efficiency principles of the circular economy.

Figures 5. PRT system in Heathrow Airport

Figure 6. PRT system in Heathrow Airport, part 2

Suncheon SkyCube PRT Ride

The Suncheon SkyCube is an eco-friendly PRT system located in the Suncheon Bay Wetlands in South Korea. It operates on a 4.6-kilometer elevated track, connecting Suncheon's 'Dream Bridge' to the Suncheon Bay Ecological Park. This system uses 40 vehicles, each accommodating six to nine passengers, and offers an on-demand service that eliminates emissions within the eco-park (Mascia et al., 2016). The SkyCube's design prioritizes minimal environmental impact by utilizing polyurethane-coated wheels to reduce noise, contributing to the preservation of the surrounding wetlands.

The system offers a green alternative to traditional public transportation systems. It excels in providing an eco-friendly transportation solution by ensuring punctuality, safety, reliability, accessibility, and privacy. The system is designed to function efficiently in urban areas, with the capability to handle short interval stops of just 1 to 2 minutes. Unlike conventional public transport that adheres to fixed schedules, PRT adapts to passenger demand. When passenger numbers are low, empty cars wait at stops, while during peak times, additional cars are dispatched to meet the increased demand. Passengers can expect minimal wait times, generally less than a minute, and enjoy a direct, non-stop journey to their destinations, akin to the experience of taxis. This ensures both privacy and efficient travel.

The cost of installing a PRT system is significantly lower, approximately one-tenth the cost of building a subway system and one-third the cost of light rail train systems, with similar capacity. Due to its automatic operation, PRT is cost-effective, with potential savings realized by not running cars during periods of low demand. As a result, PRT is an ideal option for developing new cities, new routes, or enhancing existing ones. (POSCO, 2014). Suncheon example can be an ideal model for future circular transport applications in terms of its succesful performance that it shows, especially for new city formation based on PRT.

Walkable Solar Panels Examples

Three significant examples of walkable solar panels have been selected for analysis: Solar Pavements in Tourouvre-au-Perche, France; SolaRoad in the Netherlands; and Solar Roadways in the United States. These examples were chosen for their pioneering contributions to integrating solar technology into pedestrian and vehicular surfaces. The Solar Pavements in Tourouvre-au-Perche represent a notable application of photovoltaic technology in road surfaces, marking a significant step in enhancing the sustainability of infrastructure in a rural setting (Colas Group, 2016). The SolaRoad project in the Netherlands is highlighted for its innovative approach in embedding solar panels within bike lanes, demonstrating practical and scalable integration of solar energy within urban infrastructure. (SolaRoad, 2015).Solar Roadways, an ambitious project in the United States, is recognized for its focus on developing durable solar panels for roadways and other surfaces, aiming to revolutionize energy generation and road maintenance (Brusaw, 2018). Together, these examples illustrate a diverse range of approaches to incorporating solar technology into everyday surfaces, providing valuable insights into the potential for increasing renewable energy production through innovative infrastructure solution

Solar Pavements in France

Tourouvre-au-Perche, a town in Normandy, France, has taken a significant step towards sustainable energy solutions with the inauguration of its solar road. On Wednesday, French Ecology Minister Ségolène Royal officially opened this groundbreaking kilometer-long road, marking a milestone in the integration of renewable energy into urban infrastructure. The project, which took five years to complete, represents a substantial investment in the future of eco-friendly transportation and urban planning (Colas Group, 2016).

The development of this solar road came with a hefty budget of 5.2 million dollars. The road's surface is covered with 30,000 square feet of specialized solar panels, designed to harness solar energy effectively. These panels are coated with silicone resin, providing them with enhanced durability to withstand the impact of vehicular traffic (Plummer, 2016). This coating is crucial for maintaining the integrity and functionality of the panels despite the stresses of daily use.

This initiative aligns with broader goals of reducing the carbon footprint of urban environments and promoting sustainable energy sources. By integrating solar panels into roadways, Tourouvre-au-Perche not only addresses the challenge of retrofitting street lights but also contributes to the development of renewable energy infrastructure. The solar road is a testament to France's commitment to innovative environmental solutions and serves as a model for other regions exploring similar technologies (Plummer, 2016). This French application provides an illustrative example of the installation of durable and walkable panel-paved road surfaces for the purpose of harnessing energy for urban transportation. Furthermore, the Netherlands has developed a walkable solar panel, designated 'SolaRoad', which has been implemented in Zwolle.

Figure 7. France's solar pavements

Solar Roadsways: Innovative Solar Panel Technology for Smart Highways

Solar roadways are a revolutionary idea that seeks to transform traditional asphalt and concrete roads into surfaces embedded with solar panels. The vision is to create a multifunctional infrastructure that not only supports vehicular traffic but also generates clean energy, powers streetlights, melts snow, and even charges electric vehicles on the go. The concept proposes that these solar panels be built with durable materials capable of withstanding the heavy weight and wear of vehicles while capturing sunlight to produce electricity. However, in case PRTS are to be used with much lighter body, there will not weight problem, reducing the energy cost, either (Chapman et al., 2011)

In theory, solar roadways could contribute significantly to renewable energy goals by utilizing vast stretches of road surfaces to generate power. However, in practice, the implementation has faced numerous obstacles. One of the most prominent challenges is the high cost of installation and maintenance. Solar panels that are durable enough for road use are expensive to produce, and their installation requires significant infrastructure changes. Moreover, these panels have not yet proven to be as energy-efficient as traditional, roof-mounted solar panels due to their less-than-ideal positioning and shading issues (Hyder, 2023).

Figure 8. Solar roadsways pavements

Figure 9. Solar roadsways pavements

APPLICATION EXAMPLE OF CIRCULAR INTEGRATION IN KARŞIYAKA CASE

The objective was to ascertain how the various technologies described in previous sections could be integrated in a concrete case study. The region of Karşıyaka, located within the city of Izmir in Turkey, was selected as the sample area due to its high population density. The results of the analysis indicated that the public transport systems in the region are unable to meet the demand for passenger transportation. Therefore, it was decided to test a hypothetical PRT system that would work integrated with other transport systems. Karşıyaka is a place with narrow streets and high-rise buildings, but it also has many transport system options from ferry to subway and minibuses. The centre area of the district has about 344 thousand inhabitants, where most businesses and restaurants are located. Due to the crowded neighbourhood and narrow streets, both noise and visual pollution are concentrated. As a possible solution to this new situation, a PRT system was considered due to its potential to provide a high capacity public transport with a high level of service at attractive capital and operating costs. Since the hard surfaces of the roads create heat island effect, the roads are considered to be covered with walkable solar panels and the PRT system to be powered by solar panels on the road.

Walkable solar panels are subject to pedestrian traffic, which can cause wear and tear over time. Thus, durable materials and coatings that enhance wear resistance are considered. Regular maintenance and periodic inspections can also identify and address such issues with early-warning. Safety is a critical concern when integrating solar panels into walkable surfaces to avoid slips, trips, and falls, incorporating anti-slip coatings or textures on the surface of the panels. Additionally, clear signage can alert pedestrians to the presence of solar panels.

Karşıyaka Site Details

The area selected for this study is situated in the Karşıyaka district of Izmir, located in the northern part of the city and in close proximity to the urban core. Karşıyaka has a long history of development and is currently one of the most significant urban centres in Izmir. The district's central location boasts a plethora of modern shopping centres, cafés, parks and historical landmarks.

The rationale for selecting this area is its convenient transportation infrastructure and high population density. In order to gain a deeper understanding of the project area, preliminary analyses were conducted. A daylight analysis was conducted in Karşıyaka to ascertain the potential for achieving 100% efficiency from solar panels. However, due to the height of the buildings and narrowness of the roads, it was not possible to achieve 100% efficiency from sunlight. The results of the measurements taken per square metre indicate that an average efficiency of 40 percent is probable.

In order to ensure comfortable transportation, the upper noise level is accepted as 65 dB(A), with a tolerance zone of 65-75 dB(A) and a discomfort zone of 75-120 dB(A). The noise levels produced by motor vehicles on highways are contingent upon the engine volume and the efficacy of the mufflers, with typical ranges between 72 and 92 decibels. Consequently, Karşıyaka is characterised as a noisy location due to the conventional motor traffic, including motorbikes and human density, in addition to its limited green spaces, which are unable to absorb the noise effectively.

The objective of the proposed PRT system is to mitigate both vehicle-related noise pollution and traffic congestion. Furthermore, the implementation of solar panels to meet the energy demands of PRT vehicles will contribute to the sustainability, vibrancy and liveability of the Karşıyaka environment. Two testing scenarios for the PRT system were devised, one without stops and another with stops. Concept plans were developed accordingly for each scenario.

Figure 10. PRT system without stop concept

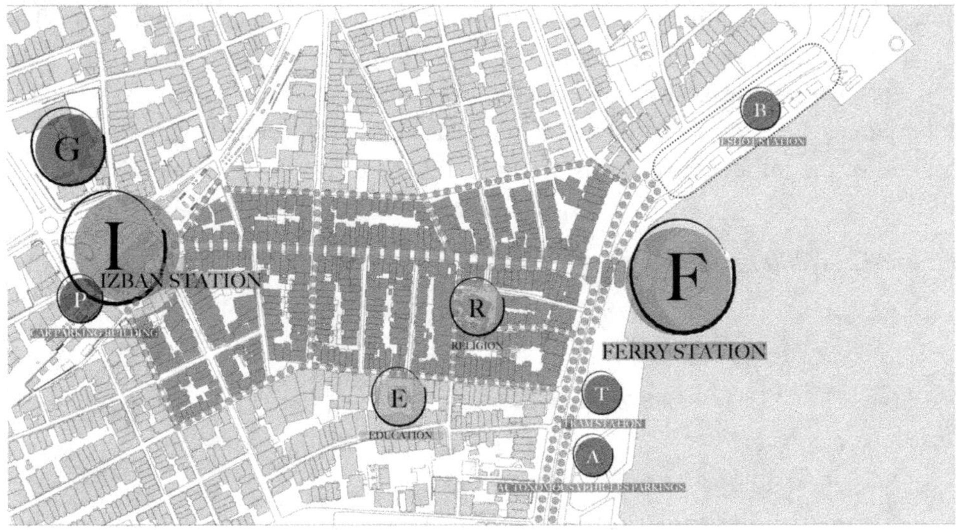

In Figure 10, the places shown as pink lines indicate the paths of PRT vehicles and their connection with other means of transportation. One of the goals was to integrate the PRT system with other transportation systems, which was also tested. Additionally, it was ensured that the roads for PRT vehicles were capable of making loops. Since the PRT system operates on non-stop schedule, a signal needs to be given, and the system, through its sensors, will recognize the person's travel demand to enter the vehicle and stop accordingly.

Figure 11. PRT system with stop concept

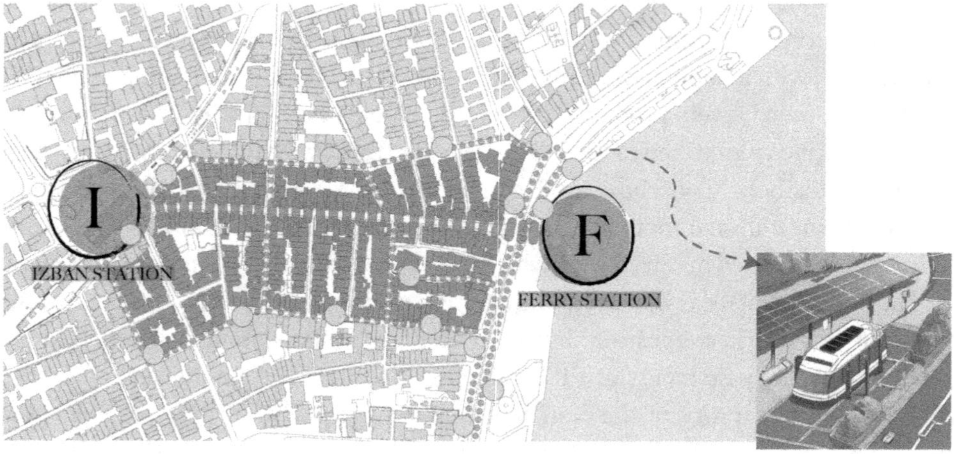

In light of the congestion that characterises the region and the resulting high demand, it would be prudent to consider the introduction of additional stops. It is, however, recognised that the construction of stops will entail additional costs. The illustration on the right depicts a prototype PRT stop. In order to implement these concepts, it was necessary to carry out the requisite energy calculations. As walkable solar panels are to be installed on the roads used by PRT vehicles, a representative design for the initial concept was developed at the urban design scale of 1/2000.

Figure 12. 1/2000 scale plan

It is hypothesised that rainwater is drained by a suitable slope, whereby the sheets positioned on top of the solar panels are tilted in a specific direction. The system is designed in such a way that the water flows in a single direction and is subsequently collected by two collection pipes. The water will be collected from both sides of the system and subsequently undergo treatment. Following purification, the water will be directed to an underground tank, where it will be available for irrigation purposes.

Energy Calculations

Energy calculations were initiated for the base assumption that one solar panel produces 400 watts of energy per meter square. Walkable solar panels vary in size from 0.5 to 1.5 square meters, with an average size of 1 square meter, as assumed. However, due to the energy transmission to the PRT system through a wireless connection, around a 40 percent energy loss is assumed. Consequently, 160 watts of energy is lost during the transfer, leaving 240 watts of usable energy per solar panel (per 1 m^2). The total area designated for solar panel installation in Karşıyaka is 18 ha which seems to be paved. Therefore, the number of solar panels that can be installed in this area is 180.000.

Figure 13. Total area of the solar pavements (the rectangle at right shows the required area for comparison)

Figure 14. Calculation of solar energy

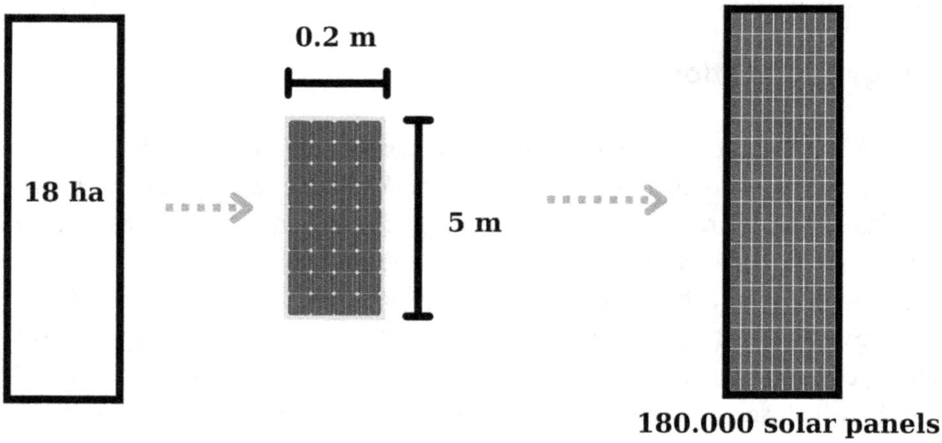

Figure 15. Solar energy generation with wireless energy transfer and efficiency of energy transfer

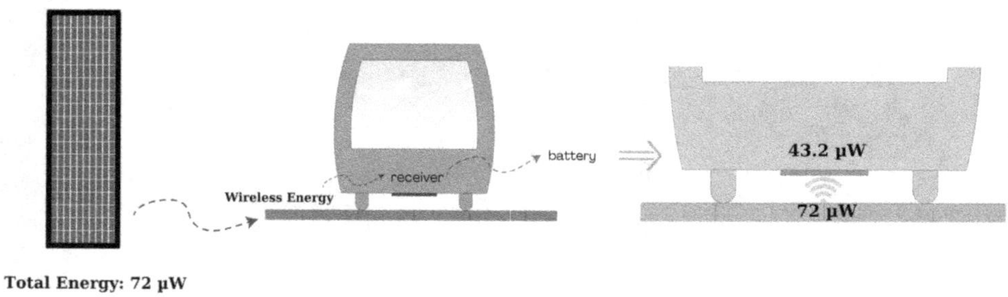

It is assumed that each panel generates 400 watts of energy, which would result in a total energy generation capacity of 72 megawatts for all panels. However, it is anticipated that a 40% loss will occur during the transport of energy through the wireless connection. Consequently, the net usable energy production of each panel is reduced to 240 watts, resulting in a total usable energy output of 43.2 megawatts.

The energy requirements of PRT vehicles can be determined concurrently or in accordance with the specified energy amount. The research conducted by Mittelman and colleagues revealed that a PRT vehicle consumes 0.08 kW of energy per hour to transport one person one kilometre. If we assume that 2,000 people will use the system in a day, the energy required for a PRT vehicle for one person amounts to 160 kW. However, since PRT vehicles will accommodate three or four people, the average energy requirement per vehicle is 40 kW. In order to accommodate 2,000 people, a total of 500 vehicles are required. Thus, the requisite energy for a single day is 20 megawatts. This indicates that the solar panels are capable of producing a surplus of energy, which is sufficient for the PRT system.

Construction Costs of Solar Panels

In accordance with the findings of Chopson (2024), the cost per square metre is estimated to be \$75.35 (1 square metre ≈ 10,764 square feet, \$7/sq. ft). The total area in question is 18 hectares, which is equivalent to 180,000 square metres. Consequently, the total cost is approximated to be approximately US\$ 13,566,000 when calculated over this area and unit price.

Fuel Consumption in Türkiye for one year

It is estimated that Turkey's annual fuel consumption is approximately 30 million metric tonnes. The price per litre of diesel fuel is 40 TL, with each tonne of diesel equivalent to approximately 1,200 litres. Therefore, Turkey's annual fuel consumption is equivalent to 36 billion litres. The total annual fuel cost is 1.44 trillion TL, which equates to approximately 40 billion USD (1 USD ≈ 36 TL).

In light of the fact that the daily fuel requirement of the PRT vehicle is 20 megawatts, it can be extrapolated that the annual fuel requirement is 7,300 megawatts. When the annual fuel consumption of Turkey and the PRT system we have designed are considered, our solar panels meet 0.004% of the total fuel requirement annually. Although this figure may appear modest, the solar panels we have designed can contribute to savings of approximately 59.4 million TL of Turkey's total annual fuel cost when the annual cost calculation is examined.

Moreover, it is anticipated that there will be an increase in carbon emissions. The combustion of diesel fuel is estimated to generate approximately 2.68 kg of CO_2 emissions per 1 litre of diesel fuel on average. Turkey's annual fuel consumption results in the emission of approximately 96.48 million tonnes of CO_2; the solar-powered PRT system we have designed has the potential to contribute to a reduction of approximately 3.95 tonnes of CO_2 emissions by replacing diesel fuel consumption. This illustrates the potential of renewable energy sources to mitigate carbon emissions.

FINDINGS, CHALLENGES AND LIMITATIONS

The study identifies several technical constraints that impact the effectiveness of the proposed integration of walkable solar panels and PRT systems. One significant challenge is the reduced energy efficiency of walkable solar panels compared to traditional solar panels. The reduction in efficiency is due to the necessity for durable materials and coatings that can withstand foot traffic, as well as safety concerns such as preventing slips, trips, and falls. Furthermore, the use of wireless energy charging introduces an additional layer of complexity, with current technology resulting in approximately 40% energy loss during transmission. This reduction in efficiency impacts the overall effectiveness of the system.

Further complications arise from the design and implementation stages of the Project; integrating the PRT system with existing public transport infrastructure presents a number of complex logistical and financial considerations. The costs associated with installing solar panels and developing the PRT infrastructure are considerable, necessitating careful planning and substantial investment. Furthermore, the energy loss associated with wireless charging and the reduced efficiency of walkable solar panels in urban settings, such as narrow streets and tall buildings, as in the Karşıyaka case, limit the overall effectiveness of the system. Maintenance requirements and varying weather conditions also affect the energy yield from the solar panels.

The daylight analysis for the Karşıyaka district reveals that achieving 100% efficiency from solar panels is unfeasible due to the area's physical characteristics. Given an estimated efficiency rate of only 40%, the total energy production is limited. This efficiency constraint, in conjunction with other environmental factors and the necessity for periodic maintenance, has an impact on the overall performance of the proposed energy solutions.

Notwithstanding these constraints, the incorporation of walkable solar panels in urban locations such as Karşıyaka demonstrates considerable potential for enhancing renewable energy generation. The study estimates that these solar panels could generate up to 43.2 megawatt of energy per day. This substantial energy output, even with reduced efficiency, demonstrates the potential for harnessing solar energy in densely populated urban environments where traditional solar installations may not be feasible.

The combination of wireless energy charging technology with PRT systems offers a promising opportunity to enhance operational efficiency and minimise energy loss. While current wireless charging technology results in a 40% energy loss, ongoing advancements are expected to enhance energy transfer efficiency over time. These developments will contribute to the creation of a more sustainable and efficient urban transport system in the future.

The study emphasises the possibility of integrating renewable energy solutions into urban transportation systems, highlighting their potential to address environmental sustainability and enhance liveability. By reducing noise and vehicle-related pollution, walkable solar panels and PRT systems can improve urban environments. However, further research and development are necessary to overcome technical limitations and optimise system efficiency. Continued innovation and investment in sustainable technologies are crucial for achieving long-term goals related to green mobility and energy efficiency.

Figure 16. Comparative analysis of solar energy and fossil fuels

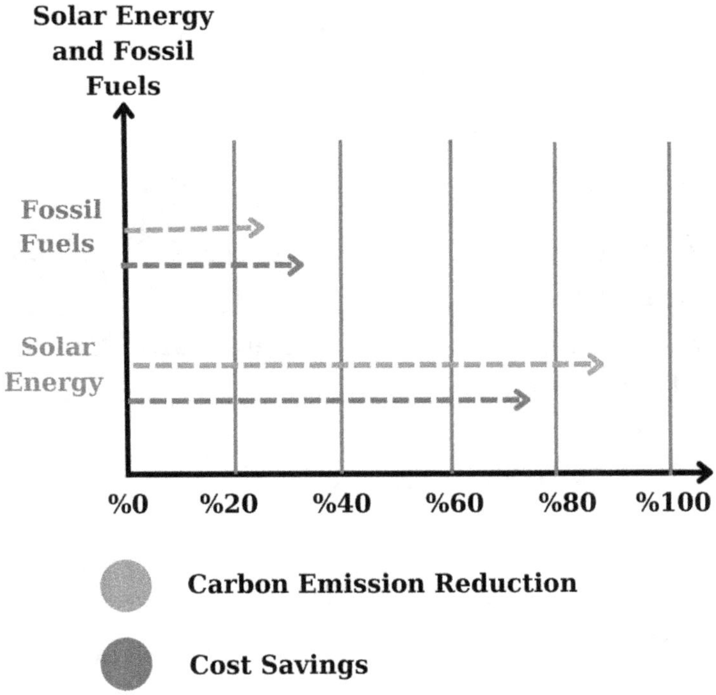

The PRT (Personal Rapid Transit) system proposed in Karşıyaka stands out as a direct response to the environmental and social problems caused by conventional transport systems. This system reflects both Spinoza's search for solutions in harmony with nature and offers a solution to Marx's critique of class inequalities. Spinoza's view that what is 'natural' will inevitably renew itself suggests that traditional systems dependent on fossil fuels should be replaced by technologies based on renewable resources such as solar energy. The proposed PRT system, integrated with innovative surface materials and solar-powered vehicles, not only ensures environmental sustainability but also promotes social inclusion. From a Marxist perspective, this system has the potential to break the class hierarchy created by traditional fossil fuel transport networks. By providing a more accessible and affordable transport system for low-income groups, it directly contributes to the reduction of social inequalities.

Figure 17. The Spinozan model of sustainability

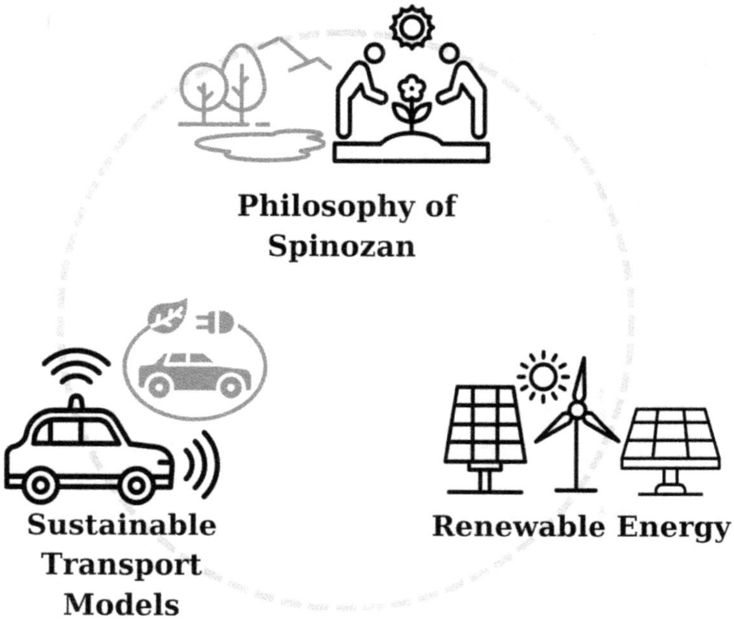

CONCLUSION

Circular transport discussed in this study, here rather referring to the use of "green" technologies and Spinozan natural solutions, offers an important solution in line with today's sustainable urbanism and environmental protection goals. Considering the environmental problems of existing transport systems, reducing fossil fuels, reusing waste and increasing energy efficiency are the cornerstones of circular transport systems. Moreover, integrating circular transport principles in urban planning has the potential to increase social welfare while minimising environmental impacts.

However, the systemic problems (*sorrows*) issued from the conventional transport systems remain to be contstant due to the status quo powers who benefit the most from the current situation. Therefore, the resoning of the existing situation required to be criticized through the lens of the Critical Theory of the Frankfurt School, that tries to see the genuine reason behind the apparent; understanding power relations, political awareness and a hegemonic shift (dismantling) in favor of people's desires is a must before initiating the circular transport and the other technical issues.

The applicability of circular transport in cities depends on various factors. In particular, transition to renewable energy sources and raising public awareness stand out as the cornerstones of this process. Technological innovations, especially autonomous electric vehicles and renewable energy systems, are considered as the basic components of circular transport. The diffusion and integration of these technologies will make a major contribution to increasing energy efficiency and reducing negative impacts on the environment. This preliminary work of ours shows through the case study that, even in strictly dense urban environment, 10% of energy requirement can be obtained from solar panels deployed around the roads and open spaces, of which the efficency can be further increased in the future.

However, the successful implementation of circular transport strategies requires strong co-operation between local authorities, national policy makers and the private sector and inclusion of the people. In particular, policy makers' adoption of circular economy principles and incentive arrangements in the field of transport will facilitate the adoption of this new approach by wider masses. In addition to policy support, raising social awareness and developing training programmes are also necessary for sustainable transport systems.

Circular transport can either directly contribute to the social and economic dynamics of cities. In addition to creating greener and more sustainable cities, this approach can also provide economic efficiency in the long term, which is not focused, here. Transport models based on circular economy principles allow for more efficient use of resources, while also contributing to the strengthening of local economies. In this sense, thus, circularity concept can be leveraged in order to gain the public support for the innovative "brave" transport.

Future research should focus on assessing how circular transport models can be implemented in different cities and their long-term impacts. Rapidly urbanising and environmentally challenged countries like Türkiye can achieve both economic development and environmental sustainability goals through innovative approaches such as circular transport. In this context, it is evident that more studies should be conducted on the feasibility and potential benefits of innovative aapproaches harnessing the concept of circular transport, to alter the existing one.

In conclusion, it can be seen that new technologies will not only ensure environmental sustainability but also contribute to economic growth. Therefore, it is of great importance that policy makers, urban planners and private sector representatives work together to promote circular transport and increase its viability. It is clear that circular transport will play a key role in creating sustainable cities and minimising environmental impacts in the future.

REFERENCES

Aggarwal, S., & Singh, A. K. (2021). Electric vehicles: The future of the transportation sector: A review. *Renewable & Sustainable Energy Reviews*, 135, 110320.

Benson, R. (2009). Shaping the public sphere: Habermas and beyond.

Bi, Z., Kan, T., Mi, C. C., Zhang, Y., Zhao, Z., & Keoleian, G. A. (2016). A review of wireless power transfer for electric vehicles: Prospects to enhance sustainable mobility. *Applied Energy*, 179, 413–425. DOI: 10.1016/j.apenergy.2016.07.003

Brusaw, S. (2018). Solar Roadways: The future of roads. [Online] Available at: https://www.solarroadways.com/

Buchanan, B. G. (2005). A (very) brief history of artificial intelligence. *AI Magazine*, 26(4), 53–53.

Calzolari, T., Genovese, A., & Brint, A. (2022). Circular economy indicators for supply chains: A systematic literature review. *Journal of Cleaner Production*, 351, 131564.

Chapman, R., Howden-Chapman, P., & Capon, A. (2016). Understanding the systemic nature of cities to improve health and climate change mitigation. *Environment International*, 94, 380–387. DOI: 10.1016/j.envint.2016.04.014 PMID: 27126780

Clairand, J.-M., Guerra-Terán, P., Serrano-Guerrero, X., González-Rodríguez, M., & Escrivá-Escrivá, G. (2019). Electric vehicles for public transportation in power systems: A review of methodologies.

Colas Group. (2016). Wattway: The solar road. [Online] Available at: https://www.wattwaybycolas.com/en/

Deakin, E. (2001). Sustainable development and sustainable transportation: Strategies for economic prosperity, environmental quality, and equity.

Diez-Jimenez, E., Fernadez-Llorca, D., Munoz, M. F., & Sotelo, M. (2020). Personal Rapid Transport System Compatible With Current Railways and Metros Infrastructure. *IEEE Transactions on Intelligent Transportation Systems*, (99), 1–11.

Dixon, T., Montgomery, J., Horton-Baker, N., & Farrell, L. (2018). Using urban foresight techniques in city visioning: Lessons from the Reading 2050 vision. *Local Economy*, 33(8), 777–799. DOI: 10.1177/0269094218800677

Doğan, E. (2022). Placement and sizing strategies for dynamic wireless charging stations on signalized intersection corridors. *Pamukkale Üniversitesi Mühendislik Bilimleri Dergisi*, 28(6), 802–811.

Duvarcı, Y. (2024). Drawing a Vision for More Sustainable Mobility: the Example of PRT System. *International Symposium on Sustainable Urban Mobility through Modular and Integrated Transport System*, 28 March 2024, Nagoya University, Nagoya.

Duvarcı, Y., & Akpinar, F. (2023). Using Personal Rapid Transit as an Effective Transport Solution in Historical Downtown Areas: A Case from Historic Kemeraltı, Izmir. *Planlama*, 33(1), 86–104.

Eames, M., Hunta, M., Dixon, T., & Britnell, J. (2013). Retrofit city futures: Visions for urban sustainability.

Electric Power Research Institute. (2021). The Electric Vehicle Maintenance Advantage. https://www.epri.com/research/products/3002020211

Enerji Atlası. (n.d.). Akaryakıt tüketimi. Enerji Atlası. Retrieved 15 January 2024, https://www.enerjiatlasi.com/akaryakit-tuketimi/

European Commission. (2020). Sustainable and Smart Mobility Strategy – putting European transport on track for the future. Retrieved from https://eur-lex.europa.eu/legal-content/EN/TXT/?uri=CELEX:52020DC0789

European Parliament. (2024). Circular economy: Definition, importance and benefits. Retrieved September 19, 2024, from https://www.europarl.europa.eu/topics/en/topic/circular-economy

Fagnant, D. J., & Kockelman, K. (2015). Preparing a nation for autonomous vehicles: Opportunities, barriers and policy recommendations. *Transportation Research Part A, Policy and Practice*, 77, 167–181. DOI: 10.1016/j.tra.2015.04.003

Goodall, W., Dovey, T., Bornstein, J., & Bonthron, B. (2017). The rise of mobility as a service. Deloitte Review. https://www2.deloitte.com/us/en/insights/focus/future-of-mobility/mobility-as-a-service.html

Gotzler, F., Neumann, F., & Adenaw, L. (2022). Assessment of personal rapid transit system configurations regarding efficiency and service quality. *Future Transportation*, 2(3), 3. Advance online publication. DOI: 10.3390/futuretransp2030041

Grabski, W., & Daszczuk, W. B. (2017). A study on cooperation of urban transport means: PRT and Light Rail. *Transport Problems*, 12(4), 5–14. DOI: 10.20858/tp.2017.12.4.1

Gramsci, A. (1997). *Hapishane defterleri (Prison Notebooks).* Belge Yayınları.

Hasan, M. H., & Van Hentenryck, P. (2021). The benefits of autonomous vehicles for community-based trip sharing.

Hossain, M. S., Kumar, L., Islam, M. M., & Selvaraj, J. (2022). A comprehensive review on the integration of electric vehicles for sustainable development. Wiley Online Library. https://onlinelibrary.wiley.com/doi/full/10.1155/2022/3868388

Hu, H., Zha, X., Li, Z., & Lv, R. (2022). Preparation and performance study of solar pavement panel based on transparent resin-concrete. *Construction & Building Materials*, 340, 127871.

Hürriyet. (n.d.). Motorin fiyatları. Bigpara. Retrieved 15 January 2024, https://bigpara.hurriyet.com.tr/akaryakit-fiyatlari/motorin/

Hyder, Z. (2023, June 26). All about Solar Roadways: The promise versus the reality. SolarReviews. https://www.solarreviews.com/blog/all-about-solar-roadways

International Energy Agency. (2020). Global EV Outlook 2020. https://www.iea.org/reports/global-ev-outlook-2020

Jaiswal, K. K., Roy Chowdhury, C., Yadav, D., Verma, R., Dutta, S., Jaiswal, K. S., Ba, S., & Karuppasamy, K. S. K. (2022). Renewable and sustainable clean energy development and impact on social, economic, and environmental health. *Energy Nexus*, 7, 100118. DOI: 10.1016/j.nexus.2022.100118

Juster, R., & Schonfeld, P. (2013). Comparative Analysis of Personal Rapid Transit as an Urban Transportation Mode. *Transportation Research Record: Journal of the Transportation Research Board*, 2350(December), 128–135. DOI: 10.3141/2350-15

KPMG. (2021). The Total Cost of Ownership for Electric Vehicles. https://home.kpmg/xx/en/home/insights/2021/04/total-cost-of-ownership-for-evs.html

Li, S., & Mi, C. C. (2014). Wireless power transfer for electric vehicle applications. *IEEE Transactions on Power Electronics*, 29(7), 3453–3461.

Liu, H., Skandalos, N., Braslina, L., Kapsalis, V., & Karamanis, D. (2023). Integrating solar energy and nature-based solutions for climate-neutral urban environments. *Solar*, 3(3), 382–415. DOI: 10.3390/solar3030022

Lobaccaro, G., Croce, S., Lindkvist, C., Munari Probst, M. C., Scognamiglio, A., Dahlberg, J., Lundgren, M., & Wall, M. (2019). A cross-country perspective on solar energy in urban planning: Lessons learned from international case studies. *Renewable & Sustainable Energy Reviews*, 108, 209–237. DOI: 10.1016/j.rser.2019.03.041

Ma, T., Yang, H., Gu, W., Li, Z., & Yan, S. (2019). Development of walkable photovoltaic floor tiles used for pavement. *Renewable Energy*, 132, 1143–1152.

Mascia, M., Hu, S., Han, K., Sun, A., North, R., & Lees-Miller, J. D. (2016). A holistic approach for performance assessment of personal rapid transit. *Research in Transportation Business & Management*, 18, 70–76. DOI: 10.1016/j.rtbm.2016.01.008

Mittelman, G., Kariv, Y., Cohen, Y., & Avineri, E. (2022). Techno-economic analysis of energy supply to personal rapid transit (PRT) systems. *Journal of Cleaner Production*, 324, 129031.

Mogge, J. (2009). The technology of Personal Rapid Transit. Retrieved from iTrans Advanced Transit

Muller, P. (2011). *PRT consulting, sustainable city enabled by personal rapid transit.* Retrieved from Sustainable City Enabled by Personal Rapid Transit.

Plummer, L. (2016) France unveils world's first solar panel road: 0.6-mile stretch to provide energy to power entire village's street lights. Daily Mail. https://www.dailymail.co.uk/sciencetech/article-4058752/France-unveils-world-s-solar-panel-road-0-6-mile-stretch-provide-energy-power-entire-village-s-street-lights.html

Poojari, S. S., & Uday, K. V. (2014). PRT systems. *International Journal of Earth Sciences & Engineering*, 7(4), 720–723.

POSCO. (2024). Korea's first personal rapid transit (PRT) Skycube. POSCO Newsroom. https://newsroom.posco.com/en/koreas-first-personal-rapid-transit-prt-skycube/

REN21. (2022). Renewables Global Status Report 2022. https://www.ren21.net/reports/global-status-report/

Road Traffic Technology. (n.d.). SolaRoad project: Krommenie. Retrieved August 28, 2024, from https://www.roadtraffic-technology.com/projects/solaroad-project-krommenie/

Robles, A. (2006). Masdar City: The soon-to-be eco city. Environment & Society Portal. Rachel Carson Center for Environment and Society. https://www.environmentandsociety.org/tools/keywords/masdar-city-soon-be-eco-city

Schmitt, R. (1988). The Materialist Dialectic. Science & Society, Vol. 52 (4) (Winter, 1988/1989), 441-456.

Solar Energy Industries Association (SEIA). (n.d.). Solar Panels. Retrieved from https://www.seia.org/solar-101/solar-panels

Solar Panels Network. (n.d.). Walkable Solar Panels: A New Era in Solar Technology. Retrieved from https://www.solarpanelsnetwork.com/walkable-solar-panels

SolaRoad. (2015). The world's first solar cycle path. [Online] Available at: https://sola-road.com/

Spinoza, B. (2012). *Ethika.* Kabalcı Yayınları.

Ultra. (2009, June 1). Corporate brochure. Archived from the original (PDF) on July 11, 2012. Retrieved August 24, 2024, from https://web.archive.org/web/20120711192906/http://www.ultraprt.net/cms/PerimeterPRT_11x17_2_side_brochure.pdf

University of Bristol. (n.d.). Heathrow personal rapid transport. Retrieved August 24, 2024, from https://www.bristol.ac.uk/research/impact/heathrow-personal-rapid-transport/

Vijayan, D. S., Koda, E., Sivasuriyan, A., Winkler, J., Devarajan, P., Kumar, R. S., Jakimiuk, A., Osinski, P., Podlasek, A., & Vaverková, M. D. (2023). Advancements in solar panel technology in civil engineering for revolutionizing renewable energy solutions—A review. *Renewable & Sustainable Energy Reviews*, 183, 113206.

White Paper. (2011). Roadmap to a Single European Transport Area – Towards a competitive and resource efficient transport system. European Commission, Brussels, 28.3.2011.

Zelyut, S. (2021). *Spinoza.* Say Yayınları.

Chapter 12
The Relationship Between COVID-19 and Urban Features in the Light of Recent Quantitative Studies

Tayfun Salihoğlu
https://orcid.org/0000-0002-9959-6961
Pamukkale University, Turkey

İhsan Tutuk
https://orcid.org/0000-0003-2373-312X
Gebze Technical University, Turkey

ABSTRACT

The COVID-19 pandemic, caused by the novel coronavirus SARS-CoV-2, has profoundly impacted urban environments globally. The virus's dependence on physical proximity for transmission and the necessity for individuals to congregate in urban spaces to meet various needs accelerated its spread. Therefore, as with many other epidemic diseases, investigating the urban factors behind contact and spread has become an important issue for the COVID-19 pandemic as well. This study investigates the relationship between COVID-19 and various urban features by synthesizing recent quantitative research. The review revealed that it is possible to conceptualise urban factors which are influencing COVID-19 spread as socio-demographic characteristics, healthcare services, infrastructure, tourism activities, economic activities, and urban conditions. Understanding these urban determinants is essential for developing targeted strategies to control the spread of COVID-19

DOI: 10.4018/979-8-3693-6695-0.ch012

and enhance urban resilience against future pandemics.

INTRODUCTION

Coronaviruses (CoV) are a large family of viruses known to cause various diseases ranging from the common cold to acute respiratory infections (Keni et al., 2020). It is known that coronaviruses lead to colds characterized by sore throat, dry cough, runny nose, malaise, and fatigue during the fall and winter months each year (COVID-19 Pandemic Assessment Report, 2020). Although coronaviruses are generally known as an RNA virus family that shows symptoms at the level of the common cold in humans, viruses such as SARS-CoV and MERS-CoV, which belong to the same family, have caused more serious diseases (Budak & Korkmaz, 2020). Indeed, these viruses infect certain respiratory cells in humans and cause lung damage, turning into a threat to human life due to the lack of specific treatment. To date, seven different coronaviruses (HCoV-OC43, HCoV-HKU1, HCoV-229E, HCoV-NL63, SARS-CoV, MERS-CoV, SARS-CoV-2) that are respiratory pathogens have been identified (Solomon & Liang, 2022). The SARS-CoV-2 virus first emerged in 2019 in the city of Wuhan, China. This new coronavirus began to spread to humans via respiratory transmission and quickly disseminated to other countries and cities. Despite being declared a “Public Health Emergency of International Concern” by the World Health Organization (WHO) in January 2020, it was reclassified as a “Pandemic” in March 2020 (Ribeiro et al., 2020). As of August 2020, more than 20 million COVID-19 cases had been confirmed worldwide, and the death toll had exceeded 761,000 (COVID-19 Pandemic Assessment Report, 2020). The virus's high transmissibility contributed to it becoming one of the deadliest pandemics in history within this short period (WHO, 2023).

During this period, the measures implemented to control the spread of the epidemic led to significant changes in the dynamics of urban areas. Schools, workplaces, and recreational facilities were closed, and strategies such as social distancing, isolation, and quarantine were implemented. However, these measures conflicted with individuals' needs for social interaction and the design and configuration of urban spaces. The virus's dependence on physical proximity for transmission and the necessity for individuals to gather in urban areas to meet various needs accelerated its spread. It has been reported that 1,430 cities in 210 countries were affected by COVID-19, with 95% of confirmed cases occurring in urban areas (WHO, 2020), which serves as evidence of how critical cities became during the pandemic period. During this time, cities faced challenges such as economic and social congestion due to the restriction measures taken to control the outbreak, as well as issues like high population density, unevenly distributed healthcare systems, inadequate housing, and declining sanitation services. This is because the COVID-19 pandemic has caused not only a global health crisis but also environmental, economic, social, and cultural crises. Figure:1 shows the main urban themes that came to the fore in cities during the COVID-19 epidemic (Sharifi & Khavarian-Garmsir, 2020).

Figure 1. Main urban themes of COVID-19

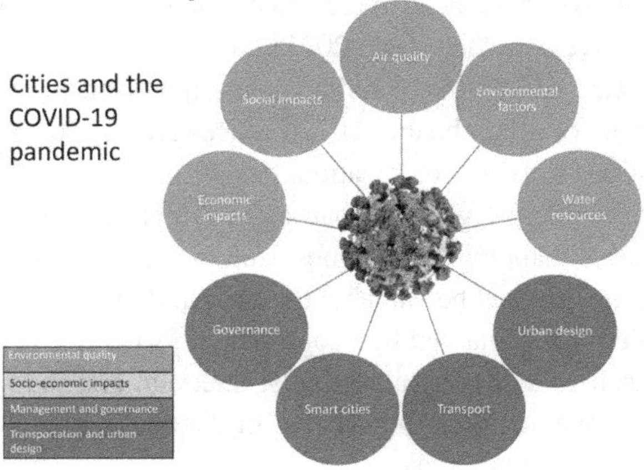

Areas with high human mobility, especially public or semi-public spaces such as schools, libraries, squares, marketplaces, and shopping centers, were closed. Transportation, trade, and industrial activities came to a standstill, leading to significant changes in daily life due to the restriction measures taken. For example, during the pandemic, urban residents preferred street shops, small neighborhood butchers, or grocery stores, which were considered risky, over shopping malls, highlighting the importance of the neighborhood concept (Negiz et al., 2021). New areas with less human density were created, providing urban residents with new movement opportunities, and the importance of urban open and green spaces increased (Özdede et al., 2021). Additionally, green spaces and open public areas were used for physical, social, and cultural activities, playing a role in mitigating the negative effects of COVID-19 on quality of life (Honey-Rosés et al., 2020; Pouso et al., 2021; cited in Mouratidis, 2021). Due to the restriction measures implemented during this period, hygiene measures in public areas were significantly tightened, and pharmacies and supermarkets became the most visited places (Glavan et al., 2022). As a result of the measures taken, individuals began to spend more time in residential areas, increasing the importance of critical spaces that connect to the outdoors, such as balconies, terraces, private courtyards, and gardens (Lekić et al., 2022). Consequently, individuals' expectations regarding their residential areas changed, highlighting the necessity for new housing planning in the future (Şolt, 2021). Furthermore, there was a significant decrease in public transportation usage during the same period, and alternative modes of transportation gained importance (Dağlı et al., 2021). Overall, it can be observed that the roles of transportation and land use, public spaces, facilities and services, housing conditions, and information and communication technology in the quality of life in cities transformed during the COVID-19 pandemic (Sharifi & Khavarian-Garmsir, 2020). During this period, the services provided by urban systems and the role of central administrations came to the forefront, emphasizing the need to reevaluate spatial policies with a sustainable, resilient, and inclusive approach (Şolt, 2021).

According to Rice (2019), the planning and design of the built environment are crucial determinants of urban health. Urban spaces are heterogeneous in terms of function distribution, density, organization, design, and usage. Consequently, the distribution of the epidemic within urban areas is not uniform (Coza, 2024). In this context, "disease mapping" techniques used to study the spread of epidemics within and between cities can be valuable tools for understanding disease patterns. Today, advanced analyses enabled by geographic information systems and spatial statistical methods help in determining spatial interventions and improving public health. Therefore, understanding the patterns of disease spread is crucial for mitigating and controlling the outbreak.

Following the COVID-19 pandemic, various spatial and spatiotemporal analyses have been conducted to identify determinants affecting disease spread and to evaluate public health policies. These studies have scientifically demonstrated that a range of urban factors, including demographic, social, economic, and environmental factors, influence the spatial distribution of COVID-19.

This study seeks to identify and elucidate prominent analytical methodologies employed in quantitative research to ascertain the relationship between the spatial dissemination of COVID-19 within urban environments and various urban components, while also conceptualizing the resultant findings.

In the subsequent section, studies compiled from high-validity sources at the scale of cities within various countries are presented. Following this, the significant variables identified in these studies are conceptualized by forming fundamental groups.

RECENT QUANTITATIVE STUDIES ON COVID-19 AND URBAN FEATURES

In recent times, numerous new scientific studies have been conducted to investigate the relationship between the COVID-19 pandemic, which has turned into a global crisis beyond a health issue, and spatial elements. These studies are generally prepared for sub-regions such as neighborhoods, districts, census units, or grid cells defined by a certain size, or for scales defined as research units such as cities or statistical regions of a country. In this context, current empirical studies in the literature that examine the relationship between the COVID-19 pandemic and spatial factors, reveal the mathematical and geographical structural model of this relationship, and scientifically prove the urban elements that directly or indirectly affect the spread of the virus have been reviewed. Within the scope of the thesis, they have been briefly summarized by focusing on the questions: "At what scale was it prepared? What variables were used? What method was applied? Which variables were statistically significant for the COVID-19 pandemic?"

Fielding-Miller et al. (2020) investigated the social elements and spatial distributions affecting the spread of the COVID-19 pandemic in the United States. In a study covering 50 states and 3024 counties, factors such as the number of agricultural workers, the proportion of the population aged 65 and over, the proportion of non-English speaking immigrant populations, the percentage of uninsured individuals, and the percentage of those living below the poverty line were defined as independent variables, while COVID-19 death rates at the county level were defined as dependent variables. In the study, areas with 1000 or more people per square mile were coded as urban (central) counties, while areas with fewer than 1000 people per square mile were coded as non-urban (rural) counties. It was statistically determined that in urban counties, agricultural workers and population density were associated with death rates, while in non-urban counties, individuals living below the poverty line and the elderly population aged 65 and over were associated with higher death rates.

Sabouri et al. (2020) focused on density in their study, investigating the risk of transmission arising from built environment elements and its impact on the COVID-19 pandemic. The study area selected was 913 urban counties in the United States, and independent variables included activity density per km^2, metropolitan population, the number of insured employees, the annual number of flights per 10,000 people, the number of intensive care beds, the number of physicians, the proportion of the Black population, the proportion of the population with at least a high school education, and the proportion of the population aged 60 and over. According to the findings obtained from the structural equation model used in the study, the metropolitan population was found to be one of the significant determinants of infection rates; metropolitan areas exhibited higher infection and mortality rates. In contrast, it was observed that the density in counties was not significantly related to infection rates, and counties with higher densities had lower mortality rates from the virus, likely due to superior healthcare systems.

Li et al. (2020) investigated the spatial distribution of the COVID-19 pandemic in Wuhan and the urban factors affecting its spread. Seven different urban indicators were used, including the proportion of the population aged 65 and over, hospital density, road density, metro density, commercial density, building footprint ratio, and land use values. It was assumed that the overall effects of these indicators were geographically evenly distributed. The findings obtained using Ordinary Least Squares (OLS) regression (which explained 64% of the variation) and the Geographically Weighted Regression (GWR) model (noted for its adjusted R^2 value of 0.822, indicating high explanatory power) showed that the density of commercial facilities, large hospitals, metro stations, high-rise buildings in certain areas, and the elderly population were factors positively associated with the pandemic.

Mollalo et al. (2020) examined the factors affecting the spatial distribution and differentiation of the COVID-19 pandemic in the United States. A total of 35 independent variables were identified based on environmental (road density, temperature, air quality, etc.), topographic (slope, elevation), socio-economic (average household income, unemployment rate, etc.), behavioral (smoking rates), and demographic (population aged 65 and over, proportion of Asian population, etc.) factors at the county level. To explain the relationship between the independent variables and the dependent variable (case numbers), five different regression models were tested. By comparing the results of these models, it was determined that four independent variables were related to case numbers. These variables were income inequality, household income, the percentage of nurses in the county, and the ratio of Black women to the total female population at the county level.

Niu et al. (2020) investigated urban factors affecting the spatial-temporal spread of COVID-19 at the city scale in China. The study included a range of variables such as population density and urban population size, as well as travel data provided by various service operators to detect inter-regional population mobility. All urban indicators were defined as independent variables, while confirmed COVID-19 cases were identified as dependent variables. A regression model was applied to determine the relationship between these variables. The findings indicated that intercity population mobility led to a higher risk of transmission, and urban population size had a significant impact on the risk of infection.

Jamshidi et al. (2020) explored the effects of meteorological parameters and urban factors on the spread of the COVID-19 pandemic. While air quality and temperature alone did not have significant implications on a global scale, at the national level, the mobility index (with a significance level of 34.32%) was found to be the most important factor influencing the growth of the COVID-19 pandemic, followed by family stay (26.14%), population (23.86%), and density (13.03%).

Nguyen et al. (2020) examined the impact of built environment characteristics at the neighborhood level on the variation of COVID-19 in the United States. Seven different built environment elements (types of family housing, building quality, greenery ratio of streets, single-lane roads, auxiliary street materials, and sidewalks) along with ten demographic characteristics (such as the proportion of Asian and Hispanic populations) were defined as independent variables, while case rates calculated at the neighborhood level (in only 20 states) were identified as dependent variables. The relationships between these variables were analyzed using regression techniques, leveraging Google Street View. The results indicated that more intense urban development (a mix of residential and commercial buildings, multiple traffic lanes), greater physical disorder, and increased walkability (which could enhance contact) were positively associated with higher case rates that could exacerbate the spread of the pandemic. Additionally, it was found that neighborhoods with a higher proportion of Black residents and individuals with the highest levels of elementary education were also associated with higher COVID-19 case rates.

Peng et al. (2020) investigated the spatial-temporal distribution of the COVID-19 outbreak using data obtained from social media applications in the city of Wuhan. A total of 729 different records were found, and it was determined that the majority of those requesting assistance were concentrated in the city center through geographic coding. Interpolation estimates were made using core density and regression analyses. According to the findings, population density and the elderly population were positively related to the spatial distribution of COVID-19 cases. Additionally, it was found that the Jiang, Jianghan, and Qiaokou regions exhibited scattered infections in the early stage, community spread in the middle stage, and full-scale epidemic distribution in the later stage, clearly reflecting the spatial-temporal spread of the outbreak.

Zhu et al. (2020) investigated the relationship between the distribution of the COVID-19 outbreak and meteorological factors in eight different cities in Brazil, Chile, Ecuador, and Peru. In the developed model, average temperature, maximum temperature, minimum temperature, absolute humidity, and average wind speed were defined as independent variables, while case numbers were defined as the dependent variable. The relationship between the dependent variable and the independent variables was analyzed using correlation analysis and a linear regression model. As a result, it was found that daily temperatures were negatively related to case numbers, except for the city of Lambayeque (Peru); in Santiago (Chile), case numbers were positively related to all meteorological factors except for absolute humidity; in Lima (Peru), case numbers were negatively related to daily minimum temperature and absolute humidity, and positively related to daily maximum temperature; in Valparaiso (Chile), case numbers were positively related to absolute humidity and average temperature; and in Antofagasta (Chile), case numbers were positively related to average temperature, maximum temperature, and absolute humidity.

Mo et al. (2020) aimed to determine the spatial-temporal distribution characteristics and changing trends of COVID-19 cases in China. Hotspot analyses were conducted using geographic coordinates, vector maps, and various national data, and the results were visualized using a grid method. According to the findings, the disease showed a tendency to spread towards the southern and northern regions, with migration movements from areas with high case rates (Hubei) increasing the risk of infection, and it was generally observed that there were sequential hotspot clusters in these regions.

Bourdin et al. (2021) investigated the urban factors causing differences in the regional distribution of the COVID-19 outbreak in Italy and the effects of restriction measures. In this context, the proportion of the population aged 65 and over at the district level, population density, total population, GDP per capita, and the number of air passengers were identified as variables to measure migration and mobility. The findings indicated that the restriction measures implemented on March 8 slowed the outbreak, and it was observed that GDP per capita, the number of passengers, population density, and total population in the district had a significant and positive relationship with COVID-19 case numbers.

Hu et al. (2021) investigated the impact of built and social conditions on the COVID-19 outbreak in Washington. Various independent variables were defined under the main headings of work, education, poverty, income, housing quality, and demographic structure, and regression techniques were used to analyze COVID-19 case numbers, death counts, and mortality rates. As a result, it was found that crowding, commuting time, and the Black population had a strong effect on COVID-19 death counts, and spatial hotspots were identified.

Kashem et al. (2021) investigated the extent to which social inequalities experienced during the pandemic period in Chicago affected COVID-19 prevalence. The study defined 15 different independent variables related to built and social environment factors (such as accessibility to open spaces, neighborhood walkability, average household size, average household income, percentage of Latinx population, percentage of Black population, percentage of population with at least a high school education, etc.) and two different dependent variables related to COVID-19 prevalence (cumulative case rates in the short and long term). According to the analysis results using Structural Equation Modeling (SEM), average household size and limited internet access played a significant role in explaining the COVID-19 spread rates in both the short and long term. Similarly, it was found that the Latinx population had a significant indirect effect during both time periods. Additionally, areas with lower education levels consistently exhibited higher case rates in both periods.

Rios & Gianmoena (2021) investigated the relationship between temperature and COVID-19 incidence at a regional level in Italy during the initial outbreak phase of 2020. The study also included various other climatic conditions (such as relative humidity, daily solar radiation, air quality), demographic factors (proportion of the population aged 0-35, proportion of the population aged over 75, population density per km^2), social factors (corruption indicators, commuting time, length of highways and railways per km^2), and health policy factors (health expenditures). The findings indicated that temperature was one of the most significant determinants reducing COVID-19 prevalence, with inter-regional temperature differences explaining a substantial portion (over 30%) of the observed variation in infection rates. However, no significant evidence was found for the other factors.

Mollalo et al. (2021) examined the factors affecting the spatial distribution of COVID-19 deaths in the United States and its relationship with 20 different diseases (such as tuberculosis, cancer, diabetes, etc.). The analysis, conducted using Global and Local Moran I statistics, revealed that the distribution of COVID-19 case fatality rates (CFR) was predominantly concentrated in urban areas of Louisiana, Connecticut, and New Jersey. Additionally, it was observed that asthma prevalence and COVID-19 death rates were significantly clustered in the same cities.

Urban & Nakada (2021) investigated the impact of social-spatial factors on the COVID-19 pandemic in São Paulo, Brazil. The study employed five different regression models: Ordinary Least Squares (OLS), Spatial Error Model (SEM), Spatial Lag Model (SLM), Geographically Weighted Regression (GWR), and Multiscale Geographically Weighted Regression (MGWR). The results indicated that Geographically Weighted Regression was the most suitable model for this analysis. In the study, 18 different independent variables were identified, including access to drinking water, population density, proportion of the population aged over 60, average household size, size of urban settlement areas, average income, proportion of households with electricity access, and access to toilets and drinking water. Strong correlations were found between COVID-19 mortality incidence and variables such as population density, average household size, and informal urban settlements. Furthermore, it was demonstrated that these four explanatory variables accounted for 85% of the deaths at the county level, and the pandemic had a more severe impact in areas inhabited by spatially vulnerable populations compared to other regions.

Han et al. (2021) conducted a study on the spatial distribution characteristics of the COVID-19 pandemic in Beijing, focusing on its relationship with environmental factors. They identified nine independent variables reflecting the intensity of human activities and the likelihood of contact: population density, distance to the Xinfadi market, distance to hospitals, distance to workplaces, distance to educational facilities, distance to traffic facilities, distance to shopping sites, distance to parks, and distance to restaurants. The findings indicated that population density and distance to the Xinfadi market were the most critical factors explaining the development of the pandemic.

In a separate study, Perone (2021) examined the factors affecting the spatial differences in COVID-19 case fatality rates across 20 regions and 107 cities in Italy. They identified 18 explanatory variables related to environmental, demographic, and health factors. The regional-level analysis included variables such as humidity, health system performance index, health expenditures, number of hospital beds, density of cars and firms, average electricity consumption per capita, number of single young adults aged 20-34, and the ratio of hospital beds before and during the pandemic. The city-level analysis included urbanization index, density, air quality metrics (e.g., PM10, PM2.5), and elevation. Common variables used at both scales included the number of physicians, the proportion of the population aged over 70, 80, and 90, temperature, the ratio of positive COVID-19 tests to the total population in the province, and the ratio of recovered individuals to normal hospital beds. Regression and clustering analyses were performed to identify the relationships between these independent variables and the dependent variable, COVID-19 case fatality rates, and to classify areas with similar mortality risks. The results showed that the health system performance index and the number of doctors had a significant negative relationship with case fatality rates at the regional level, while at the city level, the number of general practitioners and average temperature were significantly and negatively correlated. Additionally, the population aged over 70 showed a significant positive correlation at both scales.

Phiri et al. (2021) conducted a study on the district-level distribution of COVID-19 cases in Zambia, examining the environmental and socio-economic factors influencing this distribution. In their model, the case rates were defined as the dependent variable, while various other factors were treated as independent variables. The identified environmental factors included topographical features (elevation, slope) and climatic characteristics (temperature, rainfall, etc.), while socio-economic factors encompassed demographic characteristics (population density, total population, etc.), proximity to commercial and urban establishments such as airports and city centers, accessibility (distance to roads), and economic factors. The findings indicated a positive relationship between proximity to airports, urban areas, mining centers, and border regions with COVID-19 cases, while districts with temperatures above 22 °C showed a negative relationship with COVID-19 cases.

Qiu et al. (2021) investigated the impact of environmental, economic, and social factors on the COVID-19 outbreak at the provincial level in China. Their model included independent variables such as climate (temperature, relative humidity), atmospheric environmental quality (nitrogen levels, air quality index, etc.), land features (elevation), population characteristics (population density, household size), economic indicators (GDP per capita), human mobility (migration rates), and public health interventions (national health intervention data). The relationship between these independent variables and the dependent variable, the incidence rate of COVID-19 cases, was tested using a logistic regression model. The results revealed that air pollution levels (CO, PM10, and PM2.5), migration mobility, and economically more developed and densely populated cities had a positive relationship with COVID-19 cases, while humidity levels showed a negative relationship with a 15-day lag.

Aldana et al. (2021) investigated the distribution of the COVID-19 outbreak in Mexico and the spatial factors influencing this distribution. They defined variables related to individuals' health conditions (such as diabetes and obesity), as well as socio-demographic factors at the urban level (like literacy rates), economic indicators (such as GDP), mobility (number of flights), and climatic characteristics (like temperature). The effects of these variables on the COVID-19 mortality rate were analyzed using regression techniques. The results indicated that both external (with the second highest positive relationship) and internal migration percentages, the number of physicians, and GDP were positively associated with the COVID-19 mortality rate.

Başer (2021) investigated the impact of population density at the district level on the Covid-19 outbreak. The study also included demographic, economic, and social factors as independent variables. These variables include education and welfare levels, the number of healthcare workers, the ratio of male to female population, and the proportion of the population aged 65 and over. Additionally, temperature data calculated since 1941 was included in the study. The relationship between the identified independent variables and the dependent variable, which is the number of cases, was determined using a regression model. According to the findings, it was observed that the number of healthcare workers, population-weighted density, the proportion of the population aged 65 and over, and per capita income were positively related to the number of cases, while higher education levels were negatively related to the number of cases.

Sridhar (2021), in his study conducted at the state level in India, examined the socio-economic factors of the Covid-19 outbreak and the impact of urbanization on the spread of the disease. The analysis, conducted using regression techniques, revealed that urbanization, labor force participation, population density, and higher income contributed to an increase in Covid-19 prevalence. Additionally, a positive relationship was established between households living in slum areas and Covid-19 cases, while a negative relationship was found between healthcare infrastructure and parks with Covid-19 cases.

Jo et al. (2021) aimed to investigate the role of spatial structure on Covid-19 by comparing different urban density and connectivity measures during the first eight months of the outbreak in Korea. They utilized various demographic variables such as population density, the proportion of the population aged 65 and over, the ratio of female population, and the proportion of the population under 20 years old. Additionally, socio-economic variables like employment density and GDP per capita were included, along with physical variables such as park area per 1,000 people, the land area of provinces and districts, and medical accessibility. They also considered other supplementary indicators, such as the number of doctors per 1,000 people, to examine their effects on Covid-19 cases and infection rates. According to the findings obtained from the model using Ordinary Least Squares (OLS) and Negative Binomial Regression (NBR), it was observed that density had a positive relationship with the number of cases, while the number of doctors per 1,000 people and park area had a negative relationship with Covid-19 cases.

Ahmed et al. (2021) conducted a study investigating the relationship between environmental and socio-economic factors and Covid-19 infection cases across 70 different cities worldwide. The research considered environmental data such as temperature, humidity, air quality, and population density, alongside socio-economic data including GDP per capita, health expenditure per capita, and average life expectancy. These data were analyzed using statistical models like Poisson and Negative Binomial models. The results indicated that high population density triggered the rapid spread of the pandemic, and air quality (PM10) was strongly associated with the increase in Covid-19 case numbers.

In a study by Al Kindi et al. (2021) focusing on Oman, the researchers identified demographic and socio-economic factors related to Covid-19 at the city level. They utilized 24 variables, including population density, total number of households, and average household size, to approach a geographical modeling of the Covid-19 outbreak. Using Generalized Linear Models and Geographically Weighted Regression, the study found that specific demographic and social variables at various age levels were related to the pandemic, with population density, number of hospital beds, and total number of households indicating spatial interactions.

Bijari et al. (2021) explored urban factors affecting the Covid-19 outbreak in the Middle East and Central Asia, including 31 provinces of Iran at the city scale. The study selected 10 key independent variables for each city and country (such as land area, urban population, rural population, and energy consumption) and examined their relationships with confirmed Covid-19 cases. The findings revealed significant relationships suggesting that the spread of the pandemic could be linked to urbanization in the study area, as well as to fuel consumption, energy use, and emissions of greenhouse gases, $CO2$, and $NO2$.

Sharifi et al. (2021) examined the spatial factors affecting the spread of confirmed Covid-19 cases at the neighborhood level in Tehran. The study selected a total of 12 different location-based independent variables, including demographic characteristics (population density, population over 65 years old, etc.) and built environment elements (proportion of green spaces, building density, number of pharmacies, metro stations, shopping and retail centers in the neighborhood, etc.). According to the results of the analysis conducted using spatial statistical techniques, it was observed that the hotspots of Covid-19 cases were distributed in the northeast, southeast, east, and central Tehran, with some hotspots clustering closely together in certain areas. Furthermore, based on the importance levels among the independent variables, it was found that population density (R^2= 0.88), the number of pharmacies (R^2=0.64), neighborhood centers and chain stores (R^2=0.59), bus stops (R^2=0.46), building density (R^2=0.44), and green spaces (R^2=0.34) were associated with Covid-19 cases. These findings highlight the significance of spatial factors in the spread of Covid-19.

Ciupa and Suligowski (2021) investigated how the values observed during the Covid-19 pandemic in Poland varied spatially over the course of a year (March 2020 - March 2021). In the study, case and death rates were defined as dependent variables, while 24 environmental factors (mostly green spaces such as forests, orchards, meadows and pastures, recreational and leisure areas, biologically active arable land, lakes and artificial water reservoirs, rivers, ecological areas, etc.) and population density were identified as independent variables. The study included 66 cities and 380 districts, and it was found that green spaces had a negative relationship, while population density had a positive relationship with case and death rates.

Kim et al. (2021) examined the relationship between Covid-19 case and death rates and 15 different racial/ethnic inequalities in 18 counties in the United States during three different phases of the pandemic (January 21 - June 15, June 16 - August 31, and September 1 - December 18, 2020). In the early phase, it was observed that counties with higher populations of Black/African American and Hispanic individuals, where discrimination was more prevalent and population density was high, had higher case and death rates. As the pandemic progressed, it was noted that counties with higher populations of non-Hispanic white individuals, where discrimination was less prevalent and population density was lower, also had higher case and death rates. The study highlighted that such county-level inequalities changed throughout the pandemic.

Gaglioti et al. (2021) investigated socio-demographic risk factors at the county level in the United States during the Covid-19 pandemic, focusing on inequalities related to population, race, and ethnicity, as well as their effects on the pandemic. In the model created, Covid-19 case and death rates were defined as dependent variables, while factors such as the proportion of Hispanic population, proportion of Black population, proportion of Asian population, proportion of uninsured population, proportion of impoverished population, and proportion of the population aged 65 and over were identified as independent variables. The study, which applied various regression techniques, demonstrated that counties with a higher density of Black population had higher death rates, and that Black individuals faced an 18 times higher risk of death compared to the white population.

Klompmaker et al. (2021) investigated the impact of green spaces on Covid-19 incidence at the county level in the United States. They developed a plant pattern index for green spaces, which ranges from -1 to 1, where values close to 1 indicate areas with full coverage of live green vegetation, values near zero indicate areas with little live vegetation (e.g., rocks, sand), and negative values correspond to wetland areas. The study assessed the relationships between this index and various demographic (population density, proportion of Black population, proportion of Hispanic population, etc.), socioeconomic (poverty status, average household income, etc.), ecological (temperature, relative humidity, air quality), and some spatial variables (number of urban counties, number of counties under restriction orders) with Covid-19 incidence and mortality using a six-step model analyzed through negative binomial regression. The findings indicated that green spaces at the county level had a negative relationship with Covid-19 incidence, the pandemic spread more rapidly in densely populated counties, and the green spaces in these counties were associated with Covid-19 mortality.

Wu et al. (2021) examined the role of population mobility in the spread of the pandemic at the state level. They defined confirmed Covid-19 case numbers as the dependent variable, while the number of permanent residents in the states, traffic density, and the number of migrants, along with population density, were identified as independent variables. The analysis, conducted using ordinary least squares regression, revealed that migration movements from various cities to Hubei (the province where the pandemic first spread) and traffic density were critical factors in the spread of the virus.

Cutrini & Salvati (2021) investigated the spatial patterns and determinants of Covid-19 infection in Italy during the initial period of the pandemic (March 9 - June 15, 2020). They defined independent variables as the population aged 65 and over, the size of metropolitan areas, the number of air passengers per 100,000 people, the number of industrial facilities in a specific region, and the share of employees working in establishments with more than 250 employees. Covid-19 cases were identified as the dependent variable. The results, based on negative binomial regression and linear spatial models, indicated that the size of metropolitan areas (urban agglomeration), industrial regions, the concentration of large companies, and economic agglomeration in the north-south region were the most significant determinants of Covid-19 infection.

Li et al. (2021) examined the impact of built environment characteristics at the city level on the spread of Covid-19 in China during the early stages of the pandemic (January 3 - February 23, 2020). They categorized independent variables into three main headings: intercity built environment characteristics (number of trains, number of flight departures), intracity built environment characteristics (distance between train stations, travel times, activity density around train stations, and population density), and a control variable (population aged 65 and over). The relationship between these independent variables and the dependent variable, Covid-19 infection rates, was analyzed using three different regression models. The findings indicated that the number of trains and the distance between train stations had a positive relationship with Covid-19 spread, while the elderly population had a negative relationship.

Hamidi et al. (2021) investigated whether built environmental factors in high-density population areas along subway routes in New York City, USA, were associated with Covid-19 infection rates. The study included demographic and socio-economic factors (such as the proportion of Hispanic and Black populations, average household size, etc.) and utilized regression techniques. The research findings revealed a strong relationship between crowding and high infection rates, with average household size significantly impacting crowding. Additionally, it was concluded that demographic and socio-economic factors were the most important determinants of spatial variation in Covid-19 infection rates.

Li et al. (2021) investigated the impact of neighborhood-level built environment characteristics in Huanggang, China, on the Covid-19 pandemic. They focused on variables within a 1000-meter radius, including commercial amenities, medical services, ATMs, grocery stores, hair salons, shops, restaurants, clinics, pharmacies, bus stops, road length, building density, and housing prices. Using density-based clustering algorithms (DBSCAN) and structural equation modeling (SEM), the findings indicated that building density, commercial development, medical services, transportation infrastructure, and housing prices significantly affected Covid-19 case numbers.

Das et al. (2021) investigated the factors influencing the spatial distribution of Covid-19 hotspots in Kolkata, India. They developed a living environment conditions index categorized into two dimensions: socio-economic and eco-environmental. Key variables included housing conditions, household asset levels, access to basic technical infrastructure, technical facilities in housing, and gender inequality. The study found that Covid-19 hotspots were concentrated in areas with limited access to essential services (such as infrastructure), which also had higher densities of slum-type settlements compared to other regions.

In a related study, Lata et al. (2021) examined livability in Delhi, Mumbai, and Kolkata, identifying urban factors affecting Covid-19 cases using twelve key indicators (including quality of life index, health index, pollution index, and sizes of open and natural areas). Their findings revealed a negative correlation (-0.58) between Covid-19 cases and quality of life, a positive correlation (0.75) with the pollution index, and a significant positive relationship (0.71) with natural areas.

Razavi-Termeh et al. (2021) aimed to create a risk distribution map for Covid-19 in Tehran, Iran, by evaluating the existing land use structure based on socio-economic criteria. They established a spatial database using eight variables, including public transport stations, supermarkets, banks, ATMs, bakeries, pharmacies, gas stations, and hospitals. Their analysis, employing various algorithmic techniques and logistic regression models, indicated that the central and eastern regions of Tehran were at higher risk, with public transport stations and pharmacies being the most associated factors with the locations of Covid-19 patients. Additionally, land use density was found to contribute to an increase in Covid-19 prevalence.

Ruck et al. (2021) investigated the case rates in counties that had not yet reported cases during the spread of Covid-19 in the United States, using socio-economic variables at the county level. They defined nine different independent variables in the model, including education level (those with a university degree), employment (average income, etc.), ethnicity (Black, Hispanic, etc.), age (those over 65, etc.), health (number of hospitals, number of intensive care beds, etc.), political views (Democrat, Republican), clustering level (population size, population density, etc.), and the points of Covid-19 spread (Italy, Iran, China). The relationship between the dependent variable, Covid-19 death rates, and the independent variables was analyzed using regression techniques. The findings indicated that population size, population density, public transportation, political views, and the African American population were positively associated with the pandemic and explained a significant portion of the variance in death data. These results highlight the impact of socio-economic factors on the dynamics of Covid-19 spread.

Spotswood et al. (2021) investigated the impact of green spaces and a range of socio-demographic characteristics on the Covid-19 pandemic at the county level in the United States. The main variables used in the study included the amount of green space, access to parks, average income, average age, percentage of white individuals, the number of days since the first recorded case, and population density. The relationships between the independent variables and the dependent variable, Covid-19 cases, were analyzed using a negative binomial model. It was found that counties with a high density of green spaces were less affected by the pandemic, and statistically, a 4.1% increase in green spaces was associated with a 0.1% decrease in Covid-19 cases.

Sarmadi et al. (2021) examined the relationship between environmental and socio-demographic factors at the county level and the distribution of the Covid-19 pandemic in the United States. In the research model, independent variables such as air pollution, non-Hispanic Black/African American population, population density, median age, human development index, and GDP per capita were defined, and their relationships with dependent variables like case numbers, death counts, and case fatality rates were analyzed. The findings indicated that some air quality values (such as ozone) and population density were positively associated with Covid-19 variables.

Tchicaya et al. (2021) aimed to analyze the relationship between social, spatial, and health factors at the county level and cumulative Covid-19 deaths in France. In the study, independent variables such as urban population, poverty rate, gender, number of hospital beds, number of doctors, and population density were identified. Three different regression models were applied to determine the relationship between these independent variables and the dependent variable, Covid-19 mortality rate. It was found that 75% of the deaths occurring in the initial phase of the disease were concentrated in five regions, and the number of doctors and intensive care bed availability were positively associated with the mortality rate in urban areas. In the second phase, the number of doctors was also positively related to the Covid-19 mortality rate. Additionally, a 1% increase in the poverty rate was found to be associated with a 7.07% decrease in the Covid-19 mortality rate.

Silva & Alves (2021) investigated the relationship between neighborhood-level social, demographic, and economic factors and Covid-19 case and death rates in Brazil. They utilized a Poisson regression model to identify factors influencing the spread of the pandemic and to explain the temporal changes of these factors. Independent variables included age ranges, gender, crowding, education, income, and racial characteristics, while the dependent variables were the number of cases, deaths, and case fatality rates. The findings indicated that populations over 80 years old, males, and economically affluent neighborhoods during the early stages of the pandemic had higher case and death rates.

Dintrans et al. (2021) studied the temporal dynamics of Covid-19 case and death rates at the neighborhood level in Santiago, Chile, focusing on the role of spatial factors. In their study, case and death rates were defined as dependent variables, while health, socioeconomic, and demographic factors that could influence the spatial spread of the pandemic were identified as independent variables. These independent variables included the number of women, populations aged 65 and over, number of immigrants, population density, rural/urban index, size of green areas, distance to health centers, number of insured individuals, and poverty rate. Ordinary Least Squares (OLS) and spatially lagged regression methods were used to determine the relationship between the dependent and independent variables. The results showed that case numbers were positively associated with public transport usage, the population aged over 65, green areas, and distance to health centers, while death numbers were positively related to poverty rate and population density.

Tokey (2021) examined the relationship between mobility indicators at the county level in the United States and COVID-19 infection rates within a spatio-temporal framework. The study included variables defining mobility, such as the number of trips per capita and the percentage of trips outside the county, as well as socio-economic indicators like poverty rates, the proportion of the population aged 65 and over, and education levels. Additionally, data on policy stringency indices implemented during the pandemic were incorporated to assess their impact. The relationship between these independent variables and the dependent variable, COVID-19 cases, was analyzed using three different regression models. The findings indicated that counties with higher rates of individuals working from home, those experiencing greater external migration, higher population density, and higher poverty levels were positively associated with COVID-19 case rates. Furthermore, it was found that counties with a higher concentration of elderly individuals had lower case rates, and that the policy stringency index had a significant effect on reducing the spread of the virus.

Verma et al. (2021) investigated the impact of neighborhood-level mobility on COVID-19 cases in New York and Chicago using socio-economic indicators. They defined independent variables as socio-economic factors such as income status, poverty rate, and percentage of the Black population, along with mobility indicators derived from various applications, including total trips within neighborhoods and weekly distributions of travel distances. To account for the risk of exposure to the disease, they also included data on the duration of stay at home and the contact intensity index. The results indicated that contact intensity was more prevalent in low-income areas, which also had higher case rates.

Wu & Zhang (2021) focused on the spatial-temporal distribution of COVID-19 and examined the effects of cumulative social, economic, and environmental factors at the county level in Texas. The study utilized 14 explanatory variables, including income distribution, unemployed population, average temperature, average rainfall, air quality, land area size, total population, male population, Black population, population aged 20-59, population over 80, number of hospitalizations, and hospital beds per capita. The relationship between these independent variables and the dependent variable (case numbers) was analyzed using regression techniques. The results from six different models showed that in the first model, the number of hospitalizations was positively related to population density; in the second model (demographic), the population aged 20-59 was positively related, while the population over 80 was negatively related; in the third model (environmental), air quality was positively related; in the fourth model (economic), annual income was negatively related, and unemployment was positively related; in the fifth model (spatial), total land area per county was negatively related; and finally, in the sixth model (health), the number of hospital beds was positively related to COVID-19 case numbers.

Kulkarni et al. (2021) investigated the relationship between meteorological factors and the COVID-19 pandemic in India. They identified 46 different geographical locations (32 cities, 12 districts, and 2 union territories) and analyzed the effects of the selected locations' meteorological characteristics (temperature, rainfall, relative humidity, wind speed, and air pressure), population sizes (population density, total population), and geographical data (area size, elevation) on cumulative COVID-19 cases. The results indicated that the selected 46 locations accounted for 60% of COVID-19 cases in India, with high temperatures, wind speed, and rainfall showing a negative relationship, while relative humidity and air pressure were positively related to case numbers.

In the United States, Frank & Wali (2021) examined the impact of built and natural environmental conditions on COVID-19 at the county level. They identified four key components—density, urban design, accessibility, and green spaces—to control the spatial spread of the disease. Additionally, they included nine control variables at the social, demographic, and economic levels in their model. The relationship between a total of 13 variables and the dependent variables of COVID-19 infection and mortality rates was analyzed using regression techniques. The findings revealed that green spaces and housing density were negatively related to COVID-19 mortality rates, with each 1% increase associated with a decrease of 0.88 and 4.5 units in COVID-19 deaths, respectively. Furthermore, counties with higher concentrations of elderly and Black individuals had higher mortality rates, with a 1% increase in the elderly (aged 65 and over) and Black populations associated with increases of 3.55 and 0.97 units in COVID-19 mortality rates, respectively. Additionally, population density was found to be positively related to COVID-19 case rates.

Wu et al. (2021) investigated the relationship between the distribution of COVID-19 and daily activities (social activities, outdoor activities, etc.) as well as built environment characteristics such as building density, educational facilities, cultural facilities, restaurants, shopping centers, sports facilities, open spaces, and proximity to green areas across seven administrative regions in Wuhan. In the study, data on the age and location of individuals requesting help through the Weibo application were obtained to explain the impact of these built environment characteristics on the dependent variable. The population was classified into three different age groups: young, middle-aged, and elderly, and the spatial distribution densities of the patients were analyzed. The findings indicated that bus stops, middle schools, universities, and sports facilities affected the young and middle-aged groups, while building density, shopping facilities, dining facilities, distance to water bodies, distance to green areas, and distance to sports facilities were identified as built environment factors affecting the elderly population.

Bayode et al. (2022) conducted a study in Nigeria to investigate the risk factors and urban spread of COVID-19 incidence during the first quarter (March-May). In this study, COVID-19 cases were used as the dependent variable, while independent variables included socioeconomic characteristics such as population, gross domestic product (GDP), literacy rate, percentage of people living on less than one dollar ($) a day, limited access to food, improved water sources, and mobility features (availability of local and international airports). The study utilized various spatial regression models and demonstrated a strong relationship between the rate of COVID-19 and population density, with a 1% increase in population density associated with approximately a 0.006% increase in the incidence rate of COVID-19.

Aral & Bakır (2022) examined the spatial relationship and spatiotemporal patterns of the COVID-19 outbreak in Turkey between February 8 and May 28, 2021. The main independent variables used in the study included population density, elderly dependency ratio, daily GDP per capita, literacy rate, number of physicians per 1,000 people, and number of hospital beds per 100,000 people. The COVID-19 cases obtained over the 16-week period were defined as the dependent variable. The results indicated that population density and elderly dependency ratio were significant factors in the spread of the outbreak, with various meaningful spatial clusters identified.

Watts (2022) investigated the impact of demographic, socioeconomic, climate, and health factors on the pandemic in the United States during the COVID-19 period. The study identified that the main factors associated with the doubling time of infections were high population density and overcrowding. Additionally, it was found that air pollution and exposure to diabetes acted as driving forces in the spread of the pandemic, while air temperatures could slow down the growth of the outbreak.

Ronael & Baycan (2022) conducted a study investigating the spatial factors affecting the COVID-19 pandemic across Turkish cities. They utilized the Local Moran I statistic to identify significant clusters in the spatial spread of the pandemic. Subsequently, they tested the significance levels of twenty different descriptive variables, including population density, health capacity, immigrant population, active insured workers, hospital capacity, average housing sizes, air quality, and accommodation facilities, using regression methods. The results indicated that the most suitable variables were population density and hospital capacity. In the later sections of the study, various techniques such as Ordinary Least Squares (OLS), spatial lag models, spatial error models, and geographically weighted regression were employed to test the significance of these variables and whether their effects varied by region. The analyses revealed that among the twenty different variables initially defined, population density, health capacity, and annual temperature values were highly correlated with the spread of COVID-19 in Turkey. This highlights the importance of these factors in understanding the dynamics of the pandemic's spread in different urban contexts.

Mellander & Florida (2022) investigated the spatial factors associated with the spread of COVID-19 during the first wave in Sweden. Their study was conducted across 290 municipalities and focused on 34 neighborhoods within three major cities. In this research, case and death rates were defined as the dependent variables, while independent variables included population, density, average age, income, average household size, unemployment, number of nursing homes, and access to case rates from other regions with a one-week lag. The results of the regression model indicated that more than half of the variation in COVID-19 cases in regions with high infection rates could be explained by proximity factors. Additionally, the study found a significant and positive relationship between household size and COVID-19 spread at both scales of analysis. This underscores the importance of spatial dynamics and demographic factors in understanding the transmission patterns of the virus in urban settings.

Xiao et al. (2022) conducted a study using a survey method with 237 individuals living in the UK during the COVID-19 restrictions, examining the relationship between the built environment and mental health (stress, depression, etc.), while considering individuals' socio-demographic characteristics. The variables defined in the study included housing quality (such as housing size, type, proportion of homes without gardens, and access to open communal areas), socio-demographic characteristics (including Black population, White population, gender, age, education level, household income, etc.), indoor quality (natural light, comfort, indoor air quality, indoor physical activity space, etc.), and neighborhood-level activity quality (proximity to the nearest open space, proximity to the nearest stores, etc.). The analysis was performed using logistic and hierarchical regression methods. Key findings indicated that gardens and balconies had a positive impact on mental health during the COVID-19 quarantine. Specifically, individuals living in homes without gardens were more likely to feel anxious ($B = 3.240$, $p = 0.018$) compared to those living in homes with gardens. Additionally, housing and neighborhood types were significantly associated with anxiety, with individuals living in towns experiencing more anxiety ($B = 2.277$, $p = 0.002$) compared to those living in cities. These results highlight the importance of the built environment in influencing mental health outcomes during periods of social restrictions.

Alidadi and Sharifi (2022) compiled studies from the literature that analyze how built environment and human factors (social, demographic, economic) affect the spread of COVID-19 across different scales (country, state, county, city, or urban area). They found that at least 166 empirical articles had been published that included at least one case study and analyzed the effects of at least one built environment factor on the spread of COVID-19. According to the findings from these articles, the most significant urban factors associated with the COVID-19 pandemic were found to be density, land use, transportation and mobility, housing conditions, demographic characteristics, health systems, and socioeconomic factors.

Wu et al. (2022) investigated the determinants of vulnerability and resilience to COVID-19 from the perspective of the built environment and socioeconomic inequalities in Salt Lake City, USA, using 56 stations and 6 light rail systems. They included four different built environment variables defined as housing density, street connectivity, land use, and floor area ratio, as well as socioeconomic characteristics such as the proportion of minorities, income status, and education level. The analyses indicated that the built environment and minority population were the most significant factors affecting COVID-19.

Wang et al. (2022) examined variations in built environments that were measurably associated with the spread of COVID-19 in Wuhan. They defined incidence rates during different time periods as the dependent variable, while socioeconomic and demographic characteristics such as population density, income status, and the proportion of the population over 60 years old, as well as distances to hospitals, schools, markets, and bus stops, were defined as independent variables. Hierarchical linear and ridge regression models were used to explain the relationship between dependent and independent variables. The findings indicated that the distance to market areas and schools had a positive relationship with COVID-19 deaths before the quarantine period, and that income status, as well as distances to hospitals and bus stops, were negatively related to the COVID-19 pandemic in both periods.

Gaisie et al. (2022) investigated the spatial factors influencing the spread of the Covid-19 pandemic by analyzing a range of built environment elements (such as distance to train stations, parks, open spaces, playgrounds, access to major activity centers, intersection density, etc.) and socioeconomic characteristics (average household size, age level, etc.) in Melbourne. The study applied four different methods: global spatial autocorrelation (Moran I), local indicators of spatial association (LISA), multiple regression (MR), and geographically weighted regression (GWR). The findings indicated that hotspots clustered in the inner and northern regions (mostly in suburbs), while cold spots (i.e., areas with the lowest case rates) were found in the eastern and southern regions. Additionally, it was statistically proven that during the first and third phases of the pandemic, population density and distance to train stations had a positive relationship with case rates, while accessibility to the central business district had a negative relationship.

Ha and Lee (2022) examined the sociodemographic, economic, and built environment factors that influenced the spread of Covid-19 during three different time periods. The study included all 3,108 counties in the United States, defining the number of cases per 10,000 people as the dependent variable (separately for each of the three periods) and 35 different factors related to spatial and social structure as independent variables. A regression model was used to explain the relationship between the independent variables and the dependent variable of case numbers. The findings showed that during the initial phase of the pandemic, sociodemographic factors such as education level, total population, and the proportion of Black/African American population were significant; during the second phase, density and household size were important; and during the third phase, built environment factors such as the number of restaurants and supermarkets were the key determinants of Covid-19 cases. Furthermore, it was found that poverty, the number of restaurants, and the proportion of remote workers were associated with the incidence of Covid-19 in all three pandemic periods.

Xu et al. (2022) investigated the relationship between the distribution of 910 different Covid-19 cases reported through social media in Wuhan, China, and some spatial factors. Five independent variables were identified: population density, the population over 60 years old, metro stations, main roads, and commercial areas, and their relationships with the dependent variable, the number of cases, were analyzed. The study applied three different methods: Kernel Density Estimation (KDE), Global Linear Regression (OLS), and Geographically Weighted Regression (GWR). As a result, it was observed that all five selected variables were positively related to the number of cases, with the highest significance found in commercial areas (r^2=0.82). Additionally, in terms of transportation, metro stations had a stronger effect on Covid-19 cases than main roads. Demographically, it was found that a 1% increase in population density and aging rate led to an increase in Covid-19 cases by 14% and 6%, respectively.

Lin et al. (2022) investigated the relationship between Covid-19 cases at the county level in the United States and structural inequalities within the context of social health services and racial/ethnic characteristics. In this study, 22 independent indicators were identified. Some of these include: the proportion of Black/African American population, the proportion of Hispanic or Latino population, the proportion of non-Hispanic white population, the proportion of the population over 18 years old, the proportion of individuals below the poverty line, the proportion of low-income elderly population, an urban opportunity index for identifying those with high living standards, and a composite immigrant adaptation and accessibility index developed to identify immigrants, individuals living in crowded environments, and young people without health insurance.

Desmet & Wacziarg (2022) investigated the relationships between various independent variables—such as population, population density, public transportation, income status, social capital, ethnic origin, education, and votes in presidential elections—and Covid-19 case and death rates at the county level in the United States. They also examined the spatial variation of these relationships. The key findings are as follows: Density is positively related to case numbers; public transportation usage is negatively related to death rates; the population over 75 years old is negatively related to both cases and deaths; nursing home populations are positively related to both cases and deaths; average household income is positively related to case and death rates in the early stages of the pandemic; African American and Native American populations are positively related to both case and death rates; and education level is negatively related to the severity of the pandemic.

Guan et al. (2022) investigated the relationship between the geographic distribution of the Covid-19 pandemic and various factors using 17 built environment variables—such as housing density, employment density, pedestrian-focused road density, total road network length, and proximity to public transportation—and 6 socio-economic and demographic variables—such as population, population density, income status, and unemployment rate—at the county level in the United States. To examine the correlation between the independent variables and the dependent variables, which are case rates, death rates, and case fatality rates, Ordinary Least Squares (OLS) regression was applied. The findings revealed that built environment variables responded differently to infection rates, death rates, and case fatality rates, with road network density being the most influential indicator among them. Population density was positively related to Covid-19 prevalence in the first two phases of the pandemic but negatively related in later phases. Throughout the entire course of the pandemic, there was a consistently increasing positive correlation between the unemployment rate and the death rate.

Kuschel et al. (2023) investigated the relationship between urban characteristics of the 32 communes in Santiago de Chile and the Covid-19 pandemic, analyzing the impact of various demographic, educational, and employment-related factors. According to the results obtained through principal component analysis, communes with more advanced economic, social, organizational, and infrastructure resources were found to be less affected by the pandemic.

Lin et al. (2023) investigated the effects of urban parks in Beijing, China, on Covid-19 case clusters. Surveys were conducted with park users to establish criteria for park accessibility. The regression analysis, which used metrics such as the number of parks, distance to parks, park size, frequency of visits, and duration of visits, concluded that urban parks had a positive impact on visitors' health during the Covid-19 pandemic. The study found that the accessibility of urban parks was positively related to the pandemic.

Yang et al. (2023) analyzed the relationship between the spatial distribution characteristics of neighborhoods in Beijing and the Covid-19 pandemic using Geographically Weighted Regression, Binomial Logistic Regression, and Clustering Analysis models. The analysis, which used independent variables such as building density, building footprint area, amount of wetlands and green space, commercial facility density, service facility density, transportation facility density, housing prices, building age, number of buildings, and number of households, found that factors related to the urban external environment (building density, volume ratio, commercial facility density, service facility density, and transportation facility density) had a positive correlation with Covid-19 incidence. In contrast, wetlands and green spaces were negatively correlated with Covid-19 incidence. Additionally, in low-income neighborhoods, housing prices and the number of households were positively correlated with Covid-19 incidence, while in medium-income neighborhoods, the number of households showed a positive correlation with the frequency of Covid-19 outbreaks.

Daisuke et al. (2023) conducted a study in Tokyo using 53 different variables to investigate the relationship between urban factors and Covid-19 infection rates. The main independent variables used in the study included residential land use, commercial land use, green space land use, household crowding rate, proportion of people living in single-family homes, average income, proportion of the population over 65 years old, unemployment rate, population density, and school density. The analysis, which utilized Local Moran I and Multivariate Regression models, found that regions with higher densities of retail stores, restaurants, healthcare facilities, public transportation use, and remote work had higher Covid-19 infection rates. Additionally, housing crowding and the rate of remote work were identified as the most significant determinants of the pandemic's impact.

Guoqing et al. (2023) analyzed and tested the spatial relationships between Covid-19 cases and urban facilities in Wuhan. According to the findings obtained through various statistical analysis methods, Covid-19 cases exhibited a spatial distribution spreading from the city center to the urban periphery. The relationship between Covid-19 cases and facilities such as movie theaters, restaurants, and metro stations increased with distance. Additionally, urban facilities such as parking lots, residential areas, shopping malls, banks, entertainment venues, and hospitals were found to be associated with clustered Covid-19 cases.

Ahadnejad Reveshty et al. (2023) investigated the spatial factors influencing the Covid-19 outbreak in the settlements of Zanjan. Using Geographically Weighted Regression and spatial autocorrelation analyses, the findings revealed that Covid-19 case clusters were concentrated in vulnerable neighborhoods and informal settlements. It was concluded that various socio-economic and demographic factors, such as independent and labor-intensive jobs, population density, unemployment, underlying health conditions, the number of people per room, household density in residential areas, illiteracy, and gender, were associated with case rates.

İçöz & Yenilmez (2023) investigated the spatiotemporal clusters of the Covid-19 pandemic at the urban scale in Turkey. Using spatiotemporal scan statistics and Moran I measurements, the findings revealed that there were eight distinct clusters across the country, with the highest number of cases occurring in the eighth cluster, which includes 15 provinces from various regions such as the Black Sea and Eastern Anatolia. Additionally, it is suggested that these clusters could be associated with several significant factors, including socioeconomic factors and health policies.

In the research conducted to analyze the emergence, spread, or increase of the Covid-19 pandemic and to determine urban solution strategies, it has been observed that a range of variables, which could have a direct or indirect impact on the pandemic, were used. However, since the data used in these studies are generally sourced from local resources, different indicators are often employed in each study. In this context, the variables used in the studies have been color-coded to relate them to each other. This approach aims to group each dataset based on similarities in content and characteristics. The relevant visual is presented in Figure 2.

Figure 2. Current spatial studies on the COVID-19 pandemic and key variables

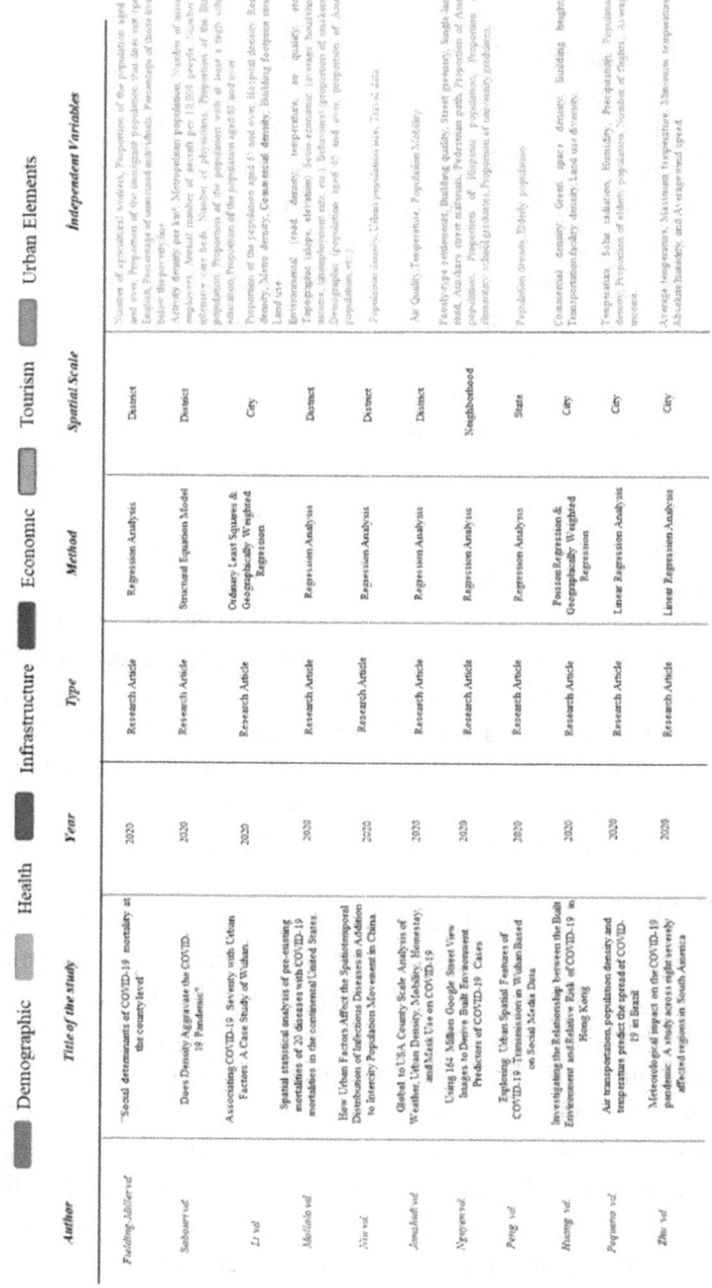

Author	Title of the study	Year	Type	Method	Spatial Scale	İndependent Variables
Fielding-Miller vd.	"Social determinants of COVID-19 mortality at the county level"	2020	Research Article	Regression Analysis	District	Number of agricultural workers, Proportion of the population aged 65 and over, Proportion of the immigrant population that does not speak English, Percentage of uninsured individuals, Percentage of those living below the poverty line
Sabouri vd.	Does Density Aggravate the COVID-19 Pandemic?	2020	Research Article	Structural Equation Model	District	Activity density per km^2, Metropolitan population, Number of insured employers, Annual number of aircraft per 10,000 people, Number of intensive care beds, Number of physicians, Proportion of the Black population, Proportion of the population with at least a high school education, Proportion of the population aged 60 and over
Li vd.	Associating COVID-19 Severity with Urban Factors: A Case Study of Wuhan.	2020	Research Article	Ordinary Least Squares & Geographically Weighted Regression	City	Proportion of the population aged 65 and over, Hospital density, Road density, Metro density, Commercial density, Building footprint ratio, Land use
Mollalo vd.	Spatial statistical analysis of pre-existing mortalities of 20 diseases with COVID-19 mortalities in the continental United States.	2020	Research Article	Regression Analysis	District	Environmental (road density, temperature, air quality, etc.) Topographic (slope, elevation) Socio-economic (average household income, unemployment rate, etc.) Behavioral (proportion of smokers), Demographic (population aged 65 and over, proportion of Asian population, etc.)
Niu vd.	How Urban Factors Affect the Spatiotemporal Distribution of Infectious Diseases in Addition to Intercity Population Movement in China	2020	Research Article	Regression Analysis	District	Population density, Urban population size, Travel data
Jamshidi vd.	Global to USA County Scale Analysis of Weather, Urban Density, Mobility, Homestay, and Mask Use on COVID-19	2020	Research Article	Regression Analysis	District	Air Quality, Temperature, Population Mobility
Nguyen vd.	Using 164 Million Google Street View Images to Derive Built Environment Predictors of COVID-19 Cases	2020	Research Article	Regression Analysis	Neighborhood	Family-type settlement, Building quality, Street greenery, Single-lane road, Auxiliary street materials, Pedestrian path, Proportion of Asian population, Proportion of Hispanic population, Proportion of elementary school graduates, Proportion of university graduates.
Peng vd.	Exploring Urban Spatial Features of COVID-19 Transmission in Wuhan Based on Social Media Data	2020	Research Article	Regression Analysis	State	Population density, Elderly population
Huang vd.	Investigating the Relationship between the Built Environment and Relative Risk of COVID-19 in Hong Kong	2020	Research Article	Poisson Regression & Geographically Weighted Regression	City	Commercial density, Green space density, Building heights, Transportation facility density, Land use diversity.
Pequeno vd.	Air transportation, population density and temperature predict the spread of COVID-19 in Brazil	2020	Research Article	Linear Regression Analysis	City	Temperature, Solar radiation, Humidity, Precipitation, Population density, Proportion of elderly population, Number of flights, Average income.
Zhu vd.	Meteorological impact on the COVID-19 pandemic: A study across eight severely affected regions in South America	2020	Research Article	Linear Regression Analysis	City	Average temperature, Maximum temperature, Minimum temperature, Absolute humidity, and Average wind speed.

Figure 3. Current spatial studies on the COVID-19 pandemic and key variables (continued)

Author	*Title of the study*	*Year*	*Type*	*Method*	*Spatial Scale*	*İndependent Variables*
Mo vd.	An analysis of spatiotemporal pattern for COIVD-19 in China based on space-time cube	2020	Research Article	Cluster Analysis	Neighborhood	Population density
Bourdin vd.	Does lockdown work? A spatial analysis of the spread and concentration of Covid-19 in Italy	2020	Research Article	Regression Analysis	District	Population ratio aged 65 and over, Population density, Total population, GDP per capita, Number of passengers traveling by airplane
Hu vd.	The role of built and social environmental factors in Covid-19 transmission: A look at America's capital city	2021	Research Article	Regression Analysis	City	Employment status, Education status, Poverty status, Income status, Housing quality
Kashem vd.	Exploring the nexus between social vulnerability, built environment, and the prevalence of COVID-19: A case study of Chicago	2021	Research Article	Structural Equation Modeling	Neighborhood	Accessibility of open spaces, Neighborhood walkability, Average household size, Average household income, Latin population ratio, Black population ratio, Proportion of population with at least a high school education
Rioc & Giannoene	On the link between temperature and regional COVID-19 severity: Evidence from Italy	2021	Research Article	Regression Analysis	District	Relative humidity, Daily solar radiation, Air quality, Proportion of population aged 0-15, Proportion of population aged 75 and over, Population density per km², **Corruption indicators, Commuting time, Length of highways and railways per km²**, Health expenditures
Mollalo vd.	Spatial statistical analysis of pre-existing mortalities of 20 diseases with COVID-19 mortalities in the continental United States	2021	Research Article	Global and Local Moran's I Analysis	City	Some diseases observed in the community (e.g., tuberculosis, cancer, etc.).
Urban & Nakada	GIS-based spatial modelling of COVID-19 death incidence in São Paulo, Brazil	2021	Research Article	Regression Analysis	Neighborhood	**Access to drinking water,** Population density, Proportion of population aged 60 and over, Average number of people per household., Size of urban settlement area., Average income, **Proportion of households with access to electricity, Proportion of individuals with access to toilets and drinking water**
Kan vd.	Spatial distribution characteristics of the COVID-19 pandemic in Beijing and its relationship with environmental factors	2021	Research Article	Regression Analysis	State	Population density, Distance to Xinfadi market, Distance to hospitals, **Distance to workplaces**, Distance to educational facilities, Distance to transportation options, **Distance to shopping centers, Distance to parks, Distance to restaurants.**
Perone	The determinants of COVID-19 case fatality rate (CFR) in the Italian regions and provinces: An analysis of environmental, demographic, and healthcare factors	2021	Research Article	Regression Analysis & Cluster Analysis	City	Humidity, Health system performance index, Health expenditures, Number of hospital beds, **Density of cars and businesses, Average electricity consumption per capita,** Number of single young people aged 20-34, Ratio of hospital beds before and during the pandemic, Number of recovered individuals during the pandemic, Number of intensive care beds, Urbanization index, Density, Air quality metrics
Phiri vd.	Spread of COVID-19 in Zambia: An assessment of environmental and socioeconomic factors using a classification tree approach	2021	Research Article	Regression Analysis	Neighborhood	Elevation, Slope, Temperature, Precipitation Amount, Population density, Total population, Proximity to airports and city centers, Proximity to mining centers and border areas, Distance to roads
Qiu vd.	A multiplicity of environmental, economic and social factor analyses to understand COVID-19 diffusion	2021	Research Article	Logistic Regression Analysis	State	Temperature, Relative humidity, Nitrogen levels, Air quality index, Elevation, Population density, Household size, GDP per capita, Migration rate, National health intervention data

Figure 4. Current spatial studies on the COVID-19 pandemic and key variables (continued)

Author	*Title of the study*	*Year*	*Type*	*Method*	*Spatial Scale*	*İndependent Variables*
Aldama vd.	Spatial epidemiological study of the distribution, clustering, and risk factors associated with early COVID-19 mortality in Mexico	2021	Research Article	Regression Analysis	City	Individuals diabetes status, Individuals obesity status, Literacy rate, GDP, Number of flights, Temperature, Air quality, Precipitation amount
Bayar	Population density index and its use for distribution of Covid-19: A case study using Turkish data	2021	Research Article	Regression Analysis	District	Education level, Welfare level, Number of healthcare workers, Ratio of male to female population, Proportion of population aged 65 and over, Population density
Sridhar	Urbanization and COVID-19 Prevalence in India	2021	Research Article	Regression Analysis	State	Urbanization rate, Labor force participation rate, Population density, Income per capita, Proportion of population living in slum areas, Health infrastructure, Access to parks
Jo vd.	Density or Connectivity: What Are the Main Causes of the Spatial Proliferation of COVID-19 in Korea?	2021	Research Article	Ordinary Least Squares & Negative Binomial Regression	City	Population density, Proportion of population aged 65 and over Proportion of female population, Proportion of population under 20 years of age, Employment density, GDP per capita, Park area per 1,000 people, Area of provinces and districts, Medical accessibility, Number of doctors per 1,000 people
Ahmed vd.	Effect of environmental and socio-economic factors on the spreading of COVID-19 at 70 cities/provinces	2021	Research Article	Positive and Negative Binomial Statistics	City	Temperature, Humidity, Air quality, Population density, GDP, Per capita health expenditure, Life expectancy
Al Kindi vd.	Demographic and socioeconomic determinants of COVID-19 across Oman - A geospatial modelling approach	2021	Research Article	Regression Analysis	City	Population density, Total number of households, Average household size, Number of hospital beds
Bijari vd.	Investigation of the urbanization contribution to the COVID 19 outbreak in Iran and the MECA countries	2021	Research Article	Regression Analysis	Country & State	Land area, Urban population, Rural population, Energy consumption
Sharifi vd.	Spatio-temporal patterns of the COVID-19 pandemic, and place-based influential factors at the neighborhood scale in Tehran	2021	Research Article	Spatial Regression Analysis	Neighborhood	Population density, Population aged 65 and over, Proportion of green spaces, Building density, Number of pharmacies, Distance to metro stations, Distance to shopping and retail centers in the neighborhood
Ghosh vd.	Association between overcrowded households, multigenerational households, and COVID-19: a cohort study	2021	Research Article	Regression Analysis	City	Obesity prevalence, Prevalence of heart disease, Proportion of the white population, Average household income
Ciupa & Suligowski	Green-Blue Spaces and Population Density versus COVID-19 Cases and Deaths in Poland	2021	Research Article	Regression Analysis	City	Forest areas, Fruit orchard areas, Pastures and meadows, Recreational and leisure areas, Lakes and artificial water reservoirs, Rivers, Ecological areas, Population density
Kim vd.	Temporal transition of racial/ethnic disparities in COVID-19 outcomes in 3108 counties of the United States: Three phases from January to December 2020	2021	Research Article	Regression Analysis	District	Black/African American population, Hispanic population, Population density
Gagliotti vd.	Population-Level Disparities in COVID-19: Measuring the Independent Association of the Proportion of Black Population on COVID-19 Cases and Deaths in US Counties	2021	Research Article	Regression Analysis	District	Proportion of Hispanic population, Proportion of Black population, Proportion of Asian population, Proportion of uninsured population, Proportion of impoverished population, Proportion of population aged 65 and over

Figure 5. Current spatial studies on the COVID-19 pandemic and key variables (continued)

Author	*Title of the study*	*Year*	*Type*	*Method*	*Spatial Scale*	*İndependent Variables*
Klompmaker vd.	County-level exposures to greenness and associations with COVID-19 incidence and mortality in the United States	2021	Research Article	Negative Binomial Regression	District	Population density, Proportion of Black population, Proportion of Hispanic population, Poverty status, Average household income, Temperature, Relative humidity, Air quality, Number of urban districts, Number of districts under restriction orders, Vegetation pattern index
Wu vd	Population migration, spread of COVID-19, and epidemic prevention and control: empirical evidence from China	2021	Research Article	Ordinary Least Squares Regression	City	Number of permanent residents in their respective provinces, Traffic density, Number of migrants, Population density
Cutrini & Salvati	Unraveling spatial patterns of COVID-19 in Italy: Global forces and local economic drivers	2021	Research Article	Negative Binomial Regression	City	Population aged 65 and over, Size of metropolitan area, Number of air passengers per 100,000 people, **Number of industrial facilities in a specific area, Share of employees in workplaces with more than 250 employees in the total number of employees.**
Li vd.	Association of built environment attributes with the spread of COVID-19 at its initial stage in China	2021	Research Article	Regression Analysis	City	Number of trains, Number of aircraft departures, Distance between train stations, Travel times **Activity density around train stations,** Population density, Number of population aged 65 and over
Hamidi vd.	Effect of environmental and socio-economic factors on the spreading of COVID-19 at 70 cities/provinces	2021	Research Article	Regression Analysis	City	Proportion of Hispanic and Black population, Average household size, Population density, Crowding rate
Li vd.	Subway Ridership, Crowding, or Population Density: Determinants of COVID-19 Infection Rates in New York City	2021	Research Article	Structural Equation Model	Neighborhood	**Commercial welfare within a 1000-meter radius, Medical services, ATMs, Grocery stores, Hairdressers, Shops, Restaurants, Clinics, Pharmacies**, Bus stops, Road length, Building density, and Housing prices
Das vd	Living environment matters: Unravelling the spatial clustering of COVID-19 hotspots in Kolkata megacity, India	2021	Research Article	Regression Analysis	City	Housing conditions, Household asset levels, **Access to basic technical infrastructure**, Technical facilities related to housing, Gender inequality
Lata vd.	Liveability of Indian Cities and Spread ofCovid-19 Case of Tier-1 Cities	2021	Research Article	İndex	City	Quality of life index, Health index, **Pollution index**, sizes of open spaces, Sizes of natural areas
Razavi-Termeh vd.	COVID-19 Risk Mapping with Considering Socio-Economic Criteria Using Machine Learning Algorithms	2021	Research Article	Logistic Regression Analysis	City	Public transportation stations, Supermarkets, Banks, Automated teller machines (ATMs), Bakeries, Pharmacies, Gas stations, and Hospitals
Ruck vd.	Early warning of vulnerable counties in a pandemic using socio-economic variables	2021	Research Article	Regression Analysis	District	Number of university students, Average income level, Black population, Hispanic population, Population aged 65 and over, Number of hospitals, number of intensive care beds, Democrats, Republicans, Population size, Population density
Spotswood vd.	Nature inequity and higher COVID-19 case rates in less-green neighbourhoods in the United States	2021	Research Article	Regression Analysis	District	Amount of green space, Access to parks, Average income status, average age, Percentage of white people, Number of days since the first recorded case, Population density
Sarmadi vd	Association of COVID-19 distribution with air quality, socio demographic factors, and comorbidities: an ecological study of US states	2021	Research Article	Regression Analysis	District	Air pollution, Non-Hispanic Black African American population, population density, median age, Human Development Index, GDP per capita

Figure 6. Current spatial studies on the COVID-19 pandemic and key variables (continued)

Author	*Title of the study*	*Year*	*Type*	*Method*	*Spatial Scale*	*İndependent Variables*
Tchicaya vd.	COVID-19 mortality with regard to healthcare services availability, health risks, and socio-spatial factors at department level in France: A spatial cross-sectional analysis.	2021	Research Article	Regression Analysis	District	Urban population ratio, Poverty rate, Gender, Number of beds, Number of intensive care beds, Number of doctors, Population density
Silva & Alves	Social inequalities and the pandemic of COVID-19: the case of Rio de Janeiro	2021	Research Article	Poisson Regression Model	Neighborhood	Age characteristics, Gender status, Crowding rate, Education status, Income status, Racial characteristics
Dintrans vd.	COVID-19 incidence and mortality in the Metropolitan Region, Chile: Time, space, and structural factors	2021	Research Article	Negative Binomial Regression	Neighborhood	Number of women, Population aged 65 and over, Population aged 80 and over, Number of immigrants, Population density, Rural-urban index, Size of green spaces, Distance to health center, Number of people with health insurance, Poverty rate
Tokey	Spatial association of mobility and COVID-19 infection rate in the USA: A county-level study using mobile phone location data	2021	Research Article	Regression Analysis	District	Number of trips per person, Percentage of trips outside the district, Poverty rate, Percentage of population aged 65 and over, Education level.
Verma vd.	Spatiotemporal contact density explains the disparity of COVID 19 spread in urban neighborhoods	2021	Research Article	Regression Analysis	Neighborhood	Income status, Poverty rate, Percentage of Black population, Total number of trips made within the neighborhood, Weekly distribution of travel distances.
Wu & Zhang	Exploration of Temporal-Spatially Varying Impacts on COVID-19 Cumulative Case in Texas Using Geographically Weighted Regression (GWR)	2021	Research Article	Spatial Regression Analysis	District	Income distribution, Unemployed population., Average temperature, Average rainfall, Air quality, Size of residential area, Total population, Male population, Percentage of Black population, Population aged 20-59, Population aged [illegible] and over, Number of hospitalized individuals, Number of beds per person.
Kulkarni vd.	Independent association of meteorological characteristics with initial spread of Covid-19 in India	2021	Research Article	Regression Analysis	City	Temperature, precipitation, Relative humidity, wind speed, Air pressure, Population density, Total population, Area sizes, Elevation
Frank & Wali	Treating two pandemics for the price of one: Chronic and infectious disease impacts of the built and natural environment	2021	Research Article	Regression Analysis	District	Density of green spaces, Housing density, Elderly population, Percentage of Black population, Population aged 65 and over.
Wu vd.	Population migration, spread of COVID-19, and epidemic prevention and control: empirical evidence from China	2021	Research Article	Kernel Density Analysis	City	Access to social activities, Access to outdoor activities, Building density, Distance to educational facilities, Distance to cultural facilities, Distance to restaurants, Distance to shopping centers, Distance to sports facilities, Distance to vacant areas, Distance to green spaces, Population by age groups (youth, middle-aged, and elderly).
Bayode vd.	Spatial variability of COVID-19 and its risk factors in Nigeria: A spatial regression method	2022	Research Article	Regression Analysis	City	Population, Gross Domestic Product (GDP), Literacy rate, Percentage of people living on less than one dollar ($1) a day, Limited access to food, Improved water source, Availability of local and international airports
Aral & Bakır	Spatiotemporal Analysis of Covid-19 in Turkey	2022	Research Article	Cluster Analysis	City	Population density, Elderly dependency ratio, GDP per capita per day, Literacy rate, Number of physicians per 1,000 people, and number of hospital beds per 100,000 people.
Watts	Macro-level drivers of SARS-CoV-2 transmission: A data-driven analysis of factors contributing to epidemic growth during the first wave of outbreaks in the United States	2022	Research Article	Cluster Analysis	State	Population density, Overcrowding, Air pollution, Temperature, Diabetes exposure rate.

Figure 7. Current spatial studies on the COVID-19 pandemic and key variables (continued)

Author	*Title of the study*	*Year*	*Type*	*Method*	*Spatial Scale*	*İndependent Variables*
Romoel & Baycan	COVID-19 mortality with regard to healthcare services availability, health risks, and socio-spatial factors at department level in France: A spatial cross-sectional analysis.	2022	Research Article	Clustering Analysis & Regression Analysis	City	Population density, Health capacity, Immigrant population, Active insured workers, Hospital capacity., OAverage housing sizes, Air quality, Number of accommodation facilities
Mellander & Florida	Social inequalities and the pandemic of COVID-19: the case of Rio de Janeiro	2022	Research Article	Regression Analysis	City & Neighborhoods	Total population, Density, Average age, Income distribution, Average household size, Unemployment, Access to nursing homes, Access to case rates with a one-week delay
Xiao vd.	COVID-19 incidence and mortality in the Metropolitan Region, Chile: Time, space, and structural factors	2022	Research Article	Logistic and Hierarchical Regression	Neighborhoods	Housing size, Housing type, Ratio of homes without gardens, Ratio of homes with access to open communal areas, Black population, White population, Gender, Age, Educational status, Household income, Proximity to the nearest open space, **Proximity to the nearest stores**
Alidadi & Sharifi	Spatial association of mobility and COVID-19 infection rate in the USA: A county-level study using mobile phone location data	2022	Research Article	Compilation & Research	Country & State & District & City & Urban Area	Density, Land use, Transportation and mobility, Housing conditions, Demographic characteristics, **Health systems**
Wu vd.	Spatiotemporal contact density explains the disparity of COVID 19 spread in urban neighborhoods	2022	Research Article	Regression Analysis	City	Housing density, Street connectivity, Land use, Building coverage ratio, Proportion of minorities, Income status, Educational status
Wang vd.	Exploration of Temporal-Spatially Varying Impacts on COVID-19 Cumulative Case in Texas Using Geographically Weighted Regression (GWR)	2022	Research Article	Geographic Regression Analysis	City	Population density, Income status, Population aged 60 and over, Distance to the hospital, Distance to the school, Distance to the supermarket, Distance to the bus stop
Gatsie vd.	Independent association of meteorological characteristics with initial spread of Covid-19 in India	2022	Research Article	Clustering Analysis & Regression Analysis	City	Distance to the train station, Distance to parks, Distance to open spaces, Distance to playgrounds, Access to major activity centers, Intersection density, Average household size, Age level
Ha & Lee	Treating two pandemics for the price of one: Chronic and infectious disease impacts of the built and natural environment	2022	Research Article	Regression Analysis	District	Educational status, Total population, Proportion of Black/African American population, Density, Household size, Number of restaurants, Number of supermarkets.
Xu vd.	Population migration, spread of COVID-19, and epidemic prevention and control: empirical evidence from China	2022	Research Article	Kernel Density Analysis & Linear and Geographically Weighted Regression	City	Population density, Population aged 60 and over, Metro stations, Main roads, Commercial areas.
Lin vd.	Spatial variability of COVID-19 and its risk factors in Nigeria: A spatial regression method	2022	Research Article	Regression Analysis	District	Proportion of Black/African American population, Proportion of Hispanic or Latino population, Proportion of non-Hispanic White population, Proportion of population aged 18 and over, Proportion of individuals below the poverty line, Proportion of low-income elderly population, Urban opportunity index, Mixed immigrant integration and accessibility index.
Desmet & Wacziarg	Spatiotemporal Analysis of Covid-19 in Turkey	2022	Research Article	Regression Analysis	District	Population, Population density, Public transportation, Income status, Social capital, Ethnicity, Education, Votes cast in presidential elections.
Guan vd.	Macro-level drivers of SARS-CoV-2 transmission: A data-driven analysis of factors contributing to epidemic growth during the first wave of outbreaks in the United States	2022	Research Article	Ordinary Least Squares Regression (OLS).	District	Housing density, Employment density, Pedestrian-oriented road density, Total road network length, Proximity to public transportation, Population density, Income status, Unemployment status

Figure 8. Current spatial studies on the COVID-19 pandemic and key variables (continued)

Author	*Title of the study*	*Year*	*Type*	*Method*	*Spatial Scale*	*İndependent Variables*
Kuschel vd.	Preparing Cities for Future Pandemics: Unraveling the Influence of Urban and Housing Variables on COVID-19 Incidence in Santiago de Chile	2023	Research Article	Principal Component Analysis	City	Population density, Number of University Graduates, Numbes of High School Graduates, **Immigrant population**, Active insured workers, **Proportion of those with access to infrastructure services**
Lin vd.	Urban park use and self-reported physical, mental, and social health during the COVID-19 pandemic: An on-site survey in Beijing, China	2023	Research Article	Survey study	City	Number of parks, Distance to parks, Size of parks, Frequency of visits, Duration of visits
Yang vd.	The impact of urban spatial environment on COVID-19: a case study in Beijing	2023	Research Article	Geographically Weighted Regression, Binomial Logistic Regression, Clustering Analysis.	Neighborhoods	Building density, Building footprint areas, Amount of wetland and green space, Density of commercial facilities, Density of service facilities, Density of transportation facilities, Housing prices, Building age, Number of buildings, Number of households
Daisuke vd.	Tokyo's COVID-19: An urban perspective on factors influencing infection rates in a global city	2023	Research Article	Local Moran's I, Multivariate Regression.	City	Residential land use, **Commercial land use**, Green space land use, Household overcrowding rate, Proportion of those living in detached houses, Average income, Proportion of the population aged 65 and over, Unemployment rate, Population density, School density
Guoqing vd.	Spatial co-location patterns between early COVID-19 risk and urban facilities: a case study of Wuhan, China	2023	Research Article	Clustering Analysis	City	Parking lots, Residential areas, Shopping malls, Banks, Entertainment venues, Hospitals, Cinemas, Restaurants, Metro stations
Ahadnejad Reveshty vd.	Spatial analysis of the factors impacting on the spread of Covid-19 in the neighborhoods of Zanjan, Iran	2023	Research Article	Geographically Weighted Regression Analysis	City	Independent and labor-intensive jobs, Population density, Unemployment, Underlying health conditions, Number of people in a room, Household density in settlements, Illiteracy, Gender, Informal settlements, Vulnerable neighborhoods
İçöz & Yenilmez	Spatio-temporal analysis of the COVID-19 pandemic in Türkiye: results of the controlled normalization	2023	Research Article	Spatio-temporal Scanning Statistical Analysis and Moran's I	City	---

CONCEPTUALISATION OF CITY FEATURES AFFECTING COVID-19 SPREAD

The exploration of features between cities with different characteristics in relation to the spatial distribution of the Covid-19 pandemic is crucial for the early detection of disease spread. This is important because identifying potential risk factors, risk levels, and high-risk areas is vital for preventing the outbreak and ensuring the sustainable development of urban health. In this context, a literature review has identified statistically proven variables that affect the spatial spread of the disease, categorized under the following headings:

- Sociodemographic Characteristics (population, gender, household size, etc.• Healthcare Services (number of hospitals, number of beds, etc.)
- Infrastructure (access to clean drinking water, etc.)
- Tourism Activities (number of accommodation facilities, number of flights, etc.)
- Economic Activities (active insured working population, export-import values, etc.)
- Urban Conditions (size of residential areas, number of buildings, etc.)

Socio-Demographic Characteristics

Various studies have been conducted to explain the changes in the number of deaths and cases following the coronavirus pandemic and to identify the demographic factors influencing these changes. Many of these studies have focused on the racial characteristics of the population and have attempted to discuss the demographic inequalities caused by the crisis. For instance, a study conducted in the United States concluded that communities with a higher African and Hispanic population were more affected by the COVID-19 pandemic (Kim et al., 2021). Similar studies have also found that the African population (Ruck et al., 2021), Black population (Nguyen et al., 2020; Ha & Lee, 2022), Latin population (Kashem et al., 2021; Lin et al., 2022), and African and Native American populations (Desmet & Wacziarg, 2022) had higher COVID-19 mortality rates compared to other communities. These findings are consistent with studies that determined that during the pandemic, the Black population had an 18 times higher mortality rate compared to the White population (Gaglioti et al., 2021) and that a 1% increase in the Black population was associated with a 0.97 increase in COVID-19 mortality rates (Frank & Wali, 2021). These results indicate that communities of different ethnic backgrounds are likely to be more affected by the pandemic due to existing economic and health infrastructure inequalities.

The varying changes of the COVID-19 pandemic between regions are fundamentally explained by different models in demographic structure. Among these models, population density plays a significant role in understanding the impact of the population on the pandemic. For instance, Jamshidi et al. (2020) found that areas with higher population density are associated with higher COVID-19 mortality rates. Similarly, a study conducted in New York and Chicago showed that in certain neighborhoods where density—and consequently the risk of contact—is higher, the case rates were also higher (Verma et al., 2021). It is generally accepted that airborne infectious diseases, such as the COVID-19 pandemic, can spread more easily in areas with higher population density due to the increased risk of physical contact. This situation is explained by the fact that higher human density and activities make cities more vulnerable. There are many studies in the relevant literature that support this explanation (Nguyen et al., 2020; Peng et al., 2020; Pequeno et al., 2020; Bourdin et al., 2021; Urban & Nakada, 2021; Han et al., 2021; Qiu et al., 2021; Başer, 2021; Sridhar, 2021; Ahmed et al., 2021; Sharifi et al., 2021; Gaisie et al., 2022; Bayode et al., 2022; Ronael & Baycan, 2022; Xu et al., 2022). However, some studies conducted by different researchers at various scales have contradicted these theories. For example, Sabouri et al. (2020) found that districts with higher density had lower COVID-19 case rates; Guan et al. (2022), who examined the spatiotemporal distribution of the pandemic, revealed that population density was negatively associated with COVID-19 prevalence (in the third phase). On the other hand, many other studies have shown that deaths related to the virus are associated with high-density areas and that human density creates favorable conditions for the spread of the pandemic.

In other studies conducted to understand the impact of demographic characteristics on the Covid-19 pandemic, it has been observed that research focused on the structure of the population and socio-demographic factors such as Gender Status (Mollalo et al., 2020; Silva & Alves, 2021), Age Status (Li et al., 2020; Peng et al., 2020; Li et al., 2021; Tokey, 2021; Frank & Wali, 2021; Başer, 2021; Dintrans et al., 2021), Education Status (Nguyen et al., 2020; Kashem et al., 2021; Başer, 2021; Ha & Lee, 2022; Desmet & Wacziarg, 2022), Household Size (Kashem et al., 2021; Urban & Nakada, 2021; Al Kindi et al., 2021; Hamidi et al., 2021; Mellander & Florida, 2022; Ahadnejad Reveshty et al., 2023), Income Status (Mollalo et al., 2020; Başer, 2021; Silva & Alves, 2021; Xiao et al., 2022; Wu et al., 2022; Yang et al., 2023), and Employment Status (Mollalo et al., 2020; Mellander & Florida, 2022; Guan et al., 2022; Daisuke et al., 2023). These studies generally used spatial statistical methods and investigated the relationship between the spatial distribution of the disease and demographic structure with the help of related explanatory variables. Although these studies differ in terms of quality and level, it has been observed that each variable has a positive or negative relationship with the Covid-19 pandemic. This situation shows that multiple variables dependent on demographic characteristics can be effective against the spread of the pandemic and should be taken into account.

Healthcare Services

Although the Covid-19 pandemic has directly or indirectly affected all factors of urban systems, it is evident that healthcare systems have faced numerous challenges following the rapid increase in medical needs. With the virus spreading to almost all regions, the need for effective healthcare services in areas such as medical supplies, facility capacity, sufficient healthcare personnel, and resource management has emerged. To meet these needs, healthcare systems have been subjected to pressure beyond their capacities, leading to disruptions in healthcare services, especially in cities with limited resources, making it difficult to prevent the spread of the pandemic. In this context, many studies have been conducted using urban healthcare infrastructure and human capital data to identify the determinants of healthcare services against the spatial distribution of the pandemic. For example, Perone (2021), who investigated the spatial distribution of the pandemic in Italy using a series of health variables, found that Covid-19 cases were lower in districts with a higher number of doctors and physicians. Another study observed that cities in Korea with a higher number of doctors per 1,000 people (Jo et al., 2021) had lower Covid-19 cases. Similar results were found in other studies indicating that health resources in India (Sridhar, 2021) and proximity to healthcare facilities in Chile (Dintrans et al., 2021) were negatively associated with Covid-19 cases. Therefore, it is evident that all factors related to healthcare services in cities are crucial intervention tools used to control the virus during the pandemic. However, there are studies that theoretically present different approaches to such research. For example, studies conducted in Wuhan found that areas with large hospitals had higher Covid-19 cases (Li et al., 2020) and that as the distance to hospitals increased, there was a significant decrease in Covid-19 cases (Wang et al., 2022). Similarly, a study by Daisuke et al. (2023) in Tokyo observed that regions with denser healthcare facilities had higher Covid-19 cases. Similar results were found in studies showing higher Covid-19 case and death rates in areas with more hospital beds (Zhang, 2021), more healthcare workers (Başer, 2021), and more physicians (Aldana et al., 2021). However, these results are likely due to the constant exposure of healthcare personnel and facilities to infected patients. In other words, regions where healthcare services and activities are conducted are also areas with a higher risk of virus transmission, leading to such outcomes.

During the pandemic, another important issue that was discussed as much as the adequacy of healthcare system capacity was the problem of access to healthcare services. Researchers approached these problems from social, economic, and spatial dimensions and highlighted the inequalities experienced in healthcare systems. For example, Arnault et al. (2021), examining the economic dimension of access to healthcare, found that individuals with low income during the pandemic could not find sufficient financial resources to address their health problems, leading them to forgo, postpone, or fail to schedule medical appointments when needed. In a study conducted in India by Das et al. (2021), it was found that Covid-19 cases were higher in regions with more restricted access to healthcare. Additionally, a study in the USA revealed that individuals with lower socio-economic status were five times more likely to experience Covid-19 deaths compared to those with higher status (Stokes et al., 2021). As a result, inadequate access to healthcare exacerbated existing social disadvantages and made urban healthcare systems more vulnerable to the Covid-19 crisis.

Infrastructure

The rapid growth of the Covid-19 pandemic has caused significant disruptions in critical infrastructure services such as electricity, water, food, transportation, and telecommunications (Carvalhaes et al., 2020). The demand for such services, which was expected before the pandemic, increased due to temporal and spatial changes during the pandemic. This led to a rise in the spread of the pandemic in cities with inadequate urban social and technical infrastructure, especially in essential facilities, service institutions, public spaces, and transportation networks. On the other hand, due to the inadequacy of urban infrastructure, city dwellers had to cope with economic, social, and psychological issues, even if these did not directly impact the pandemic. Researchers aiming to uncover the impact of the spatial distribution of inequalities in urban infrastructure systems on the spread of the Covid-19 pandemic have obtained significant findings. For example, Kuschel et al. (2023), examining 32 communes in Santiago de Chile, concluded that communes with more developed infrastructure resources were less affected by the pandemic. Das et al. (2021) found that regions with more limited infrastructure services in India had higher Covid-19 cases. A similar approach by Sridhar (2021) observed that regions with more developed health infrastructure had lower Covid-19 cases. Additionally, studies have identified transportation infrastructure (Li et al., 2021) and information technology infrastructure (Mouratidis, 2021) as critical determinants in combating the Covid-19 pandemic. Consequently, it is evident that the availability of appropriate infrastructure and resources, particularly health information platforms and services in urban areas, is crucial in responding to public health crises like the Covid-19 pandemic.

Tourism Activities

According to data from the United Nations World Tourism Organization (UNWTO), global tourism mobility decreased by 74% in 2020 compared to the previous year (Yu et al., 2023). The restrictions implemented during the pandemic, which lasted for about two years, brought tourism activities to a halt. This is because tourism involves a series of spatial movements by tourists (Yu et al., 2023). However, during a pandemic like Covid-19, such mobility activities caused the virus to be transported to different regions, contributing to the spread of the disease. Therefore, the spatial reflections and changes in travel policies and how they affected the spread of the pandemic became a highly interesting research area. In this context, although there are a limited number of studies in the literature investigating the spatial connection between tourism activities and the Covid-19 pandemic, noteworthy results have been obtained. These studies focus on some activities in the tourism industry and the mobility in these activities, such as proximity to airports (Phiri et al., 2021; Bayode et al., 2022), tourist arrivals and departures, incoming and outgoing flights (Pequeno et al., 2020), and the number of tourism facilities (Jamshidi et al., 2020; Ronael & Baycan, 2022) (Saputra, 2023). For example, a study in Brazil found that the first case was detected in a tourist who arrived by air from Italy, and the infected individual subsequently transmitted the virus to family and close contacts, spreading the pandemic to different regions (Tchicaya et al., 2021). Another study by Pequeno et al. (2020) in 27 different cities in Brazil found a positive relationship between regions with high numbers of flights related to tourism activities and higher Covid-19 cases. Similar results were observed in other studies that used air travel and proximity to airports as independent variables (Aldana et al., 2021; Phiri et al., 2021), showing higher Covid-19 cases and deaths in these regions. These results indicate that economic dynamics should be considered to understand the impact of mobility caused by tourism activities on the Covid-19 pandemic. This is supported by studies showing that air transportation is typically preferred by high-income individuals (Tchicaya et al., 2021). Additionally, research conducted during the early stages of the pandemic found that communities with high income were likely more affected by the pandemic due to tourism or economic reasons (Sridhar, 2021; Silva & Alves, 2021).

Economic Activities

The fact that Covid-19 spreads through contact accelerated the transmission of the virus in commercial units and work environments where people gather. On the other hand, the measures taken to prevent this led to a global economic downturn (Ronael & Baycan, 2022). Every country was affected by this recession, but the distribution and magnitude of these effects were related to the industrial structures of spatial economic networks. Especially during periods of lockdown measures, limiting interregional mobility caused regions whose economies heavily rely on retail sales, warehousing, and food services to be more severely impacted by the economic crisis (Kim et al., 2022). Although these measures were effective in controlling the spread of the pandemic among communities, they resulted in long-term economic costs. For example, a study on the impact of two months of lockdown measures on the labor market in Madrid, Spain, found that the accommodation sector experienced a 31.8% decline, retail trade 11.3%, personal services 5.6%, and businesses related to cultural activities 2.5% (OECD, 2020). In other words, disruptions in the international supply chain pressured local production networks, resource expenditures, and labor markets, leading to a decrease in income and employment, thus increasing poverty. During this period, many firms and businesses engaged in commercial activities, such as stores, restaurants, and shops, closed, causing urban spaces to lose their economic vitality. In this context, measures taken to prevent further spread of the pandemic and regain economic vitality were insufficient due to the global spread of the pandemic and its transformation into a multi-layered crisis (health crisis, internal economic crisis, capital reduction, etc.). For example, cities with natural, social, and cultural identities, typically located in coastal or mountainous regions and whose economies depend on local restaurants, shops, and other businesses for cultural activities, were found to be more affected by the Covid-19 pandemic due to the decline in tourism revenue (OECD, 2020). This is because the lack of sectoral diversity and digital infrastructure in such settlements made the communities living there more vulnerable to the pandemic. In addition to the aforementioned impacts of the Covid-19 pandemic on global economic activities, significant research has been conducted to reveal its effects on spatial distribution. For example, a study by Sharifi et al. (2021) in Tehran found that areas with higher concentrations of shops had higher Covid-19 cases. Cutrini & Salvati (2021), investigating the spatial distribution of Covid-19 cases in Italy, concluded that regions where industrial facilities and large companies were clustered had higher Covid-19 cases. Similar results were found in studies identifying higher Covid-19 cases in areas with more intense commercial activities in Huanggang (Li et al., 2021), higher commercial density in Wuhan (Li et al., 2020), and more retail stores and restaurants in Tokyo (Daisuke et al., 2023).

Urban Conditions

Considering the scope and impact of the Covid-19 pandemic, identifying the built and natural environmental features that contribute to the spread of the virus has become a significant area of study, particularly for urban researchers. This is because investigating the spatial distribution characteristics of a contagious disease that has turned into a pandemic forms the foundation for formulating spatial measures against the disease. These studies, often conducted through data mining, web-based mapping, or temporal-spatial analyses, have identified the effects of Covid-19 in different urban areas and the morphological, functional, and spatial patterns of these effects.

Significant findings in the relevant literature indicate that certain environmental factors, such as housing conditions, land use, urban density, transportation, and urban-rural differences, play a critical role in understanding the dynamics of the spatial spread of the Covid-19 pandemic (Sharifi et al., 2021; Hu et al., 2021; Coza & Njjar, 2024, Gaisie et al., 2022; Yang et al., 2023). For example, a study by Li et al. (2020) found higher Covid-19 cases in cities with more intense mixed land use, including public facilities, hospitals, roads, and metro stations. Similarly, a study in Tehran by Razavi-Termeh et al. (2021) identified a higher risk of transmission in areas with denser land use. These results indicate that land uses arising from more intensive urban activities increase the risk of contact and exposure (Huang et al., 2020; Guan et al., 2022; Alidadi & Sharifi, 2022; Daisuke et al., 2023; Guoqing et al., 2023). Many empirical studies in the literature show higher cases and deaths in cities with more urban activities (Sharifi et al., 2021; Li et al., 2021; Yang et al., 2023). This is because the risk of transmission varies significantly spatially. For example, a study in the US by Sabouri et al. (2020) found that metropolitan areas were more affected by the Covid-19 pandemic than many counties in the country. A study by Lin et al. (2022) using 22 independent indicators also found that urban areas with higher social and economic density had higher case rates compared to rural areas. Although such regions offer better living services (such as healthcare), problems likely arising from socio-spatial inequalities make some communities more vulnerable to the pandemic. This is because during the spread of the pandemic, humans are considered carriers; transportation networks are channels, and economic and social activities are driving forces (Xie et al., 2021). Thus, these complex dynamics have led to social, economic, and spatial impacts due to imbalances in living conditions in the course and outcomes of the pandemic (McGowan & Bambra, 2022). For instance, a study using the Geographically Weighted Regression Model by Ahadnejad Reveshty et al. (2023) found higher clusters of Covid-19 cases in neighborhoods with socio-economically disadvantaged individuals. Similar results are seen in studies revealing that communities living in slum neighborhoods and with lower living standards have higher risks of Covid-19 infection, hospitalization, confirmed diagnosis, and death (Das et al., 2021; Sridhar, 2021). Many studies also provide findings that housing quality and living conditions are among the strongest determinants affecting the spread of Covid-19 (Nguyen et al., 2020; Xiao et al., 2022; Ahadnejad Reveshty et al., 2023; Daisuke et al., 2023). For example, a study in Washington analyzed the relationship between housing conditions and the spatial spread of the pandemic using 12 indicators (such as house size and crowding rate) and found that communities living in more crowded environments were more exposed to the pandemic (Hu et al., 2021). This is because living in an overcrowded housing type is considered riskier than living in a housing type with more extensive usage areas (Hamidi et al., 2021; Alidadi & Sharifi, 2022). Moreover, studies in the literature show that crowded environments such as public transportation also accelerate the spread of the pandemic (Dintrans et al., 2021; Wu et al., 2022; Daisuke et al., 2023). Therefore, crowding is interpreted as an indicator of increasing density in living environments and correspondingly high infectious risk at the regional level (Rubin et al., 2020). On the other hand, some findings indicate that the need for open and green public spaces increased during the pandemic period (Mouratidis & Yiannakou, 2022). This is because restriction measures during this period caused physical and mental health deteriorations, and communities living in areas with insufficient access to green spaces were more exposed to such health problems (Sharifi et al., 2021). Spatial analyses have also found a negative relationship between proximity to green open spaces and increased green space amounts and Covid-19 cases (Ciupa & Suligowski, 2021; Wu et al., 2021; Frank & Wali, 2021; Dintrans et al., 2021; Klompmaker et al., 2021; Yang et al., 2023; Lin et al., 2023). For example, a county-level study in the US statistically proved that a 4.1% increase in green spaces was associated with a 0.1% decrease in Covid-19 cases (Spotswood et al., 2021). Furthermore, there are many studies in the literature investigating the effects of environmental factors such as temperature, precipitation, humidity, and air quality on the pandemic (Rios & Gianmoena, 2021; Zhu et al., 2020; Pequeno et al., 2020; Ronael & Baycan 2022; Watts 2022). For example, Phiri et al. (2021) found that counties with temperatures above 22°C had a negative relationship with Covid-19 cases. A study in India also found that high temperature, wind speed, and precipitation were negatively correlated, while relative humidity and air pressure were positively correlated with case numbers (Kulkarni et al., 2021). Similar results are seen in studies revealing the relationship of air quality (Wu & Zhang, 2021; Ahmed et al., 2021), air pollution values (Qiu et al., 2021), temperature (Perone, 2021), and minimum, maximum, and average temperatures, absolute humidity, and average wind speed in various cities with Covid-19 cases and deaths (Rios & Gianmoena, 2021).

CONCLUSION AND EVALUATION

Literature reviews and field studies have revealed that factors associated with the spread of COVID-19 can be systematically categorized into six key domains, which we have designated as follows: Socio-Demographic, Healthcare Services, Infrastructure, Economic Activities, Tourism Services, and Urban Conditions.

Socio-Demographic factors encompass aspects such as population density, age distribution, and socioeconomic status, all of which influence susceptibility to and transmission of the virus. Healthcare Services include variables related to the availability and accessibility of medical facilities, the quality of healthcare infrastructure, and the capacity for testing and treatment. Infrastructure Services cover elements like transportation systems, public utilities, and essential services that impact both mobility and the management of the pandemic. Economic Activities involve factors such as employment patterns, business operations, and economic resilience that affect both the spread of the virus and societal responses. Tourism Services pertain to travel and tourism dynamics, which can introduce or amplify virus transmission across regions. Finally, Urban Conditions relate to the design and characteristics of urban spaces, including housing quality, public space utilization, and environmental factors.

Advanced analytical techniques employing contemporary geographic information systems (GIS) and spatial statistical methods are utilized to dissect these factors. These analyses help identify critical risk factors, evaluate distribution levels of the virus, and examine clustering tendencies that contribute to the spatial complexity of the pandemic's spread. By leveraging these sophisticated tools, researchers can develop geographic models that illustrate the pandemic's progression and pinpoint areas of heightened risk.

Thus, a comprehensive investigation into the spatial patterns and fundamental determinants of COVID-19 facilitates a deeper understanding of the pandemic's evolution. This approach enables the identification of key risk areas and supports the development of targeted interventions to mitigate the impact of the pandemic.

ACKNOWLEDGMENT

This study was developed within the scope of the TUBITAK-SOBAG project numbered 122K710 and titled 'Investigation of the Relationship of the COVID 19 Epidemic with the Urban and Environmental Components at the Regional and Local Level.

REFERENCES

Agudo-Sierra, E., & Llaguno-Munitxa, M. (2022). Nature-based Solutions for Industrial Logistic Parks. *IOP Conference Series. Earth and Environmental Science*, 1122(1), 012007. DOI: 10.1088/1755-1315/1122/1/012007

Ahadnejad Reveshty, M., Heydari, M. T., & Tahmasebimoghaddam, H. (2023, September). Spatial analysis of the factors impacting the spread of COVID-19 in the neighborhoods of Zanjan, Iran. *Spatial Information Research*, 32(2), 151–164. DOI: 10.1007/s41324-023-00550-0

Ahmed, J., Jaman, M. H., Saha, G., & Ghosh, P. (2021). Effect of environmental and socio-economic factors on the spreading of COVID-19 at 70 cities/provinces. *Heliyon*, 7(5, e06979), 1–8. DOI: 10.1016/j.heliyon.2021.e06979 PMID: 33969235

Al Kindi, K. M., Al-Mawali, A., Akharusi, A., Alshukaili, D., Alnasiri, N., Al Awadhi, T., Charabi, Y., & El Kenawy, A. M. (2021). Demographic and socioeconomic determinants of COVID-19 across Oman - A geospatial modelling approach. *Geospatial Health*, 16(985), 145–160. DOI: 10.4081/gh.2021.985 PMID: 34000790

Aldana, R. R., Gomez-Verjan, J. C., Bello-Chavolla, O. Y., & García Peña, C. (2021, July 21). Spatial epidemiological study of the distribution, clustering, and risk factors associated with early COVID-19 mortality in Mexico. *PLoS One*, 16(7), e0254690. PMID: 34293014

Alidadi, M., & Sharifi, A. (2022). Effects of the built environment and human factors on the spread of COVID-19: A systematic literature review. *The Science of the Total Environment*, 850(158056), 1–13. DOI: 10.1016/j.scitotenv.2022.158056 PMID: 35985590

Aral, N., & Bakır, H. (2022, January). Spatiotemporal analysis of COVID-19 in Turkey. *Sustainable Cities and Society*, 76, 103421. DOI: 10.1016/j.scs.2021.103421 PMID: 34646730

Arun, K., & Özmutlu, S. Y. (2021). Evaluation of Turkey's Logistics Performance Index with a Strategic Perspective. *Nevşehir Hacı Bektaş Veli University Journal of Social Sciences*, 13(3), 1315–1327.

Başer, O. (2021). Population density index and its use for distribution of Covid-19: A case study using Turkish data. *Health Policy (Amsterdam)*, 125(2), 148–154. DOI: 10.1016/j.healthpol.2020.10.003 PMID: 33190934

Bayode, T., Popoola, A., Akogun, O., Siegmund, A., Magidimisha-Chipungu, H., & Ipingbemi, O. (2022). Spatial variability of COVID-19 and its risk factors in Nigeria: A spatial regression method. *Applied Geography (Sevenoaks, England)*, 138(102621), 1–8. DOI: 10.1016/j.apgeog.2021.102621 PMID: 34880507

Bijari, N. B., Mahdinia, M. H., & Daneshvar, M. R. (2021). Investigation of the urbanization contribution to the COVID-19 outbreak in Iran and the MECA countries. *Environment, Development and Sustainability*, 23(12), 17964–17985. DOI: 10.1007/s10668-021-01423-y PMID: 33880075

Bourdin, S., Jeanne, L., Nadou, F., & Noiret, G. (2021, March 3). Does lockdown work? A spatial analysis of the spread and concentration of COVID-19 in Italy. *Sustainable Cities and Society*.

Bowersox, D. J., Closs, D. J., Cooper, M. B., & Bowersox, J. C. (2013). *Supply chain logistics management* (4th ed.). McGraw-Hill Education.

Budak, F., & Korkmaz, Ş. (2020). A General Assessment of the COVID-19 Pandemic Process: The Case of Turkey. [SAYOD]. *Journal of Social Research and Management*, 1, 62–79.

Çelikok, K., & Talih, Ö. (2023). International Transportation Projects and Turkey from the Perspective of Transportation Economics. *Eskişehir Osmangazi University Journal of Social Sciences*, 24(2), 400–423.

Christopher, M. (2016). *Logistics & supply chain management* (5th ed.). Pearson Education Limited.

Ciupa, T., & Suligowski, R. (2021). Green-Blue Spaces and Population Density versus COVID-19 Cases and Deaths in Poland. *International Journal of Environmental Research and Public Health*, 18(6636), 1–17. DOI: 10.3390/ijerph18126636 PMID: 34203108

2020. COVID-19 Pandemic Assessment Report. Ankara: Turkish Academy of Sciences. Retrieved March 1, 2023, from https://www.tuba.gov.tr/files/images/2020/kovidrap

Coyle, J. J., Langley, C. J., Novack, R. A., & Gibson, B. J. (2020). *Supply chain management: A logistics perspective* (10th ed.). Cengage Learning.

Coza, H. (2024). The Impact of the COVID-19 Pandemic on Architectural Space Needs. *Kent Akademisi*, 17(1), 60–71. DOI: 10.35674/kent.1365028

Coza, H., & Njjar, M. S. (2024). The Survivability of Microorganisms on Kitchen Surfaces. *Online Journal of Art and Design.*, 12(1), 148–152.

Cutrini, E., & Salvati, L. (2021, August 18). Unraveling spatial patterns of COVID-19 in Italy: Global forces and local economic drivers. *Regional Science Policy & Practice*, 13(S1), 73–108. DOI: 10.1111/rsp3.12465 PMID: 38607853

Dağlı, E., Aydın, M. M., & Çoruh, E. (2021). Examination of Changes in Traffic Flows Due to COVID-19 Restrictions on Main Arteries in Urban Areas: The Case of Antalya. İDEALKENT, 12(34), 1199-1225

Daisuka, M., Alidadi, M., & Sharifi, A. (2023, October). Tokyo's COVID-19: An urban perspective on factors influencing infection rates in a global city. *Sustainable Cities and Society*, 97, 104743. DOI: 10.1016/j.scs.2023.104743 PMID: 37397232

Das, A., Ghosh, S., Das, K., Basu, T., Dutta, I., & Das, M. (2021, February 31). Living environment matters: Unraveling the spatial clustering of COVID-19 hotspots in Kolkata megacity, India. *Sustainable Cities and Society*, 65, 102577. DOI: 10.1016/j.scs.2020.102577 PMID: 33163331

Desmet, K., & Wacziarg, R. (2022, January). JUE Insight: Understanding spatial variation in COVID-19 across the United States. *Journal of Urban Economics*, 127, 103332. DOI: 10.1016/j.jue.2021.103332 PMID: 33723466

Dintrans, P. V., Castillo, C., Fuente, F. d., & Maddaleno, M. (2021, May 6). COVID-19 incidence and mortality in the Metropolitan Region, Chile: Time, space, and structural factors. *PLoS One*, 16(5), e0250950. PMID: 33945538

Erduman, Y., Eren, O., & Gül, S. (2020). Import content of Turkish production and exports: A sectoral analysis. *Central Bank Review*, 20(4), 155–168. DOI: 10.1016/j.cbrev.2020.07.001

Fielding-Miller, R. K., Sundaram, M. E., & Brouwer, K. (2020, October 14). Social determinants of COVID-19 mortality at the county level. *PLoS One*, 15(10), 1–11. DOI: 10.1371/journal.pone.0240151 PMID: 33052932

Frank, L. D., & Wali, B. (2021, October). Treating two pandemics for the price of one: Chronic and infectious disease impacts of the built and natural environment. *Sustainable Cities and Society*, 73, 103089. DOI: 10.1016/j.scs.2021.103089 PMID: 34155475

Gaglioti, A. H., Li, C., Douglas, M. D., Baltrus, P. T., Blount, M. A., Zahidi, R., Caplan, L. S., Willock, R. J., Fasuyi, O. B., & Mack, D. H. (2021). Population-Level Disparities in COVID-19: Measuring the Independent Association of the Proportion of Black Population on COVID-19 Cases and Deaths in US Counties. *Journal of Public Health Management and Practice*, 27(3), 268–277. DOI: 10.1097/PHH.0000000000001354 PMID: 33762542

Gaisie, E., Oppong-Yeboah, N. Y., & Cobbinah, P. B. (2022, June). Geographies of infections: Built environment and COVID-19 pandemic in metropolitan Melbourne. *Sustainable Cities and Society*, 81, 103838. DOI: 10.1016/j.scs.2022.103838 PMID: 35291308

Glavan, O. L., Nikolić, N., Folić, B., Vitoseviç, B., Mitroviç, A., & Kosanoviç, S. (2022). COVID-19 and City Space: Impact and Perspectives. Sustainability, 14(1885)

Guan, C., Tan, J., Hall, B., Liu, C., Li, Y., & Cai, Z. (2022, March 15). The Effect of the Built Environment on the COVID-19 Pandemic at the Initial Stage: A County-Level Study of the USA. *Sustainability (Basel)*, 14(3417), 1–17. DOI: 10.3390/su14063417

Guoqing, Z., Bin, M., Hui, L., Xin, Z., Min, X., Siyu, C., & Juan, W. (2024). Spatial co-location patterns between early COVID-19 risk and urban facilities: a case study of Wuhan, China. Front. Public Health, 11(1293888.), 1-15.

Ha, J., & Lee, S. (2022, December). Do the determinants of COVID-19 transmission differ by epidemic wave? Evidence from U.S. counties. *Cities (London, England)*, 131, 103892. DOI: 10.1016/j.cities.2022.103892 PMID: 35942406

Hamidi, S., & Hamidi, I. (2021, May). Subway ridership, crowding, or population density: Determinants of COVID-19 infection rates in New York City. *American Journal of Preventive Medicine*, 60(5), 614–620. DOI: 10.1016/j.amepre.2020.11.016 PMID: 33888260

Han, Y., Yang, L., Jia, K., Li, J., Feng, S., Chen, W., Zhao, W., & Pereira, P. (2021). Spatial distribution characteristics of the COVID-19 pandemic in Beijing and its relationship with environmental factors. *The Science of the Total Environment*, 761(144257), 1–11. DOI: 10.1016/j.scitotenv.2020.144257 PMID: 33352341

Hu, M., Roberts, J. D., Azevedo, G. P., & Milner, D. (2021, February). The role of built and social environmental factors in COVID-19 transmission: A look at America's capital city. *Sustainable Cities and Society*, 65, 102580. DOI: 10.1016/j.scs.2020.102580

İçöz, C., & Yenilmez, İ. (2023). Spatio-temporal analysis of the COVID-19 pandemic in Türkiye: Results of the controlled normalization. *Spat. Inf. Res.*, 31(1), 39–50. DOI: 10.1007/s41324-022-00476-z

Jamshidi, S., Baniasad, M., & Niyogi, D. (2020). Global to USA County Scale Analysis of Weather, Urban Density, Mobility, Homestay, and Mask Use on COVID-19. *International Journal of Environmental Research and Public Health*, 17(7847), 1–17. DOI: 10.3390/ijerph17217847 PMID: 33114771

Jo, Y., Hong, A., & Sung, H. (2021). Density or Connectivity: What Are the Main Causes of the Spatial Proliferation of COVID-19 in Korea? *International Journal of Environmental Research and Public Health*, 18(5084), 1–16. DOI: 10.3390/ijerph18105084 PMID: 34065031

Kashem, S. B., Fırıncı, D. M., González, S. R., & Lee, C. A. (2021). Exploring the nexus between social vulnerability, built environment, and the prevalence of COVID-19: A case study of Chicago. *Sustainable Cities and Society*, 75(103261), 1–9. DOI: 10.1016/j.scs.2021.103261 PMID: 34580620

Keni, R., Alexander, A., Ganesh, P. N., Mudgal, J., & Nandakumar, K. (2020). Mayıs 28). COVID-19: Emergence, Spread, Possible Treatments, and Global Burden. *Frontiers in Public Health*, 8(216).

Kim, H., Bell, M. L., & Zanobetti, A. (2021, January). Temporal transition of racial/ethnic disparities in COVID-19 outcomes in 3108 counties of the United States: Three phases from January to December 2020. *The Science of the Total Environment*, 791, 148167. DOI: 10.1016/j.scitotenv.2021.148167 PMID: 34118681

Klompmaker, J. O., Hart, J. E., Holland, I., Sabath, M., Wu, X., Laden, F., & James, P. (2021, August). County-level exposures to greenness and associations with COVID-19 incidence and mortality in the United States. *Environmental Research*, 199, 111331. DOI: 10.1016/j.envres.2021.111331 PMID: 34004166

Kulkarni, H., Khandait, H., Narlawar, U. W., Rathod, P., & Mamtani, M. (2021, April). 10). Independent association of meteorological characteristics with initial spread of Covid-19 in India. *The Science of the Total Environment*, 764(142801), 1–8. PMID: 33148430

Kuschel, K., Carrasco, R., Idrovo-Aguirre, B. J., Duran, C., & Contreras-Reyes, J. E. (2023, August 11). Preparing Cities for Future Pandemics: Unraveling the Influence of Urban and Housing Variables on COVID-19 Incidence in Santiago de Chile. *Health Care*, 11(2259), 1–14. PMID: 37628457

Lata, K., Thapa, K., & Rajput, A. S. (2021). Liveability of Indian Cities and Spread ofCovid-19 Case of Tier-1 Cities. İndian Journal of Public, 67(3), 365-382.

Lekić, O. Z., Nikolić, N., Folić, B. L., & Vitoseviç, B. (2022, February). COVID-19 and City Space: Impact and Perspectives. Sustainability, 14(1885), 1-19.

Li, S., Ma, S., & Zhang, J. (2021, April). Association of built environment attributes with the spread of COVID-19 at its initial stage in China. *Sustainable Cities and Society*, 67, 102752. DOI: 10.1016/j.scs.2021.102752 PMID: 33558840

Li, X., Zhou, L., Jia, T., Peng, R., Fu, X., & Zou, Y. (2020). Associating COVID-19 Severity with Urban Factors: A Case Study of Wuhan. *International Journal of Environmental Research and Public Health*, 17(6712), 1–20. DOI: 10.3390/ijerph17186712 PMID: 32942626

Li, Y., Liu, X., Han, Z., & Dou, J. (2020, January 7). Spatial proximity-based geographically weighted regression model for landslide susceptibility assessment: A case study of Qingchuan area, China. *Applied Sciences (Basel, Switzerland)*, 10(1107), 1107. DOI: 10.3390/app10031107

Lin, D., Sun, Y., Yang, Y., Han, Y., & Xu, C. (2023, January). Urban park use and self-reported physical, mental, and social health during the COVID-19 pandemic: An on-site survey in Beijing, China. *Urban Forestry & Urban Greening*, 79, 127804. DOI: 10.1016/j.ufug.2022.127804 PMID: 36447619

Lin, Q., Paykin, S., Halpern, D., Martinez-Cardoso, A., & Kolak, M. (2022, March 4). Assessment of Structural Barriers and Racial Group Disparities of COVID-19 Mortality With Spatial Analysis. *JAMA Network Open*, 5(3), 1–18. DOI: 10.1001/jamanetworkopen.2022.0984 PMID: 35244703

Mellander, C., & Florida, R. (2022). The geography of COVID-19 in Sweden. *The Annals of Regional Science*, 68(1), 125–150. DOI: 10.1007/s00168-021-01071-0 PMID: 34316091

Mo, C., Tan, D., Mai, T., Bei, C., Qin, J., Pang, W., & Zhang, Z. (2020, Nisan 2). An analysis of spatiotemporal pattern for COIVD-19 in China based on space time cube. Journol of Medıcal Vırology, s. 1-9.

Mollalo, A., Rivera, K. M., & Vahabi, N. (2021, April). Spatial statistical analysis of pre-existing mortalities of 20 diseases with COVID-19 mortalities in the continental United States. *Sustainable Cities and Society*, 67, 102738. DOI: 10.1016/j.scs.2021.102738 PMID: 33532175

Mollalo, A., Vahedi, B., & Rivera, K. M. (2020). GIS-based spatial modeling of COVID-19 incidence rate in the continental United States. *The Science of the Total Environment*, 728(13884), 1–8. DOI: 10.1016/j.scitotenv.2020.138884 PMID: 32335404

Mouratidis, K. (2021, December). How COVID-19 Reshaped Quality of Life in Cities: A Synthesis and Implications for Urban Planning. *Land Use Policy*, 111(105772), 105772. DOI: 10.1016/j.landusepol.2021.105772 PMID: 34566233

Negiz, N., Yavuzçehre, P. S., & Yalçın, Ö. (2021, October 10). Cities and Citizens During the COVID-19 Process: Observations on Experiences and Expectations (Findings from the Lakes Region). *Journal of Hacettepe University Faculty of Economics and Administrative Sciences*, 39, 135–153.

Nguyen, Q. C., Huang, Y., Kumar, A., Duan, H., Keralis, J. M., Dwivedi, P., & Tasdizen, T. (2020). Using 164 Million Google Street View Images to Derive Built Environment Predictors of COVID-19 Cases. *International Journal of Environmental Research and Public Health*, 17(6359), 6359. DOI: 10.3390/ijerph17176359 PMID: 32882867

Niu, D., Wang, L., Li, W., & Ma, Y. (2022). An International Comparative Study on the Resilience of Urban Communities after COVID-19 Pandemic: A One Year Case Study between Lanzhou, China and Sarajevo, Bosnia and Herzegovina. *International Journal of Environmental Research and Public Health*, 19(14458), 14458. DOI: 10.3390/ijerph192114458 PMID: 36361337

Özdede, S., Kalonya, D. H., & Aygün, A. (2021). Rethinking the Need for Urban Green Space per Capita in the Post-Pandemic Period. *Idealkent*, 12, 362–388.

Peng, Z., Wang, R., Liu, L., & Wu, H. (2020, June 19). Exploring Urban Spatial Features of COVID-19 Transmission in Wuhan Based on Social Media Data. *ISPRS International Journal of Geo-Information*, 9(402), 1–14. DOI: 10.3390/ijgi9060402

Perone, G. (2021, February 10). The determinants of COVID-19 case fatality rate (CFR) in the Italian regions and provinces: An analysis of environmental, demographic, and healthcare factors. *The Science of the Total Environment*, 755, 142523. DOI: 10.1016/j.scitotenv.2020.142523 PMID: 33022464

Phiri, D., Salekin, S., Nyirenda, V. R., Simwanda, M., Ranagalage, M., & Murayama, Y. (2021). Spread of COVID-19 in Zambia: An assessment of environmental and socioeconomic factors using a classification tree approach. *Scientific African*, 12(e00827), 1–12. DOI: 10.1016/j.sciaf.2021.e00827 PMID: 34250321

Qiu, J., Li, R., Han, D., Shao, Q., Han, Y., Luo, X., & Wu, Y. (2021, December). A multiplicity of environmental, economic and social factor analyses to understand COVID-19 diffusion. *One Health*, 13, 100335. DOI: 10.1016/j.onehlt.2021.100335 PMID: 34632042

Razavi-Termeh, S. V., Sadeghi-Niaraki, A., Farhangi, F., & Choi, S.-M. (2021, September 14). COVID-19 risk mapping with considering socio-economic criteria using machine learning algorithms. *International Journal of Environmental Research and Public Health*, 18(9657), 9657. DOI: 10.3390/ijerph18189657 PMID: 34574582

Rice, L. (2019). A health map for architecture: The determinants of health and wellbeing in buildings. Designing for Health & Wellbeing: Home, City [within Vernon]. *Society*, •••, 155–184.

Rios, V., & Gianmoena, L. (2021). On the link between temperature and regional COVID-19 severity: Evidence from Italy. *Regional Science Policy & Practice*, 13(S1), 109–137. DOI: 10.1111/rsp3.12472 PMID: 38607900

Ronael, M., & Baycan, T. (2022, September 10). Place-based factors affecting COVID-19 incidences in Turkey. *Asia-Pacific Journal of Regional Science*, 6(3), 1053–1086. DOI: 10.1007/s41685-022-00257-4

Ruck, D. J., Bentley, R. A., & Borycz, J. (2021, May). Early warning of vulnerable counties in a pandemic using socio-economic variables. *Economics and Human Biology*, 41, 100988. DOI: 10.1016/j.ehb.2021.100988 PMID: 33636583

Sabouri, S., Hamidi, S., & Ewing, R. (2020). Does Density Aggravate the COVID-19 Pandemic? *Journal of the American Planning Association*, 86(4), 495–509. DOI: 10.1080/01944363.2020.1777891

Sarmadi, M., Moghanddam, V. K., Dickerson, A. S., & Martelletti, L. (2021, April). Association of COVID-19 distribution with air quality, sociodemographic factors, and comorbidities: An ecological study of US states. *Air Quality, Atmosphere & Health*, 14(4), 455–465. DOI: 10.1007/s11869-020-00949-w PMID: 33078068

Sharifi, A., & Khavarian-Garmsir, A. R. (2020). The COVID-19 pandemic: Impacts on cities and major lessons for urban planning, design, and management. *The Science of the Total Environment*, 749(142391), 2. DOI: 10.1016/j.scitotenv.2020.142391 PMID: 33370924

Sharifi, A., Lak, A., Badr, S., Zali, A., Maher, A., & Mostafavi, E. (2021). Eylül). Spatio-temporal patterns of the COVID-19 pandemic, and place-based influential factors at the neighborhood scale in Tehran. *Sustainable Cities and Society*, 72(103034), 1–13.

Silva, J., & Alves, M. R. (2021). Social inequalities and the pandemic of COVID-19: The case of Rio de Janeiro. *Epidemiology and Community Health*, 75(10), 975–979. DOI: 10.1136/jech-2020-214724 PMID: 33811129

Solomon, M., & Liang, C. (2022, October 2). Human coronaviruses: The emergence of SARS-CoV-2 and management of COVID-19. *Virus Research*, 319(198882), 198882. DOI: 10.1016/j.virusres.2022.198882 PMID: 35934258

Şolt, B. H. (2021, January 5). Changes Experienced in Cities and Urban Services During the COVID-19 Pandemic. *International Journal of Management and Social Research*, 8(15), 39–47.

Spotswood, E. N., Benjamin, M., Stoneburner, L., Wheeler, M. M., Beller, E. E., Balk, D., & McDonald, R. I. (2021, January 11). Nature inequity and higher COVID-19 case rates in less-green neighborhoods in the United States. *Nature Sustainability*, 4(12), 1092–1098. DOI: 10.1038/s41893-021-00781-9

Sridhar, K. S. (2021). Urbanization and COVID-19 Prevalence in India. *Regional Science Policy & Practice*, •••, 1–13.

Tchicaya, A., Lorentz, N., Leduc, K., & Lanchy, G. (2021, September 17). COVID-19 mortality with regard to healthcare services availability, health risks, and socio-spatial factors at department level in France: A spatial cross-sectional analysis. *PLoS One*, 16(9), e0257060. DOI: 10.1371/journal.pone.0256857 PMID: 34534226

Tokey, A. I. (2021). Eylül). Spatial association of mobility and COVID-19 infection rate in the USA: A county-level study using mobile phone location data. *Journal of Transport & Health*, 22(101135). PMID: 34277349

Urban, R. C., & Nakada, L. Y. (2021, April 1). GIS-based spatial modelling of COVID-19 death incidence in São Paulo, Brazil. *Environment and Urbanization*, 33(1), 229–238. DOI: 10.1177/0956247820963962 PMID: 38603029

Verma, R., Yabe, T., & Ukkusuri, S. V. (2021, May 26). Spatiotemporal contact density explains the disparity of COVID-19 spread in urban neighborhoods. *Scientific Reports*, 11(10952), 1–11. DOI: 10.1038/s41598-021-90483-1 PMID: 34040093

Wang, J., Zeng, F., Tang, H., Wang, J., & Xing, L. (2022, January). Correlations between the urban built environmental factors and the spatial distribution at the community level in the reported COVID-19 samples: A case study of Wuhan. *Cities (London, England)*, 129, 103932. DOI: 10.1016/j.cities.2022.103932 PMID: 35975194

Watts, M. J. (2022, September). Macro-level drivers of SARS-CoV-2 transmission: A data-driven analysis of factors contributing to epidemic growth during the first wave of outbreaks in the United States. *Spatial and Spatio-temporal Epidemiology*, 43, 100539. DOI: 10.1016/j.sste.2022.100539 PMID: 36460448

WHO. (2020). Coronavirus Disease 2019 (COVID-19) Situation Report. World Health Organization. Retrieved April 03, 2023, from https://www.who.int/docs/default-source/coronaviruse/situation-reports/20200229-sitrep-40-covid-19.pdf?sfvrsn=849d0665_2

WHO. (March 22, 2023). WHO Coronavirus (COVID-19) Dashboard. Retrieved from WHO Health Emergency Dashboard: https://covid19.who.int/

Wu, X., & Zhang, J. (2021, March 4). Exploration of Temporal-Spatially Varying Impacts on COVID-19 Cumulative Case in Texas Using Geographically Weighted Regression (GWR). Environmental Science and Pollution Reserch, 1-21.

Wu, Y., Hu, Z., Su, M., Xie, L., Zhang, A., Lin, X., & Nie, Y. (2021). Population migration, spread of COVID-19, and epidemic prevention and control: Empirical evidence from China. *BMC Public Health*, 21(529), 1–12. PMID: 33731053

Wu, Y., Hu, Z., Su, M., Xie, L., Zhang, A., Lin, X., & Nie, Y. (2021). Population migration, spread of COVID-19, and epidemic prevention and control: Empirical evidence from China. *BMC Public Health*, 21(529), 1–12. PMID: 33731053

Wu, Y., Xiao, W., & Wei, Y. D. (2022). Neighborhood, built environment and resilience in transportation during the COVID-19 pandemic. Transportation Research Part, 110(103428), 1-15.

Xiao, J., Zhao, J., Luo, Z., Liu, F., & Greenwood, D. (2022, September). The impact of built environment on mental health: A COVID-19 lockdown perspective. *Health & Place*, 77, 102889. DOI: 10.1016/j.healthplace.2022.102889 PMID: 36027740

Xu, G., Jiang, Y., Wang, S., Qin, K., Ding, J., Liu, Y., & Lu, B. (2022, January). Spatial disparities of self-reported COVID-19 cases and influencing factors in Wuhan, China. *Sustainable Cities and Society*, 76, 103785. DOI: 10.1016/j.scs.2021.103485 PMID: 34722132

Yang, Z., Li, J., Huang, X., Zhang, A., Lu, Y., Zhao, X., & Yang, X. (2023). The impact of urban spatial environment on COVID-19: A case study in Beijing. *Frontiers in Public Health*, •••, 1287999. PMID: 38259769

Zhu, L., Liu, X., Huang, H., Avellán-Llaguno, R. D., Lazo, M. M., Gaggero, A., & Zhu, Y. G. (2020, November). Meteorological impact on the COVID-19 pandemic: A study across eight severely affected regions in South America. *The Science of the Total Environment*, 744, 140881. DOI: 10.1016/j.scitotenv.2020.140881 PMID: 32674022

Compilation of References

2020. COVID-19 Pandemic Assessment Report. Ankara: Turkish Academy of Sciences. Retrieved March 1, 2023, from https://www.tuba.gov.tr/files/images/2020/kovidrap

Abduljabbar, R. L., Liyanage, S., & Dia, H. (2021). The Role of Micro-mobility in Shaping Sustainable Cities: A Systematic Literature Review''. *Transportation Research Part D, Transport and Environment*, 92, 102734. DOI: 10.1016/j.trd.2021.102734

Abenoza, R. F., Cats, O., & Susilo, Y. O. (2017). Travel satisfaction with public transport: Determinants, user classes, regional disparities and their evolution. *Transportation Research Part A, Policy and Practice*, 95, 64–84. DOI: 10.1016/j.tra.2016.11.011

Abroshan, H., Luckana, R., & Wheat, D. (2021). Planning a dedicated bus lane to decrease traffic congestion considering policy implementation. In *39th International System Dynamics Conference*.

Abu-Alsaad, H. A., & Al-Taie, R. R. K. (2024). Smart Parking System Using IoT. 2024 16th International Conference on Electronics, Computers and Artificial Intelligence (ECAI), 1–7. DOI: 10.1109/ECAI61503.2024.10607419

Abusaada, H., & Elshater, A. (2021). 'Effect of people on placemaking and affective atmospheres in city streets'. *Ain Shams Engineering Journal*, 12(3), 3389–3403. DOI: 10.1016/j.asej.2021.04.019

Adenekan, O. A., Ezeigweneme, C., & Chukwurah, E. G. (2024). The evolution of smart cities: Integrating technology, governance, and sustainable development. *International Journal of Applied Research in Social Sciences*, 6(5), 891–902. DOI: 10.51594/ijarss.v6i5.1131

Aditya, A., Anwarul, S., Tanwar, R., & Koneru, S. K. V. (2023). An IoT assisted Intelligent Parking System (IPS) for Smart Cities. *Procedia Computer Science*, 218, 1045–1054. DOI: 10.1016/j.procs.2023.01.084

Afifa, K. A., Hussain, N., Ashraf, M. H., & Saleem, M. Z. (2024). Air Pollution and Climate Change as Grand Challenges to Sustainability''. *The Science of the Total Environment*, 928, 172370. Advance online publication. DOI: 10.1016/j.scitotenv.2024.172370 PMID: 38604367

Agbali, M., Trillo, C., Ibrahim, I., Arayıcı, Y., & Fernando, T. (2019). Are smart innovation ecosystems really seeking to meet citizens' needs? insights from the stakeholders' vision on smart city strategy implementation. *Smart Cities*, 2(2), 307–327. DOI: 10.3390/smartcities2020019

Aggarwal, S., & Singh, A. K. (2021). Electric vehicles: The future of the transportation sector: A review. *Renewable & Sustainable Energy Reviews*, 135, 110320.

Agudo-Sierra, E., & Llaguno-Munitxa, M. (2022). Nature-based Solutions for Industrial Logistic Parks. *IOP Conference Series. Earth and Environmental Science*, 1122(1), 012007. DOI: 10.1088/1755-1315/1122/1/012007

Ahadnejad Reveshty, M., Heydari, M. T., & Tahmasebimoghaddam, H. (2023, September). Spatial analysis of the factors impacting the spread of COVID-19 in the neighborhoods of Zanjan, Iran. *Spatial Information Research*, 32(2), 151–164. DOI: 10.1007/s41324-023-00550-0

Ahmed, J., Jaman, M. H., Saha, G., & Ghosh, P. (2021). Effect of environmental and socio-economic factors on the spreading of COVID-19 at 70 cities/provinces. *Heliyon*, 7(5, e06979), 1–8. DOI: 10.1016/j.heliyon.2021.e06979 PMID: 33969235

Ajchariyavanich, C., Limpisthira, T., Chanjarasvichai, N., Jareonwatanan, T., Phongphanpanya, W., Wareechuensuk, S., Srichareonkul, S., Tachatanitanont, S., Ratanamahatana, C., Prompoon, N., & Pipattanasomporn, M. (2019). Park King: An IoT-based Smart Parking System. 2019 IEEE International Smart Cities Conference (ISC2), 729–734. DOI: 10.1109/ISC246665.2019.9071721

Akbar, K. F., Hale, W., & Headley, A. D. (2003). Assessment of scenic beauty of the roadside vegetation in northern England. *Landscape and Urban Planning*, 63(3), 139–144. DOI: 10.1016/S0169-2046(02)00185-8

Akyüz, E. (2021). *''Urban Transportation: Pros and Cons of Elektric Vehicles''*. Hagia Sophia 3. İnternational Conference on Multidisciplinary Scienific Studies, İstanbul, 685-696.

Al Kindi, K. M., Al-Mawali, A., Akharusi, A., Alshukaili, D., Alnasiri, N., Al Awadhi, T., Charabi, Y., & El Kenawy, A. M. (2021). Demographic and socioeconomic determinants of COVID-19 across Oman - A geospatial modelling approach. *Geospatial Health*, 16(985), 145–160. DOI: 10.4081/gh.2021.985 PMID: 34000790

Albalate, D., ve Bel, G. (2010). What shapes local public transportation in Europe? Economics, mobility, institutions, and geography. Transportation Research Part E: Logistics and Transportation Review, 46(5), 775–790. 11 Mayıs 2012 DOI: 10.1016/j.tre.2009.12.010

Aldana, R. R., Gomez-Verjan, J. C., Bello-Chavolla, O. Y., & García Peña, C. (2021, July 21). Spatial epidemiological study of the distribution, clustering, and risk factors associated with early COVID-19 mortality in Mexico. *PLoS One*, 16(7), e0254690. PMID: 34293014

Alidadi, M., & Sharifi, A. (2022). Effects of the built environment and human factors on the spread of COVID-19: A systematic literature review. *The Science of the Total Environment*, 850(158056), 1–13. DOI: 10.1016/j.scitotenv.2022.158056 PMID: 35985590

Alnour, M., Awan, A., & Hossain, M. E. (2024). Towards a green transportation system in Mexico: The Role of Renewable Energy and Transport Public-Private Partnership to Curb Emissions. *Journal of Cleaner Production*, 442, 140984. Advance online publication. DOI: 10.1016/j.jclepro.2024.140984

Alsnih, R., ve Hensher, D. (2003). The mobility and accessibility expectations of seniors in an aging population. Transportation Research Part A: Policy and Practice, 37(10), 903–916. 22 Aralık 2014, DOI: 10.1016/S0965-8564(03)00073-9

Ampuni, A., Fonataba, S., Fitrianto, A., & Wang, G. (2019). Smart Parking System With Automatic Cashier Machine Utilize the IoT Technology. 2019 International Conference on ICT for Smart Society (ICISS), 1–4. DOI: 10.1109/ICISS48059.2019.8969793

Andersson, I. (2016). 'Green cities' going greener? Local environmental policy-making and place branding in the 'Greenest City in Europe.'. *European Planning Studies*, 24(6), 1197–1215. DOI: 10.1080/09654313.2016.1152233

Antonson, H., Mårdh, S., Wiklund, M., & Blomqvist, G. (2009). Effect of surrounding landscape on driving behaviour: A driving simulator study. *Journal of Environmental Psychology*, 29(4), 493–502. DOI: 10.1016/j.jenvp.2009.03.005

Aral, N., & Bakır, H. (2022, January). Spatiotemporal analysis of COVID-19 in Turkey. *Sustainable Cities and Society*, 76, 103421. DOI: 10.1016/j.scs.2021.103421 PMID: 34646730

Ardıçoğlu, R. (2023). Bicycle Network Planning in Urban Transportation: The Case of Elazığ. *Urban Academy*, 16(1), 194–223.

Arshad, K., Hussain, N., Ashraf, M. H., & Saleem, M. Z. (2024). Air pollution and climate change as grand challenges to sustainability. *The Science of the Total Environment*, •••, 172370. PMID: 38604367

Arthurs, P., Gillam, L., Krause, P., Wang, N., Halder, K., & Mouzakitis, A. (2022). A taxonomy and survey of edge cloud computing for intelligent transportation systems and connected vehicles. *IEEE Transactions on Intelligent Transportation Systems*, 23(7), 6206–6221. DOI: 10.1109/TITS.2021.3084396

Arun, K., & Özmutlu, S. Y. (2021). Evaluation of Turkey's Logistics Performance Index with a Strategic Perspective. *Nevşehir Hacı Bektaş Veli University Journal of Social Sciences*, 13(3), 1315–1327.

Ashworth, G., & Kavaratzis, M. (2007). Beyond the logo: Brand management for cities. *Journal of Brand Management*, 16(8), 520–531. DOI: 10.1057/palgrave.bm.2550133

Ayaz, Ç. E. (2023). A Strategic Overview on Urban Mobility Management: A Qualitative Analysis on Istanbul Sustainable Urban Mobility Plan. *Eskişehir Osmangazi University Journal of Social Sciences*, 24(1), 41–55.

Aydın, M. M. (2023). A charging station planning study to prepare for the utilization of electric vehicles on university campuses: The case of Ondokuz Mayıs University. *International Journal of Engineering and Applied Sciences*, 15(4), 184–204. DOI: 10.24107/ijeas.1394018

Aydın, M. M., & Bıyık, B. (2023). Şerit yoğunluk tespit ve bilgilendirme sistemlerinin tasarımdan işletmeye genel altyapısının belirlenmesi: Örnek bir sistem tasarımı. *Gazi University Journal of Science Part C: Design and Technology*, 11(4), 1092–1107.

Aydın, M. M., & Keleş, M. A. (2024). Trafikte acil durum müdahale araçlarının geçişi esnasındaki karmaşa problemleri için AUS tabanlı bir çözüm önerisi geliştirilmesi. *Journal of the Institute of Science and Technology*, 14(1), 294–311.

Aydin, M., Potoglou, D., & Cipcigan, L. (2023). Working principle and performance evolution of camera-based intelligent signalized intersections: Samsun city, Türkiye example. *Scientific Journal of Silesian University of Technology.Series Transport*, 121, 5–17.

Aygün, A., & Torlak, S. E. (2020). An Examination on Climate Change Resilience on Denizli''. *Nigde Omer Halisdemir University Journal of Engineering Sciences*, 9(1), 648–664.

Babalık-Sutcliffe, E. (2009). "Clean Transport: Innovative Solutions to the Creation of a More Sustainable Urban Transport System". Prospects for Research in Transport and Logistics on a Regional: Global Perspective (pp. 89-94). İstanbul: Dogus University Publications.

Badhoutiya, A., & Saxena, A. (2022). SPS: A framework Of Smart Parking System Using IoT Technology. 2022 6th International Conference On Computing, Communication, Control And Automation (ICCUBEA, 1–4. DOI: 10.1109/ICCUBEA54992.2022.10011086

Bahga, A., & Madisetti, V. K. (2013). Cloud-based information technology framework for data driven intelligent transportation systems. *Journal of Transportation Technologies*, 03(02), 131–141. DOI: 10.4236/jtts.2013.32013

Balasubramanian, S., Irulappan, C., & Kitchley, J. L. (2022). 'Aesthetics of urban commercial streets from the perspective of cognitive memory and user behavior in urban environments'. *Frontiers of Architectural Research*, 11(5), 949–962. DOI: 10.1016/j.foar.2022.03.003

Bandung, Y., Aldiansyah, M. F. N., Dwi Putra, M. R., & Syiraaj, M. I. I. (2023). Design and Implementation of IoT-Based Smart Parking System in Campus Area. 2023 10th International Conference on ICT for Smart Society (ICISS), 1–6. DOI: 10.1109/ICISS59129.2023.10292040

Banister, D. (2008). The sustainable mobility paradigm. *Transport Policy*, 15(2), 73–80. DOI: 10.1016/j.tranpol.2007.10.005

Barriga, J. J., Sulca, J., León, J. L., Ulloa, A., Portero, D., Andrade, R., & Yoo, S. G. (2019). Smart Parking: A Literature Review from the Technological Perspective. *Applied Sciences (Basel, Switzerland)*, 21(21), 4569. DOI: 10.3390/app9214569

Baruah, R., Siddique, A. R., & Santhi, H. (2023). Unıfıed payment system for Indıan transportatıon. [IRJET]. *International Research Journal of Engineering and Technology*, 10(5), 1540–1548.

Başer, O. (2021). Population density index and its use for distribution of Covid-19: A case study using Turkish data. *Health Policy (Amsterdam)*, 125(2), 148–154. DOI: 10.1016/j.healthpol.2020.10.003 PMID: 33190934

Bastian, O., Grunewald, K., Syrbe, R.-U., Walz, U., & Wende, W. (2014). Landscape Services: The Concept and Its Practical Relevance. *Landscape Ecology*, 29(9), 1463–1479. DOI: 10.1007/s10980-014-0064-5

Baydar, T. (2007). İnsan ilişkileri ve davranış. 11 Haziran 2013, https://www.istanbul.edu.tr/yuksekokullar/sosyalbilimler/iuwebson/turgutbaydar07guz/davranis.pdf

Bayode, T., Popoola, A., Akogun, O., Siegmund, A., Magidimisha-Chipungu, H., & Ipingbemi, O. (2022). Spatial variability of COVID-19 and its risk factors in Nigeria: A spatial regression method. *Applied Geography (Sevenoaks, England)*, 138(102621), 1–8. DOI: 10.1016/j.apgeog.2021.102621 PMID: 34880507

Benson, R. (2009). Shaping the public sphere: Habermas and beyond.

Bibri, S. (2019). The anatomy of the data-driven smart sustainable city: Instrumentation, datafication, computerization and related applications. *Journal of Big Data*, 6(1), 59. Advance online publication. DOI: 10.1186/s40537-019-0221-4

Bibri, S. E., Krogstie, J., & Kärrholm, M. (2020). "Compact city planning and development: Emerging practices and strategies for achieving the goals of sustainability". *Developments in the Built Environment*, 4(20), 1–20. DOI: 10.1016/j.dibe.2020.100021

Bibri, S., Alahi, A., Sharifi, A., & Krogstie, J. (2023). Environmentally sustainable smart cities and their converging ai, iot, and big data technologies and solutions: An integrated approach to an extensive literature review. *Energy Informatics*, 6(1), 9. Advance online publication. DOI: 10.1186/s42162-023-00259-2 PMID: 37032812

Bifulco, F., Amitrano, C., & Tregua, M. (2014). Driving smartization through intelligent transport. *The China Business Review*, 13(4), 243.

Bijari, N. B., Mahdinia, M. H., & Daneshvar, M. R. (2021). Investigation of the urbanization contribution to the COVID-19 outbreak in Iran and the MECA countries. *Environment, Development and Sustainability*, 23(12), 17964–17985. DOI: 10.1007/s10668-021-01423-y PMID: 33880075

Billones, R., Guillermo, M., Lucas, K., Era, M., Dadios, E., & Fillone, A. (2021). Smart region mobility framework. *Sustainability (Basel)*, 13(11), 6366. DOI: 10.3390/su13116366

Bitam, S., & Mellouk, A. (2012). ITS-cloud: Cloud computing for intelligent transportation system. *2012 IEEE Global Communications Conference (GLOBECOM)*. https://doi.org/DOI: 10.1109/GLOCOM.2012.6503418

Bıyık, Y., & Civelekoğlu, G. (2018). Investigation of Carbon Footprint Change Originated from Transportation Sector''. *Bilge International Journal of Science and Technology Research*, 2(2), 157–166.

Bi, Z., Kan, T., Mi, C. C., Zhang, Y., Zhao, Z., & Keoleian, G. A. (2016). A review of wireless power transfer for electric vehicles: Prospects to enhance sustainable mobility. *Applied Energy*, 179, 413–425. DOI: 10.1016/j.apenergy.2016.07.003

Bjertnæs, G. H. (2019). Efficient taxation of fuel and road use. *CESifo Working Paper* No. 8019

Björklund, M. (2011). Influence From The Business Environment on Environmental Purchasing -Drivers and Hinders of Purchasing Green Transportation Services. *Journal of Purchasing and Supply Management*, 17(1), 11–22. DOI: 10.1016/j.pursup.2010.04.002

Black, W. R. (2010). *Sustainable transportation: problems and solutions*. Guilford Press.

Blumentrath, C., & Tveit, M. S. (2014). "Visual characteristics of roads: A literature review of people's perception and Norwegian design practice," Transportation Research Part A. *Policy and Practice, Elsevier*, 59(C), 58–71.

Boarnet, M. G. (2010). Planning, climate change, and transportation: Thoughts on policy analysis. *Transportation Research Part A, Policy and Practice*, 44(8), 587–595. DOI: 10.1016/j.tra.2010.03.001

Boerema, S. T., Van Velsen, L., Vollenbroek-Hutten, M. M. R., ve Hermens, H. J. (2016). Value-based design for the elderly: An application in the field of mobility aids. Assistive Technology, 29(2), 1–9. 25 Haziran 2013, DOI: 10.1080/10400435.2016.1208303

Bolund, P., & Hunhammar, S. (1999). 'Ecosystem services in urban areas'. *Ecological Economics*, 29(2), 293–301. DOI: 10.1016/S0921-8009(99)00013-0

Bourdin, S., Jeanne, L., Nadou, F., & Noiret, G. (2021, March 3). Does lockdown work? A spatial analysis of the spread and concentration of COVID-19 in Italy. *Sustainable Cities and Society*.

Bowersox, D. J., Closs, D. J., Cooper, M. B., & Bowersox, J. C. (2013). *Supply chain logistics management* (4th ed.). McGraw-Hill Education.

Braiterman, J. (2011). City Branding through New Green Spaces. In *Palgrave Macmillan UK eBooks* (pp. 70–81). https://doi.org/DOI: 10.1057/9780230294790_9

Brusaw, S. (2018). Solar Roadways: The future of roads. [Online] Available at: https://www.solarroadways.com/

Buchanan, B. G. (2005). A (very) brief history of artificial intelligence. *AI Magazine*, 26(4), 53–53.

Budak, F., & Korkmaz, Ş. (2020). A General Assessment of the COVID-19 Pandemic Process: The Case of Turkey. [SAYOD]. *Journal of Social Research and Management*, 1, 62–79.

Bulkeley, H., Schroeder, H., Janda, K., Zhao, J., Armstrong, A., Chu, S. Y., & Ghosh, S. (2009). ''Cities and Climate Change: The role of İnstitutions, Governance and Urban Planning''.*5th Urban Research Symposium: Cities and Climate Change*, Marseille, France, 1-92.

Buosi, M., Cilloni, M., Corradi, A., Rolt, C., Dias, J., Foschini, L., Montanari, R., & Zito, P. (2018). A crowdsensing campaign and data analytics for assisting urban mobility pattern determination. In *Proceedings of the 2018 IEEE Symposium on Computers and Communications (ISCC)* (pp. 00224-00229), Natal, Brazil. DOI: 10.1109/ISCC.2018.8538483

Burden, D. (2006). *22 Benefits of Urban Street Trees*. Glatting Jackson and Walkable Communities, Inc.

Burns, P. C. (1999). Navigation and the mobility of older drivers. The Journals of Gerontology Series B: Psychological Sciences and Social Sciences, 54B(1), S49–S55. 24 Ocak 2018, DOI: 10.1093/geronb/54B.1.S49

Çalışkan, D. Z., Kurt, Ü., & Timur, M. C. (2017). ''Econometric Analysis of Climate Change and Transportation Sector Relation: The Case of Turkey''.*International Congress of Energy, Economy and Policy*, İstanbul.

Calzolari, T., Genovese, A., & Brint, A. (2022). Circular economy indicators for supply chains: A systematic literature review. *Journal of Cleaner Production*, 351, 131564.

Camata, S., Sartori, S., Bettini, M., & Conti, D. (2022). Innovation and sustainable development of smart cities: A systematic review to identify research trends and gaps. [RISUS]. *Journal on Innovation and Sustainability*, 13(4), 32–44. DOI: 10.23925/2179-3565.2022v13i4p32-44

Carmona, M., Heath, T., Oc, T., & Tiesdell, S. (2010). *Public Places Urban Spaces:The Dimensions of Urban Design*. Published Elseiver Ltd.

Cartiere, C., & Willis, S. (2008). *The Practice of Public Art*. Routledge. DOI: 10.4324/9780203926673

Çelikok, K., & Talih, Ö. (2023). International Transportation Projects and Turkey from the Perspective of Transportation Economics. *Eskişehir Osmangazi University Journal of Social Sciences*, 24(2), 400–423.

Çelikyay, H. H., & Bayraktar, H. K. (2023). *"Climate Friendly Cities and Resilience Strategies in Climate Policy Papers"*. HUMANITAS –. *The International Journal of Social Sciences (Islamabad)*, 11, 87–107.

Čepeliauskaitė, G., Keppner, B., Simkute, Z., Stasiškienė, Ž., Leuser, L., Kalniņa, I., Kotoviča, N., Andiņš, J., & Muiste, M. (2021). Smart-mobility services for climate mitigation in urban areas: Case studies of Baltic countries and Germany. *Sustainability (Basel)*, 13(8), 4127. DOI: 10.3390/su13084127

Chan, C., & Marafa, L. M. (2014b). Rebranding Hong Kong "Green": The potential for connecting city branding with green resources. *World Leisure Journal*, 56(1), 62–80. DOI: 10.1080/04419057.2013.876587

Chang, Y.C. (2013). Factors affecting airport access mode choice for elderly air passengers. Transportation Research Part E: Logistics and Transportation Review, 57, 105–112. 9 Eylül 2014, DOI: 10.1016/j.tre.2013.01.010

Chapman, R., Howden-Chapman, P., & Capon, A. (2016). Understanding the systemic nature of cities to improve health and climate change mitigation. *Environment International*, 94, 380–387. DOI: 10.1016/j.envint.2016.04.014 PMID: 27126780

Chatfield, A. T., & Reddick, C. G. (2016). Smart city implementation through shared vision of social innovation for environmental sustainability: A case study of Kitakyushu, Japan. *Social Science Computer Review*, 34(6), 757–773. DOI: 10.1177/0894439315611085

Choo, S., Sohn, D., ve Park, M. (2016). Mobility characteristics of the elderly: A case for Seoul Metropolitan Area. KSCE Journal of Civil Engineering, 20(3), 1023–1031. 20 Şubat 2018, DOI: 10.1007/s12205-016-0651-x

Chourabi, H., Nam, T., Walker, S., Gil-García, J. R., Mellouli, S., Nahon, K., Pardo, T. A., & Scholl, H. J. (2012). Understanding smart cities: An integrative framework. In *Proceedings of the 2012 45th Hawaii International Conference on System Sciences* (pp. 2289-2297), Maui, HI, United States. DOI: 10.1109/HICSS.2012.615

Christopher, M. (2016). *Logistics & supply chain management* (5th ed.). Pearson Education Limited.

Ciupa, T., & Suligowski, R. (2021). Green-Blue Spaces and Population Density versus COVID-19 Cases and Deaths in Poland. *International Journal of Environmental Research and Public Health*, 18(6636), 1–17. DOI: 10.3390/ijerph18126636 PMID: 34203108

Clairand, J.-M., Guerra-Terán, P., Serrano-Guerrero, X., González-Rodríguez, M., & Escrivá-Escrivá, G. (2019). Electric vehicles for public transportation in power systems: A review of methodologies.

Cleave, E., & Arku, G. (2020). Immigrant attraction through place branding? Evidence of city-level effectiveness from Canada's London. *Cities (London, England)*, 97, 102502. DOI: 10.1016/j.cities.2019.102502

Colas Group. (2016). Wattway: The solar road. [Online] Available at: https://www.wattwaybycolas.com/en/

Conlon, M., Foote, P., O'Malley, K., & Stuart, D. (2001). Successful arterial street limited-stop express bus service in Chicago. *Transportation Research Record: Journal of the Transportation Research Board*, 1760(1), 74–80. DOI: 10.3141/1760-10

Connors, B. T. (2019). Do we still need to pay?. *The PayTech Book: The Payment Technology Handbook for Investors, Entrepreneurs and FinTech Visionaries*, 21-22.

Copaja-Alegre, M., & Esponda-Alva, C. (2019). Tecnología e innovación hacia la ciudad inteligente. avances, perspectivas y desafíos. *Bitácora Urbano Territorial*, 29(2), 59–70. DOI: 10.15446/bitacora.v29n2.68333

Corbett, J. (2001). Torsten Hägerstrand: Time geography. CSISS Classics. UC Santa Barbara: Center for Spatially Integrated Social Science. 24 Haziran 2018, https://escholarship.org/uc/item/2t75b8sj

Çoruh, E., Aydın, M. M., Dağlı, E., Öztürk, Ö. F., & Yıldırım, M. S. (2020). Kapalı otobüs durağı uygulamalarının kullanıcı beklentileri ve memnuniyeti doğrultusunda incelenmesi: Gümüşhane ve Antalya örneği. *Gümüşhane Üniversitesi Fen Bilimleri Dergisi*, 10(4), 1027–1038.

Coutts, G., & Jokela, T. (2008). *Art, Community and Environment: EducationalPerspectives*. Intellect Books. DOI: 10.2307/j.ctv36xw54v

Coyle, J. J., Langley, C. J., Novack, R. A., & Gibson, B. J. (2020). *Supply chain management: A logistics perspective* (10th ed.). Cengage Learning.

Coza, H. (2024). The Impact of the COVID-19 Pandemic on Architectural Space Needs. *Kent Akademisi*, 17(1), 60–71. DOI: 10.35674/kent.1365028

Coza, H., & Njjar, M. S. (2024). The Survivability of Microorganisms on Kitchen Surfaces. *Online Journal of Art and Design.*, 12(1), 148–152.

Crankshaw, N. (2009). *Creating Vibrant Public Spaces; Streetscape Design in Commercial and Historic Districts*. Islandpress.

Çubukçu, K. M. (2015). *Planlama ve coğrafyada temel istatistik ve mekansal istatistik.* Nobel Akademik Yayıncılık Eğitim Danışmanlık Tic. Ltd. Şti.

Cullen, I. (1976). Reviews. *Lund Studies in Geography*, 44, 961–972.

Cutrini, E., & Salvati, L. (2021, August 18). Unraveling spatial patterns of COVID-19 in Italy: Global forces and local economic drivers. *Regional Science Policy & Practice*, 13(S1), 73–108. DOI: 10.1111/rsp3.12465 PMID: 38607853

Dağlı, E., Aydın, M. M., & Çoruh, E. (2021). Examination of Changes in Traffic Flows Due to COVID-19 Restrictions on Main Arteries in Urban Areas: The Case of Antalya. İDEALKENT, 12(34), 1199-1225

Dağlı, E., Büber, M., & Taspınar, Y. S. (2022). Detection of accident situation by machine learning methods using traffic announcements: The case of metropol Istanbul. *International Journal Of Applied Mathematics Electronics And Computers*, 10(3), 61–67. DOI: 10.18100/ijamec.1145293

Daisuka, M., Alidadi, M., & Sharifi, A. (2023, October). Tokyo's COVID-19: An urban perspective on factors influencing infection rates in a global city. *Sustainable Cities and Society*, 97, 104743. DOI: 10.1016/j.scs.2023.104743 PMID: 37397232

Daldanise, G. (2020). From Place-Branding to Community-Branding: A Collaborative Decision-Making Process for cultural heritage Enhancement. *Sustainability (Basel)*, 12(24), 10399. DOI: 10.3390/su122410399

Dankovtsev, V. I., Toutova, N. V., Vorozhtsov, A. S., & Andreev, I. A. (2022). Multicriteria application placement in fog computing. *2022 Intelligent Technologies and Electronic Devices in Vehicle and Road Transport Complex (TIRVED), 31 15*, 1-4. https://doi.org/DOI: 10.1109/TIRVED56496.2022.9965545

Das, A., Ghosh, S., Das, K., Basu, T., Dutta, I., & Das, M. (2021, February 31). Living environment matters: Unraveling the spatial clustering of COVID-19 hotspots in Kolkata megacity, India. *Sustainable Cities and Society*, 65, 102577. DOI: 10.1016/j.scs.2020.102577 PMID: 33163331

Dastgerdi, A. S., & De Luca, G. (2019). Strengthening the city's reputation in the age of cities: An insight in the city branding theory. *City. Territory and Architecture*, 6(1), 2. Advance online publication. DOI: 10.1186/s40410-019-0101-4

DBSA. (2021). "How Green Transport Can be a Catalyst for Economic Development and Growth *[online]*". (18.07.2024), https://www.dbsa.org/article/how-green-transport-can-be-catalyst-economic-development-and-growth

De Oña, J., Estévez, E., & De Oña, R. (2020). Perception of public transport quality of service among regular private vehicle users in Madrid, Spain. *Transportation Research Record: Journal of the Transportation Research Board*, 2674(2), 213–224. DOI: 10.1177/0361198120907095

De Vos, J., Waygood, E. O. D., & Letarte, L. (2020). Modeling the desire for using public transport. *Travel Behaviour & Society*, 19, 90–98. DOI: 10.1016/j.tbs.2019.12.005

Deakin, E. (2001). Sustainable development and sustainable transportation: Strategies for economic prosperity, environmental quality, and equity.

Deakin, M.; Mitchell, G.; Nijkamp, P.; Vreeker, R. (2007) Sustainable UrbanDevelopment: Volume 2; The Environmental Assessment Methods, Routledge Press; p365

Dee, C. (2001). *Form and Fabric in Landscape Architecture*. Taylor & Francis.

Del Castillo, J. M., & Benitez, F. G. (2013). Determining a public transport satisfaction index from user surveys. *Transportmetrica A: Transport Science*, 9(8), 713–741. DOI: 10.1080/18128602.2011.654139

Demirbaş, M., & Aydın, R. (2020). The Biggest Threat of the 21st Century: Global Climate Change. *Ecological Life Sciences*, 15(4), 163–179.

Desmet, K., & Wacziarg, R. (2022, January). JUE Insight: Understanding spatial variation in COVID-19 across the United States. *Journal of Urban Economics*, 127, 103332. DOI: 10.1016/j.jue.2021.103332 PMID: 33723466

Development Bureau - Urban Renewal Strategy Review. (n.d.). Wanchai DAS final report. Retrieved from https://www.ursreview.gov.hk/eng/doc/Wanchai%20DAS%20final%20report_080410.pdf

Development Bureau. (2007, December 20). Press Release: Secretary for Development talks about Wan Chai old area revitalization project [:]https://www.devb.gov.hk/tc/publications_and_press_releases/press/index_id_3836.html

Dhar, S., & Shukla, P. R. (2015). Low Carbon Scenarios for Transport in India: Co-benefits Analysis. *Energy Policy*, 81, 186–198. DOI: 10.1016/j.enpol.2014.11.026

Dheeven, T. A., Kumar, P. M., Venkatesh, V., & Sailaja, K. A. I. (2024). IoT based sensor enabled vehicle parking system. *Measurement. Sensors*, 100953, 100953. Advance online publication. DOI: 10.1016/j.measen.2023.100953

Dickerson, A. E., Molnar, L. J., Eby, D. W., Adler, G., Bedard, M., Berg-Weger, M.,Trujillo, L. (2007). Transportation and aging: A research agenda for advancing safe mobility. The Gerontologist, 47(5), 578–590. 25 Ocak 2018, DOI: 10.1093/geront/47.5.578

Diez-Jimenez, E., Fernadez-Llorca, D., Munoz, M. F., & Sotelo, M. (2020). Personal Rapid Transport System Compatible With Current Railways and Metros Infrastructure. *IEEE Transactions on Intelligent Transportation Systems*, (99), 1–11.

Dimitrova, V. (2024). Education in the era of smart cities: transformation and opportunities. *Environment Technology Resources Proceedings of the International Scientific and Practical Conference*, 2, 352-357. DOI: 10.17770/etr2024vol2.8084

Ding, A., Huang, X., & Fu, C. (2017). Air Pollution and Weather Interaction in East Asia''. *Environmental Sciences (Ruse)*. Advance online publication. DOI: 10.1093/acrefore/9780199389414.013.536

Dintrans, P. V., Castillo, C., Fuente, F. d., & Maddaleno, M. (2021, May 6). COVID-19 incidence and mortality in the Metropolitan Region, Chile: Time, space, and structural factors. *PLoS One*, 16(5), e0250950. PMID: 33945538

Dixon, T., Montgomery, J., Horton-Baker, N., & Farrell, L. (2018). Using urban foresight techniques in city visioning: Lessons from the Reading 2050 vision. *Local Economy*, 33(8), 777–799. DOI: 10.1177/0269094218800677

Doğan, E. (2022). Placement and sizing strategies for dynamic wireless charging stations on signalized intersection corridors. *Pamukkale Üniversitesi Mühendislik Bilimleri Dergisi*, 28(6), 802–811.

Doğan, S., & Tüzer, M. (2011). Combating Global Climate Change: General Approaches and International Efforts''. *Journal of Sociology Conferences*, 44, 157–194.

Drobne, S., & Bogataj, M. (2015). Optimal allocation of public service centers in the central places of functional regions. *IFAC-PapersOnLine*, 48(3), 2362–2367. DOI: 10.1016/j.ifacol.2015.06.441

Drottenborg, H., (1999). Aesthetics and Safety in Traffic Environments.

Dupre, K. (2019). Trends and gaps in place-making in the context of urban development and tourism. *Journal of Place Management and Development*, 12(1), 102–120. DOI: 10.1108/JPMD-07-2017-0072

Duvarcı, Y. (2024). Drawing a Vision for More Sustainable Mobility: the Example of PRT System. *International Symposium on Sustainable Urban Mobility through Modular and Integrated Transport System*, 28 March 2024, Nagoya University, Nagoya.

Duvarcı, Y., & Akpinar, F. (2023). Using Personal Rapid Transit as an Effective Transport Solution in Historical Downtown Areas: A Case from Historic Kemeraltı, Izmir. *Planlama*, 33(1), 86–104.

Dzhambov, A., Hartig, T., Markevych, I., Tilov, B., & Dimitrova, D. (2018). 'Urban residential greenspace and mental health in youth: Different approaches to testing multiple pathways yield different conclusions'. *Environmental Research*, 160, 47–59. DOI: 10.1016/j.envres.2017.09.015 PMID: 28961469

Eames, M., Hunta, M., Dixon, T., & Britnell, J. (2013). Retrofit city futures: Visions for urban sustainability.

Eckhoff, D., & Wagner, I. (2018). Privacy in the smart city-applications, technologies, challenges, and solutions. *IEEE Communications Surveys and Tutorials*, 20(1), 489–516. DOI: 10.1109/COMST.2017.2748998

Egebjerg, U. (2002). *Beautiful Roads - A Handbook of Road Architecture*. Danish Road Directorate.

Elassy, M., Al-Hattab, M., Takruri, M., & Badawi, S. (2024). Intelligent transportation systems for sustainable smart cities. *Transportation Engineering*, 100252.

Electric Power Research Institute. (2021). The Electric Vehicle Maintenance Advantage. https://www.epri.com/research/products/3002020211

El-Geneidy, A. M., & Surprenant-Legault, J. (2010). Limited-stop bus service: An evaluation of an implementation strategy. *Public Transport (Berlin)*, 2(4), 291–306. DOI: 10.1007/s12469-010-0030-9

Elker, C. (1999). Çağdaş uaşım politikaları. II. Ulaşım ve Trafik Kongresi- Sergisi, 175–184. 8 Şubat 2016, http://arsiv.mmo.org.tr/pdf/11164.pdf

Enerji Atlası. (n.d.). Akaryakıt tüketimi. Enerji Atlası. Retrieved 15 January 2024, https://www.enerjiatlasi.com/akaryakit-tuketimi/

Environmental Protection Department. (2018, November 2). *Solutions for Pollution Problems*https://www.epd.gov.hk/epd/english/greenrestaurant/solutions/solutions_water.html

Erçetin, C. (2023). Environmental Sustainability and Cycling as a Transport Mode: Best Pracices". *Environment, City and Climate Journal.*, 2(3), 238–269.

Ercolano, J. M. (2011). Limited-stop bus operations: An evaluation. *Transportation Research Record: Journal of the Transportation Research Board*, (994), 24–29.

Erdoğan, S. (2020). *''Energy, Environment and Greenhouse Gas''*. Çankırı Karatekin University Faculty of Economics and Administrative Sciences Journal, 10(1), 277-303.

Erduman, Y., Eren, O., & Gül, S. (2020). Import content of Turkish production and exports: A sectoral analysis. *Central Bank Review*, 20(4), 155–168. DOI: 10.1016/j.cbrev.2020.07.001

Eroğlu, F. (2017). *Davranış bilimleri (15.Baskı)*. Beta Yayın Evi.

Erwee, C., Chipungu, L., & Magidimisha-Chipungu, H. (2021). Rebranding Umhlanga as an intelligent city. In *Springer eBooks* (pp. 391–437). https://doi.org/DOI: 10.1007/978-3-030-69698-6_77

Etukudoh, E. A., Queen, Z., Nwokediegwu, S., Umoh, A. A., Ibekwe, K. I., & Adefemi, A. (2024). Solar power integration in urban areas: A review of design innovations and efficiency enhancements. *World Journal of Advanced Research and Reviews*, 21(1), 1383–1394. DOI: 10.30574/wjarr.2024.21.1.0168

European Commission. (2017). Clean transport, Urban transport. http://ec. europa.eu/transport/themes/urban/urban_mobility/urban_mobility_actions/public_transport_en. Accessed 27th January 2017.

European Commission. (2020). Sustainable and Smart Mobility Strategy – putting European transport on track for the future. Retrieved from https://eur-lex.europa.eu/legal-content/EN/TXT/?uri=CELEX:52020DC0789

European Parliament. (2024). Circular economy: Definition, importance and benefits. Retrieved September 19, 2024, from https://www.europarl.europa.eu/topics/en/topic/circular-economy

European Telecommunications Standards Institute (2010). Intelligent transport systems (ITS): communications architecture, v1.1.1, European Standard (Telecommunications series), September 2010

Fagnant, D. J., & Kockelman, K. (2015). Preparing a nation for autonomous vehicles: Opportunities, barriers and policy recommendations. *Transportation Research Part A, Policy and Practice*, 77, 167–181. DOI: 10.1016/j.tra.2015.04.003

Febrer, N., Folkvord, F., & Lupiañez-Villanueva, F. (2021). Cost-effectiveness assessment of internet of things in smart cities. *Frontiers in Digital Health*, 3, 662874. Advance online publication. DOI: 10.3389/fdgth.2021.662874 PMID: 34713138

Feldman, O. (2012). The GEH measure and quality of the highway assignment models. *Association for European Transport and Contributors*, 1-18.

Fernandes, E., Salgado, A., & Bernardino, J. (2020). Big data streaming platforms to support real-time analytics. *Proceedings of the 15th International Conference on Software Technologies*, 426-433. https://doi.org/DOI: 10.5220/0009817304260433

Fielding-Miller, R. K., Sundaram, M. E., & Brouwer, K. (2020, October 14). Social determinants of COVID-19 mortality at the county level. *PLoS One*, 15(10), 1–11. DOI: 10.1371/journal.pone.0240151 PMID: 33052932

Flink, A. C.; Olka K.; Searns, R. M. (2001)Trails for the Twenty-First Century:Planning, Design, and Management Manual for Multi-Use Trails, Island Press;p88,89.

Florek, M., Hereźniak, M., & Augustyn, A. (2021b). Measuring the effectiveness of city brand strategy. In search for a universal evaluative framework. *Cities (London, England)*, 110, 103079. DOI: 10.1016/j.cities.2020.103079

Fox, G. C., Kamburugamuve, S., & Hartman, R. D. (2012). Architecture and measured characteristics of a cloud based internet of things. *2012 International Conference on Collaboration Technologies and Systems (CTS)*, 6–12. DOI: 10.1109/CTS.2012.6261020

Fraifer, M., & Fernstrom, M. (2016). Smart car parking system prototype utilizing CCTV nodes: A proof of concept prototype of a novel approach towards IoT-concept based smart parking. 2016 IEEE 3rd World Forum on Internet of Things (WF-IoT), 649–654. DOI: 10.1109/WF-IoT.2016.7845458

Frank, L. D., & Wali, B. (2021, October). Treating two pandemics for the price of one: Chronic and infectious disease impacts of the built and natural environment. *Sustainable Cities and Society*, 73, 103089. DOI: 10.1016/j.scs.2021.103089 PMID: 34155475

Gagan, S., & Ravi, H. K. (2024). ParkNest - The Smart Parking System Using IoT. *2024 International Conference on Smart Systems for Applications in Electrical Sciences (ICSSES)*, 1–6. DOI: 10.1109/ICSSES62373.2024.10561362

Gaglioti, A. H., Li, C., Douglas, M. D., Baltrus, P. T., Blount, M. A., Zahidi, R., Caplan, L. S., Willock, R. J., Fasuyi, O. B., & Mack, D. H. (2021). Population-Level Disparities in COVID-19: Measuring the Independent Association of the Proportion of Black Population on COVID-19 Cases and Deaths in US Counties. *Journal of Public Health Management and Practice*, 27(3), 268–277. DOI: 10.1097/PHH.0000000000001354 PMID: 33762542

Gaisie, E., Oppong-Yeboah, N. Y., & Cobbinah, P. B. (2022, June). Geographies of infections: Built environment and COVID-19 pandemic in metropolitan Melbourne. *Sustainable Cities and Society*, 81, 103838. DOI: 10.1016/j.scs.2022.103838 PMID: 35291308

Gandhi, M. (2023). https://pub.aimind.so/adaptive-traffic-control-systems-a-comprehensive-review-part-2-64a0a48a985e

García-Ayllón, S., & Miralles, J. (2015). New strategies to improve governance in territorial management: Evolving from "Smart Cities" to "Smart Territories". *Procedia Engineering*, 118, 3–11. DOI: 10.1016/j.proeng.2015.08.396

Geng, J., Cao, J., Jia, H., Zhu, Z., Fang, H., Gao, C., Ji, C., Jia, G., Han, G., & Zhou, X. (2024). Heter-train: A distributed training framework based on semi-asynchronous parallel mechanism for heterogeneous intelligent transportation systems. *IEEE Transactions on Intelligent Transportation Systems*, 25(1), 959–972. DOI: 10.1109/TITS.2023.3286400

Gergin, E. D. (2024). Cities Resistant to Climate Change: Examples of Good Local Governance Practice in The World and In Turkey''. *Erzurum Technical University Social Sciences Institute Journal*, 18, 94–111.

Gerosa, A., & Tartari, M. (2021). The bottom-up place branding of a neighborhood: Analyzing a case of selective empowerment. *Space and Culture*, 120633122110323. Advance online publication. DOI: 10.1177/12063312211032355

Giuliano, G., & Hu, H. ve Lee, K. (2003). Travel patterns of the elderly: The role of land use. 24 Ocak 2018, https://www.metrans.org/sites/default/files/research-project/00-08_Final_0_0.pdf

Glavan, O. L., Nikolić, N., Folić, B., Vitoseviç, B., Mitroviç, A., & Kosanoviç, S. (2022). COVID-19 and City Space: Impact and Perspectives. Sustainability, 14(1885)

Global Carbon Project. (2022). ''*Carbon dioxide emissions 2022 [online]*''. (15.08.2024). https://www.globalcarbonproject.org/

Global Street Design Guide. (2014). National Association of City Transportation Officials.

Goodall, W., Dovey, T., Bornstein, J., & Bonthron, B. (2017). The rise of mobility as a service. Deloitte Review. https://www2.deloitte.com/us/en/insights/focus/future-of-mobility/mobility-as-a-service.html

Google. (2015). The Google geocoding API. 1 Mart 2017, https://developers.google.com/maps/documentation/geocoding/

Google. (2017). Distance matrix API. 1 Mart 2017, https://developers.google.com/maps/documentation/distance-matrix/start

Gotzler, F., Neumann, F., & Adenaw, L. (2022). Assessment of personal rapid transit system configurations regarding efficiency and service quality. *Future Transportation*, 2(3), 3. Advance online publication. DOI: 10.3390/futuretransp2030041

Grabski, W., & Daszczuk, W. B. (2017). A study on cooperation of urban transport means: PRT and Light Rail. *Transport Problems*, 12(4), 5–14. DOI: 10.20858/tp.2017.12.4.1

Gramsci, A. (1997). *Hapishane defterleri (Prison Notebooks)*. Belge Yayınları.

Grzybowska, H., Wijayaratna, K., Shafiei, S., Amini, N., & Travis Waller, S. (2022). Ramp metering strategy implementation: A case study review. *Journal of Transportation Engineering. Part A, Systems*, 148(5), 03122002. DOI: 10.1061/JTEPBS.0000641

Guan, C., Tan, J., Hall, B., Liu, C., Li, Y., & Cai, Z. (2022, March 15). The Effect of the Built Environment on the COVID-19 Pandemic at the Initial Stage: A County-Level Study of the USA. *Sustainability (Basel)*, 14(3417), 1–17. DOI: 10.3390/su14063417

Guoqing, Z., Bin, M., Hui, L., Xin, Z., Min, X., Siyu, C., & Juan, W. (2024). Spatial co-location patterns between early COVID-19 risk and urban facilities: a case study of Wuhan, China. Front. Public Health, 11(1293888.), 1-15.

Hacıoğlu, E., Eren, İ., & Polatoğlu, Ç. (2021). Enhancing socio-spatial activities in urban areas through urban furniture: Case of Bagdat Avenue-Istanbul. In *Contemporary Issues in Architecture and Urban Planning* (pp. 70–95). Architectural & Urban Forms.

Hagerstrand, T. (1970). What about people in regional science. Papers Of The Regional Science Association, 24, 7–21. 24 Haziran 2014, https://link.springer.com/content/pdf/10.1007/BF01936872.pdf

Ha, J., & Lee, S. (2022, December). Do the determinants of COVID-19 transmission differ by epidemic wave? Evidence from U.S. counties. *Cities (London, England)*, 131, 103892. DOI: 10.1016/j.cities.2022.103892 PMID: 35942406

Hakak, S., Gadekallu, T. R., Maddikunta, P. K. R., Ramu, S. P., Parimala, M., De Alwis, C., & Liyanage, M. (2023). Autonomous Vehicles in 5G and beyond: A Survey. *Vehicular Communications*, 39, 100551. DOI: 10.1016/j.vehcom.2022.100551

Hamidi, S., & Hamidi, I. (2021, May). Subway ridership, crowding, or population density: Determinants of COVID-19 infection rates in New York City. *American Journal of Preventive Medicine*, 60(5), 614–620. DOI: 10.1016/j.amepre.2020.11.016 PMID: 33888260

Han, Y., Yang, L., Jia, K., Li, J., Feng, S., Chen, W., Zhao, W., & Pereira, P. (2021). Spatial distribution characteristics of the COVID-19 pandemic in Beijing and its relationship with environmental factors. *The Science of the Total Environment*, 761(144257), 1–11. DOI: 10.1016/j.scitotenv.2020.144257 PMID: 33352341

Hasan, M. H., & Van Hentenryck, P. (2021). The benefits of autonomous vehicles for community-based trip sharing.

Hasan, U., Whyte, A., & Jassmi, H. A. (2021). Public bus transport service satisfaction: Understanding its value to urban passengers towards improved uptake. *Transactions on Transport Sciences*, 12(1), 25–37. DOI: 10.5507/tots.2021.002

Heidrich, O., Reckien, D., Olazabal, M., Foley, A., Salvia, M., de Gregorio Hurtado, S., Orru, H., Flacke, J., Geneletti, D., Pietrapertosa, F., Hamann, J. P., Tiwary, A., Feliu, E., & Dawson, R. J. (2016). National climate policies across Europe and their impacts on cities strategies''. *Journal of Environmental Management*, 168, 36–45. DOI: 10.1016/j.jenvman.2015.11.043 PMID: 26696604

He, J., Zeng, Z., & Li, Z. (2010). Benefit evaluation framework of ıntelligent transportation systems. *Journal of Transportation Systems Engineering and Information Technology*, 10(1), 81–87. DOI: 10.1016/S1570-6672(09)60025-8

Héran, F., & Ravalet, E. P. J. (2008). *La consommation d'espace-temps des divers modes de déplacement en milieu urbain: Application au cas de l'île de France.* PREDIT.

Herrera-Quintero, L. F., Vega-Alfonso, J., Bermudez, D., Marentes, L. A., & Banse, K. (2019). ITS for Smart Parking Systems, towards the creation of smart city services using IoT and cloud approaches. *2019 Smart City Symposium Prague (SCSP)*, 1–7. DOI: 10.1109/SCSP.2019.8805705

Herstein, R., & Berger, R. (2013). Much more than sports: sports events as stimuli for city rebranding. *Journal of Business Strategy/˜the œJournal of Business Strategy, 34*(2), 38–44. https://doi.org/DOI: 10.1108/02756661311310440

Hess, D. B. (2009). Access to public transit and its influence on ridership for older adults in two U. S. Cities. Journal of Transport and LandUse 2, 2(1), 3–27. 2 Mayıs 2013, https://www.jtlu.org/index.php/jtlu/article/view/11

Hightower, A., Ziedan, A., Crossland, C., & Brakewood, C. (2022). Current Practices and potential rider benefits of fare capping policies in the USA. *Transportation Research Record: Journal of the Transportation Research Board*, 2676(10), 376–390. DOI: 10.1177/03611981221089572

Holden, E., Linnerud, K., & Banister, D. (2013). Sustainable passenger transport: Back to Brundtland''. *Transportation Research Part A, Policy and Practice*, 54, 67–77. DOI: 10.1016/j.tra.2013.07.012

Hossain, M. S., Kumar, L., Islam, M. M., & Selvaraj, J. (2022). A comprehensive review on the integration of electric vehicles for sustainable development. Wiley Online Library. https://onlinelibrary.wiley.com/doi/full/10.1155/2022/3868388

Hu, X., Wang, J., ve Wang, L. (2013). Understanding the travel behavior of elderly people in the developing country: a case study of Changchun, China. Procedia - Social and Behavioral Sciences, 96(Cictp), 873–880. 4 Ocak 2015, DOI: 10.1016/j.sbspro.2013.08.099

Huang, R., Chang, B., Tsai, Y., & Liang, Y. (2017). Mobile edge computing-based vehicular cloud of cooperative adaptive driving for Platooning autonomous self driving. *2017 IEEE 7th International Symposium on Cloud and Service Computing (SC2)*. https://doi.org/DOI: 10.1109/SC2.2017.13

Hu, H., Zha, X., Li, Z., & Lv, R. (2022). Preparation and performance study of solar pavement panel based on transparent resin-concrete. *Construction & Building Materials*, 340, 127871.

Hu, M., Roberts, J. D., Azevedo, G. P., & Milner, D. (2021, February). The role of built and social environmental factors in COVID-19 transmission: A look at America's capital city. *Sustainable Cities and Society*, 65, 102580. DOI: 10.1016/j.scs.2020.102580

Hürriyet. (n.d.). Motorin fiyatları. Bigpara. Retrieved 15 January 2024, https://bigpara.hurriyet.com.tr/akaryakit-fiyatlari/motorin/

Hussain, B., Zefreh, M. M., & Torok, A. (2018). Designing the appropriate data collection method for public transport passenger satisfaction analysis. *IJTTE. International Journal for Traffic and Transport Engineering*, 8(2), 177–183. DOI: 10.7708/ijtte.2018.8(2).03

Hyder, Z. (2023, June 26). All about Solar Roadways: The promise versus the reality. SolarReviews. https://www.solarreviews.com/blog/all-about-solar-roadways

İçöz, C., & Yenilmez, İ. (2023). Spatio-temporal analysis of the COVID-19 pandemic in Türkiye: Results of the controlled normalization. *Spat. Inf. Res.*, 31(1), 39–50. DOI: 10.1007/s41324-022-00476-z

IEA. (2021). *''Global Energy Review 2021 [online]''*. (07.07.2024), https://www.iea.org/reports/global-energy-review-2021

IEA. (2022). *''Uses of Energy' [online]'*. (10.07.2024), https://www.iea.org/energy-system

INRIX. (2024). INRIX 2023 Traffic Scorecard Report. https://inrix.com/scorecard-city-2023/?city=Istanbul&index=5

Insch, A., & Bowden, B. (2016). Possibilities and limits of brand repositioning for a second-ranked city: The case of Brisbane, Australia's "New World City", 1979–2013. *Cities (London, England)*, 56, 47–54. DOI: 10.1016/j.cities.2016.03.003

International Energy Agency. (2020). Global EV Outlook 2020. https://www.iea.org/reports/global-ev-outlook-2020

Ipingbemi, O. (2010). Travel characteristics and mobility constraints of the elderly in Ibadan, Nigeria. Journal of Transport Geography, 18(2), 285–291. 25 Ocak 2018, https://doi.org/DOI: 10.1016/j.jtrangeo.2009.05.011

Irnazarov, F., & Kayumova, M. (2017). Toward smart city development in Central Asia: A comparative assessment. *Central Asian Affairs*, 4(1), 51–82. DOI: 10.1163/22142290-00401003

Isa, N., Mohamed, A., & Yusoff, M. (2015). Implementation of dynamic traffic routing for traffic congestion: A review. In *Soft Computing in Data Science: First International Conference, SCDS 2015, Putrajaya, Malaysia, September 2-3, 2015* [Springer Singapore.]. *Proceedings*, 1, 174–186.

Istanbul Metropolitan Municipality. (2024). İstanbul akıllı şehir stratejik planı, projeler [Istanbul smart city strategic plan, projects]. Istanbul Metropolitan Municipality. https://akillisehir.istanbul/tr/projects

İzmir Büyükşehir Belediyesi. (2018). İzmir ulaşım ana planı saha araştırmaları bilgi notu (2). 31 Ocak 2018, https://www.izmir.bel.tr/YuklenenDosyalar/file/HIZMETREHBERI/idari/Upi_SahaArastirma.pdf

Ja'afar, N. H., Sulaiman, A. B., & Shamsuddin, S. (2012) The Contribution of Landscape Features on Traditional Streets in Malaysia,Procedia - Social and Behavioral Sciences,Volume 50,Pages 643-656,ISSN 1877-0428.

Jabbar, W. A., Tiew, L. Y., & Ali Shah, N. Y. (2024). Internet of things enabled parking management system using long range wide area network for smart city. Internet of Things and Cyber-Physical Systems, 82–98. DOI: 10.1016/j.iotcps.2023.09.001

Jaiswal, K. K., Roy Chowdhury, C., Yadav, D., Verma, R., Dutta, S., Jaiswal, K. S., Ba, S., & Karuppasamy, K. S. K. (2022). Renewable and sustainable clean energy development and impact on social, economic, and environmental health. *Energy Nexus*, 7, 100118. DOI: 10.1016/j.nexus.2022.100118

Jamshidi, S., Baniasad, M., & Niyogi, D. (2020). Global to USA County Scale Analysis of Weather, Urban Density, Mobility, Homestay, and Mask Use on COVID-19. *International Journal of Environmental Research and Public Health*, 17(7847), 1–17. DOI: 10.3390/ijerph17217847 PMID: 33114771

Jara-Díaz, S. R., & Gschwender, A. (2009). The effect of financial constraints on the optimal design of public transport services. *Transportation*, 36(1), 65–75. DOI: 10.1007/s11116-008-9182-8

Jara-Díaz, S., Fielbaum, A., & Gschwender, A. (2017). Optimal fleet size, frequencies and vehicle capacities considering peak and off-peak periods in public transport. *Transportation Research Part A, Policy and Practice*, 106, 65–74. DOI: 10.1016/j.tra.2017.09.005

Jen, W., Tu, R., & Lu, T. (2011). Managing passenger behavioral intention: An integrated framework for service quality, satisfaction, perceived value and switching barriers. *Transportation*, 38(3), 321–342. DOI: 10.1007/s11116-010-9306-9

Ji, Y. G. (2012). *Advances in Affective and Pleasurable Design (Advances in HumanFactors and Ergonomics Series)*. CRC Press.

Joshi, M., Vaidya, A., & Deshmukh, M. (2018). Sustainable transport solutions for the concept of smart city. In Gautam, A., De, S., Dhar, A., Gupta, J. G., & Pandey, A. (Eds.), *Sustainable energy and transportation: Technologies and policies* (pp. 21–42). Springer., DOI: 10.1007/978-981-10-7509-4_3

Jo, Y., Hong, A., & Sung, H. (2021). Density or Connectivity: What Are the Main Causes of the Spatial Proliferation of COVID-19 in Korea? *International Journal of Environmental Research and Public Health*, 18(5084), 1–16. DOI: 10.3390/ijerph18105084 PMID: 34065031

Juster, R., & Schonfeld, P. (2013). Comparative Analysis of Personal Rapid Transit as an Urban Transportation Mode. *Transportation Research Record: Journal of the Transportation Research Board*, 2350(December), 128–135. DOI: 10.3141/2350-15

Kabakuş, N., & Eyüboğlu, M. (2023). *''Use of Environmentally Friendly Transportation Vehicles''*. In 3 th International Congress of Engineering and Natural Sciences, Ankara 258-270.

Kaffash, S., Nguyen, A. T., & Zhu, J. (2021). Big data algorithms and applications in intelligent transportation system: A review and bibliometric analysis. *International Journal of Production Economics*, 231, 107868. DOI: 10.1016/j.ijpe.2020.107868

Kalašová, A., Harantová, V., & Čulík, K. (2019). Public transport as a part of shared economy. The Archives of Automotive Engineering – Archiwum Motoryzacji, 85(3), 49-56. DOI: 10.14669/AM.VOL85.ART4

Kaplan, H., & Kaynak, Z. (2011). *''Green transportation types and regulations in urban transportation: an analysis in terms of authority and responsibilities of local governments''*. 2nd Public Transportation Week Transist, IV. Transportation Symposium and Exhibition, Haliç Congress Center, Istanbul, 01-02 December, 160-162.

Karaçoban, E. (2023). Akıllı kentler ve bölgesel kalkınma: TR32 Bölgesi kalkınma stratejilerini yeniden düşünmek [Smart cities and regional development: rethinking TR32 regional development strategies] [Unpublished MSc thesis]. Pamukkale University, Denizli, Turkey.

Karakaya, E., Akkoyun, G., & Hiçyılmaz, B. (2023). *''Carbon Pricing for Greenhouse Gas Emission Reduction: Carbon Tax or Emission Trading?''*. Journal of Research in Economics. *Politics & Finance*, 8(4), 813–841.

Kashem, S. B., Fırıncı, D. M., González, S. R., & Lee, C. A. (2021). Exploring the nexus between social vulnerability, built environment, and the prevalence of COVID-19: A case study of Chicago. *Sustainable Cities and Society*, 75(103261), 1–9. DOI: 10.1016/j.scs.2021.103261 PMID: 34580620

Kavaratzis, M. (2004). From city marketing to city branding: Towards a theoretical framework for developing city brands. *Place Branding*, 1(1), 58–73. DOI: 10.1057/palgrave.pb.5990005

Kavaratzis, M., & Hatch, M. J. (2013). The dynamics of place brands. *Marketing Theory*, 13(1), 69–86. DOI: 10.1177/1470593112467268

Kazi, S., Bagasrawala, M., Shaikh, F., & Sayyed, A. (2018). Smart e-ticketing system for public transport bus. *International Conference on Smart City and Emerging Technology (ICSCET)* (pp. 1-7). IEEE. DOI: 10.1109/ICSCET.2018.8537302

Keleş, R. (1998). *Dictionary of Urban Science Terms* (2nd ed.). İmge Bookstore Publications.

Keni, R., Alexander, A., Ganesh, P. N., Mudgal, J., & Nandakumar, K. (2020). Mayıs 28). COVID-19: Emergence, Spread, Possible Treatments, and Global Burden. *Frontiers in Public Health*, 8(216).

Khazraeian, S., & Hadi, M. (2019). Intelligent transportation systems in future smart cities. *Sustainable Interdependent Networks II: From Smart Power Grids to Intelligent Transportation Networks*, 109-120.

Khondaker, B., & Kattan, L. (2015). Variable speed limit: An overview. *Transportation Letters*, 7(5), 264–278. DOI: 10.1179/1942787514Y.0000000053

Kim, H., Bell, M. L., & Zanobetti, A. (2021, January). Temporal transition of racial/ethnic disparities in COVID-19 outcomes in 3108 counties of the United States: Three phases from January to December 2020. *The Science of the Total Environment*, 791, 148167. DOI: 10.1016/j.scitotenv.2021.148167 PMID: 34118681

Kim, T. H., Ramos, C., & Mohammed, S. (2017). Smart city and IoT. *Future Generation Computer Systems*, 76, 159–162. DOI: 10.1016/j.future.2017.03.034

Klompmaker, J. O., Hart, J. E., Holland, I., Sabath, M., Wu, X., Laden, F., & James, P. (2021, August). County-level exposures to greenness and associations with COVID-19 incidence and mortality in the United States. *Environmental Research*, 199, 111331. DOI: 10.1016/j.envres.2021.111331 PMID: 34004166

Kocagöz, E., & İğde, Ç. S. (2022). Which Factors Affect the Intention to Buy an Electric Vehicle? A Consumer Study''. *Journal of Social Sciences of Kahramanmaraş Sütçü İmam University*, 19, 104–120.

Korol, O., Shushunova, N., Lopatkin, D., Zanin, A., & Шушунова, T. (2018). Application of high-tech solutions in ecodevelopment. Matec Web of Conferences, 251, 06002. DOI: 10.1051/matecconf/201825106002

KPMG. (2021). The Total Cost of Ownership for Electric Vehicles. https://home.kpmg/xx/en/home/insights/2021/04/total-cost-of-ownership-for-evs.html

Kuhail, M. A., Ahmad, B., & Rottinghaus, C. (2018). Smart resident: a personalized transportation guidance system. In *Proceedings of the 2018 IEEE 5th International Congress on Information Science and Technology (CiSt)* (pp. 547-551), Marrakech, Morocco. DOI: 10.1109/CIST.2018.8596632

Kulkarni, H., Khandait, H., Narlawar, U. W., Rathod, P., & Mamtani, M. (2021, April). 10). Independent association of meteorological characteristics with initial spread of Covid-19 in India. *The Science of the Total Environment*, 764(142801), 1–8. PMID: 33148430

Kuschel, K., Carrasco, R., Idrovo-Aguirre, B. J., Duran, C., & Contreras-Reyes, J. E. (2023, August 11). Preparing Cities for Future Pandemics: Unraveling the Influence of Urban and Housing Variables on COVID-19 Incidence in Santiago de Chile. *Health Care*, 11(2259), 1–14. PMID: 37628457

Kwan, S. C., & Hashim, J. H. (2016). A review on co-benefits of mass public transportation in climate change mitigation. *Sustainable Cities and Society*, 22, 11–18. DOI: 10.1016/j.scs.2016.01.004

Lacmanović, I., Radulović, B., & Lacmanović, D. (2010). Contactless payment systems based on RFID technology. *In The 33rd International Convention MIPRO* (pp. 1114-1119). IEEE.

Lai, W. T., & Chen, C. F. (2011). Behavioral intentions of public transit passengers: The roles of service quality, perceived value, satisfaction and involvement. *Transport Policy*, 18(2), 318–325. DOI: 10.1016/j.tranpol.2010.09.003

Lata, K., Thapa, K., & Rajput, A. S. (2021). Liveability of Indian Cities and Spread ofCovid-19 Case of Tier-1 Cities. İndian Journal of Public, 67(3), 365-382.

Le Dantec, C. A., Asad, M., Misra, A., & Watkins, K. E. (2015). Planning with crowdsourced data: rhetoric and representation in transportation planning. *In Proceedings of the 18th ACM conference on computer supported cooperative work & social computing* (pp. 1717-1727).

Lee, W. L. (2019, February 13). Illegal driving in Wan Chai causes congestion. District councilors advocate strengthening video law enforcement [ ]. *Hk01.com*https://www.hk01.com/18 /293958/ - - #

Lekić, O. Z., Nikolić, N., Folić, B. L., & Vitoseviç, B. (2022, February). COVID-19 and City Space: Impact and Perspectives. Sustainability, 14(1885), 1-19.

Levine, J., ve Garb, Y. (2002). Congestion pricing's conditional promise: promotion of accessibility or mobility? Transport Policy, 9(3), 179–188. October 1 Ocak 2018, DOI: 10.1016/S0967-070X(02)00007-0

Li, H. R. (2016). Study on Green Transportation System of International Metropolises''. *Procedia Engineering*, 137, 762–771. DOI: 10.1016/j.proeng.2016.01.314

Li, H., Chen, Y., Li, K., Wang, C., & Chen, B. (2023). Transportation internet: A sustainable solution for intelligent transportation systems. *IEEE Transactions on Intelligent Transportation Systems*, 24(12), 15818–15829. DOI: 10.1109/TITS.2023.3270749

Li, J. Q., Song, M. K., Li, M., & Zhang, W. B. (2009). Planning for bus rapid transit in single dedicated bus lane. *Transportation Research Record: Journal of the Transportation Research Board*, 2111(1), 76–82. DOI: 10.3141/2111-10

Lin, D., Sun, Y., Yang, Y., Han, Y., & Xu, C. (2023, January). Urban park use and self-reported physical, mental, and social health during the COVID-19 pandemic: An on-site survey in Beijing, China. *Urban Forestry & Urban Greening*, 79, 127804. DOI: 10.1016/j.ufug.2022.127804 PMID: 36447619

Lin, D., Yan, M., Kong, L., Quan, R., & Guan, Y. L. (2024). A framework of real-time intelligent transportation system based on hybrid fog-cloud computing. *IEEE Communications Magazine*, 62(1), 126–132. DOI: 10.1109/MCOM.013.2200616

Lin, Q., Paykin, S., Halpern, D., Martinez-Cardoso, A., & Kolak, M. (2022, March 4). Assessment of Structural Barriers and Racial Group Disparities of COVID-19 Mortality With Spatial Analysis. *JAMA Network Open*, 5(3), 1–18. DOI: 10.1001/jamanetworkopen.2022.0984 PMID: 35244703

Li, S., Ma, S., & Zhang, J. (2021, April). Association of built environment attributes with the spread of COVID-19 at its initial stage in China. *Sustainable Cities and Society*, 67, 102752. DOI: 10.1016/j.scs.2021.102752 PMID: 33558840

Li, S., & Mi, C. C. (2014). Wireless power transfer for electric vehicle applications. *IEEE Transactions on Power Electronics*, 29(7), 3453–3461.

Litman, T. A. (2016). *Evaluating Transportation Land Use Impacts: Considering the Impacts, Benefits and Costs of Different Land Use Development Patterns* (Vol. 1). Victoria Transport Policy Institute.

Liu, H., Skandalos, N., Braslina, L., Kapsalis, V., & Karamanis, D. (2023). Integrating solar energy and nature-based solutions for climate-neutral urban environments. *Solar*, 3(3), 382–415. DOI: 10.3390/solar3030022

Liu, L., Feng, J., Mu, X., Pei, Q., Lan, D., & Xiao, M. (2023). Asynchronous deep reinforcement learning for collaborative task computing and on-demand resource allocation in vehicular edge computing. *IEEE Transactions on Intelligent Transportation Systems*, 24(12), 15513–15526. DOI: 10.1109/TITS.2023.3249745

Li, X., Zhou, L., Jia, T., Peng, R., Fu, X., & Zou, Y. (2020). Associating COVID-19 Severity with Urban Factors: A Case Study of Wuhan. *International Journal of Environmental Research and Public Health*, 17(6712), 1–20. DOI: 10.3390/ijerph17186712 PMID: 32942626

Li, Y., Liu, X., Han, Z., & Dou, J. (2020, January 7). Spatial proximity-based geographically weighted regression model for landslide susceptibility assessment: A case study of Qingchuan area, China. *Applied Sciences (Basel, Switzerland)*, 10(1107), 1107. DOI: 10.3390/app10031107

Lobaccaro, G., Croce, S., Lindkvist, C., Munari Probst, M. C., Scognamiglio, A., Dahlberg, J., Lundgren, M., & Wall, M. (2019). A cross-country perspective on solar energy in urban planning: Lessons learned from international case studies. *Renewable & Sustainable Energy Reviews*, 108, 209–237. DOI: 10.1016/j.rser.2019.03.041

Lorenz, A., Madeja, N., & Leyh, C. (2023). A framework for assessing the sustainability impact of intelligent transport systems in the smart city context. In *Proceedings of the 18th Conference on Computer Science and Intelligence Systems (FedCSIS)* (pp. 161-169), Warsaw, Poland. DOI: 10.15439/2023F6002

Louafi, S., Abdou, S., & Reiter, S. (2017). Effect of vegetation cover on thermal and visual comfort of pedestrians in urban spaces in hot and dry climate. *Nature & Technology*, (17), 30B.

Lu, R., Lin, X., Zhu, H., & Shen, X. (2009). SPARK: A New VANET-Based Smart Parking Scheme for Large Parking Lots. *IEEE INFOCOM*, 2009, 1413–1421. Advance online publication. DOI: 10.1109/INFCOM.2009.5062057

Macdonald, E. (2005). Street-facing Dwelling Units and Livability: The Impacts of Emerging Building Types in Vancouver ' s NewHigh-density Residential Neighbourhoods. *Journal of Urban Design*, 10(1), 13–38. DOI: 10.1080/13574800500062320

Martinez, L. M., Pritchard, J. P., & Crist, P. (2024). Shared mobility's role in sustainable mobility: *Past, Present, and Future.Annual Review of Environment and Resources*, •••, 49. DOI: 10.1146/annurev-environ-121522-041916

Mascia, M., Hu, S., Han, K., Sun, A., North, R., & Lees-Miller, J. D. (2016). A holistic approach for performance assessment of personal rapid transit. *Research in Transportation Business & Management*, 18, 70–76. DOI: 10.1016/j.rtbm.2016.01.008

Ma, T., Yang, H., Gu, W., Li, Z., & Yan, S. (2019). Development of walkable photovoltaic floor tiles used for pavement. *Renewable Energy*, 132, 1143–1152.

Mathew, S., Pulugurtha, S. S., & Mane, A. S. (2020). Effect of toll roads on travel time reliability within its vicinity: A case study from the state of North Carolina. *Transportation Letters*, 12(9), 604–612. DOI: 10.1080/19427867.2019.1671043

Matijosaitiene, I., Stankevičė, I., & Ribeiro, J. (2014). Road Landscape as a Product: Does it Satisfy Consumers' Aesthetic Needs? *The Baltic Journal of Road and Bridge Engineering*, 9(4), 297–305. DOI: 10.3846/bjrbe.2014.36

Ma, X., Chau, C. K., & Lai, J. H. K. (2021). 'Critical factors influencing the comfort evaluation for recreational walking in urban street environments'. *Cities (London, England)*, 116, 103286. DOI: 10.1016/j.cities.2021.103286

May, T., & Crass, M. (2007). Sustainability in Transport: Implications for Policy Makers. *Transportation Research Record: Journal of the Transportation Research Board*, 2017(1), 1–9. DOI: 10.3141/2017-01

Medeiros, T. C., Soares, E., & Campos, C. A. (2022). Elasticity of intelligent transportation system applications in a fog computing environment. *2022 International Wireless Communications and Mobile Computing (IWCMC)*, 560-565. https://doi.org/DOI: 10.1109/IWCMC55113.2022.9824638

Medeiros, T. C., Soares, E., & Vieira Campos, C. A. (2021). An intelligent transportation system application using mobile edge computing. *2021 IEEE Symposium on Computers and Communications (ISCC)*, 1-6. https://doi.org/DOI: 10.1109/ISCC53001.2021.9631498

Mellander, C., & Florida, R. (2022). The geography of COVID-19 in Sweden. *The Annals of Regional Science*, 68(1), 125–150. DOI: 10.1007/s00168-021-01071-0 PMID: 34316091

Mendiratta, S., Dey, D., & Rani Sona, D. (2017). Automatic car parking system with visual indicator along with IoT. *2017 International Conference on Microelectronic Devices, Circuits and Systems (ICMDCS)*, 1–3. DOI: 10.1109/ICMDCS.2017.8211549

Mercado, R., ve Páez, A. (2009). Determinants of distance traveled with a focus on the elderly: a multilevel analysis in the Hamilton CMA, Canada. Journal of Transport Geography, 17(1), 65–76. 2 Eylül 2012, DOI: 10.1016/j.jtrangeo.2008.04.012

Metz, D. H. (2000). Mobility of older people and their quality of life. Transport Policy, 7(2), 149–152. 5 Eylül 2014, DOI: 10.1016/S0967-070X(00)00004-4

Metz, D. H. (2003). Transport policy for an ageing population. Transport Reviews, 23(4), 375–386. 21 Nisan 2014, DOI: 10.1080/0144164032000048573

Miller, J. S., Hoel, L. a., Goswami, A., ve Ulmer, J. M. (2006). Assessing the utility of private information in transportation planning studies: A case study of trip generation analysis. Socio-Economic Planning Sciences, 40(2), 94–118. 1 Haziran 2012, DOI: 10.1016/j.seps.2004.10.003

Miller, D., Merrilees, B., & Yakimova, R. (2013). Corporate Rebranding: An integrative review of major enablers and barriers to the rebranding process. *International Journal of Management Reviews*, 16(3), 265–289. DOI: 10.1111/ijmr.12020

Mittelman, G., Kariv, Y., Cohen, Y., & Avineri, E. (2022). Techno-economic analysis of energy supply to personal rapid transit (PRT) systems. *Journal of Cleaner Production*, 324, 129031.

Mo, C., Tan, D., Mai, T., Bei, C., Qin, J., Pang, W., & Zhang, Z. (2020, Nisan 2). An analysis of spatiotemporal pattern for COIVD-19 in China based on space time cube. Journol of Medıcal Vırology, s. 1-9.

Mogaji, E., & Nguyen, N. P. (2024). Evaluating the emergence of contactless digital payment technology for transportation. *Technological Forecasting and Social Change*, 203, 123378. DOI: 10.1016/j.techfore.2024.123378

Mogge, J. (2009). The technology of Personal Rapid Transit. Retrieved from iTrans Advanced Transit

Mohammadzadeh, Z. (2023). Smart city healthcare delivery innovations: a systematic review of essential technologies and indicators for developing nations. DOI: 10.21203/rs.3.rs-3297004/v1

Mohapatra, H. (2021). Socio-technical Challenges in the Implementation of Smart City. 2021 International Conference on Innovation and Intelligence for Informatics, Computing, and Technologies (3ICT), 57–62. DOI: 10.1109/3ICT53449.2021.9581905

Mohapatra, H., & Dalai, A. K. (2022). IoT based V2I framework for accident prevention. *2022 2nd International Conference on Artificial Intelligence and Signal Processing (AISP), 0*, 1-4. https://doi.org/DOI: 10.1109/AISP53593.2022.9760623

Mohapatra, H., Rath, A. K., & Panda, N. (2022). IoT infrastructure for the accident avoidance: An approach of smart transportation. *International Journal of Information Technology : an Official Journal of Bharati Vidyapeeth's Institute of Computer Applications and Management*, 14(2), 761–768. DOI: 10.1007/s41870-022-00872-6

Mokonyama, M., & Venter, C. (2013). Incorporation of customer satisfaction in public transport contracts–A preliminary analysis. *Research in Transportation Economics*, 39(1), 58–66. DOI: 10.1016/j.retrec.2012.05.024

Molina, M. (2020). What is an intelligent system? *arXiv preprint arXiv*:2009.09083.

Mollalo, A., Rivera, K. M., & Vahabi, N. (2021, April). Spatial statistical analysis of pre-existing mortalities of 20 diseases with COVID-19 mortalities in the continental United States. *Sustainable Cities and Society*, 67, 102738. DOI: 10.1016/j.scs.2021.102738 PMID: 33532175

Mollalo, A., Vahedi, B., & Rivera, K. M. (2020). GIS-based spatial modeling of COVID-19 incidence rate in the continental United States. *The Science of the Total Environment*, 728(13884), 1–8. DOI: 10.1016/j.scitotenv.2020.138884 PMID: 32335404

Mondale, T. (2000). Transportation—a major player in smart growth. [Institute of Transportation Engineers]. *ITE Journal*, 70(11), 39–43.

Montella, A., Chiaradonna, S., Mihiel, A. C. D. S., Lovegrove, G., Nunziante, P., & Rella Riccardi, M. (2022). 'Sustainable complete streets design criteria and case study in Naples, Italy'. *Sustainability (Basel)*, 14(20), 13142. DOI: 10.3390/su142013142

Morelli, J. (2011). Environmental Sustainability: A Definition for Environmental Professionals. *Journal of Environmental Sustainability*, 1(1), 1–10. DOI: 10.14448/jes.01.0002

Mouratidis, K. (2021, December). How COVID-19 Reshaped Quality of Life in Cities: A Synthesis and Implications for Urban Planning. *Land Use Policy*, 111(105772), 105772. DOI: 10.1016/j.landusepol.2021.105772 PMID: 34566233

Mouwen, A. (2015). Drivers of customer satisfaction with public transport services. *Transportation Research Part A, Policy and Practice*, 78, 1–20. DOI: 10.1016/j.tra.2015.05.005

Mugion, R. G., Toni, M., Raharjo, H., Di Pietro, L., & Sebathu, S. P. (2018). Does the service quality of urban public transport enhance sustainable mobility? *Journal of Cleaner Production*, 174, 1566–1587. DOI: 10.1016/j.jclepro.2017.11.052

Muller, P. (2011). *PRT consulting, sustainable city enabled by personal rapid transit.* Retrieved from Sustainable City Enabled by Personal Rapid Transit.

Muniba, M., & Yu, B. (2023). Does innovative city pilot policy stimulate the chinese regional innovation: An application of did model. *International Journal of Environmental Research and Public Health*, 20(2), 1245. DOI: 10.3390/ijerph20021245 PMID: 36673993

Musa, W. (2017) The impact of smart city initiatives on cities' local economic development [Unpublished MSc thesis]. Fort Hays State University, Hays, KS, United States.

Naderi, J.,R. (2003), 'Landscape Design in Clear Zone: Effect of Landscape Variables on Pedestrian Health and Driver Safety', *Transportation Research Record Journal of the Transportation Research Board 1851(1851):119-130*

Nasiboglu, E., Kuvvetli, U., & Ozkilcik, M. ve Eliiyi, U. (2012). Origin-destination matrix generation using smart card data: Case study for Izmir. 2012 IV International Conference "Problems of Cybernetics and Informatics" (PCI), 1–4. 15 Haziran 2014, DOI: 10.1109/ICPCI.2012.6486315

Negiz, N., Yavuzçehre, P. S., & Yalçın, Ö. (2021, October 10). Cities and Citizens During the COVID-19 Process: Observations on Experiences and Expectations (Findings from the Lakes Region). *Journal of Hacettepe University Faculty of Economics and Administrative Sciences*, 39, 135–153.

Neirotti, P., Marco, A. D., Cagliano, A. C., Mangano, G., & Scorrano, F. (2014). Current trends in Smart City initiatives: Some stylised facts. *Cities (London, England)*, 38, 25–36. DOI: 10.1016/j.cities.2013.12.010

Newbold, K. B., Scott, D. M., Spinney, J. E. L., Kanaroglou, P., ve Páez, A. (2005). Travel behavior within Canada's older population: A cohort analysis. Journal of Transport Geography, 13(4), 340–351. 10 Mayıs 2015, DOI: 10.1016/j.jtrangeo.2004.07.007

Nguyen, Q. C., Huang, Y., Kumar, A., Duan, H., Keralis, J. M., Dwivedi, P., & Tasdizen, T. (2020). Using 164 Million Google Street View Images to Derive Built Environment Predictors of COVID-19 Cases. *International Journal of Environmental Research and Public Health*, 17(6359), 6359. DOI: 10.3390/ijerph17176359 PMID: 32882867

Ni, A., Zhang, C., Hu, Y., Lu, W., & Li, H. (2020). Influence mechanism of the corporate image on passenger satisfaction with public transport in China. *Transport Policy*, 94, 54–65. DOI: 10.1016/j.tranpol.2020.04.014

Niu, D., Wang, L., Li, W., & Ma, Y. (2022). An International Comparative Study on the Resilience of Urban Communities after COVID-19 Pandemic: A One Year Case Study between Lanzhou, China and Sarajevo, Bosnia and Herzegovina. *International Journal of Environmental Research and Public Health*, 19(14458), 14458. DOI: 10.3390/ijerph192114458 PMID: 36361337

Oladimeji, D., Gupta, K., Kose, N. A., Gundogan, K., Ge, L., & Liang, F. (2023). Smart transportation: An overview of technologies and applications. *Sensors (Basel)*, 23(8), 3880. DOI: 10.3390/s23083880 PMID: 37112221

Olawole, M. O., ve Aloba, O. (2014). Mobility characteristics of the elderly and their associated level of satisfaction with transport services in Osogbo, Southwestern Nigeria. Transport Policy, 35, 105–116. 10 Aralık 2017 DOI: 10.1016/j.tranpol.2014.05.018

Oliveira, J., Silva, M., Santos, S., Costa, A., & Clemente, T. (2021). Multidimensional sorting framework of cities regarding the concept of sustainable and smart cities with an application to Brazilian capitals. *Sustainable Cities and Society*, 74, 103193. DOI: 10.1016/j.scs.2021.103193

Önder, H. G., ve Akdemir, F. (2022). ''*Rebuilding Sustainable Transportation Infrastructure in the Period of Pandemic: Micromobility Trends and Turkey*''. İdealkent, 13(2), 748-770.

Our World in Data. (2020). ''Cars, planes, trains: where do CO_2 emissions from transport come from *[online]*''. (12.07.2024), https://ourworldindata.org/co2-emissions-from-transport

Ouyang, X., & Xu, M. (2022). Promoting Green Transportation under the Belt and Road Initiative: Locating Charging Stations Considering Electric Vehicle Users' Travel Behavior''. *Transport Policy*, 116, 58–80. DOI: 10.1016/j.tranpol.2021.11.023

Özdede, S., Kalonya, D. H., & Aygün, A. (2021). Rethinking the Need for Urban Green Space per Capita in the Post-Pandemic Period. *Idealkent*, 12, 362–388.

Özden, A., & Kun, S. B. K. (2024). *Evaluation of the Bicycle and E-Scooter Infrastructure in Turkey in terms of Urban Transportation*. Pamukkale University Journal of Engineering Sciences., DOI: 10.5505/pajes.2024.09633

Öztürk, K. (2002). Global Climatic Changes and Their Probable Effect upon Turkey''. *Gazi University Gazi Faculty of Education Journal*, 22(1), 47–65.

Paiva, S., Ahad, M. A., Tripathi, G., Feroz, N., & Casalino, G. (2021). Enabling technologies for urban smart mobility: A recent trends, opportunities and challenges. *Sensors (Basel)*, 21(6), 2143. DOI: 10.3390/s21062143 PMID: 33803903

Paricio, A., & Lopez-Carmona, M. A. (2019). Urban traffic routing using weighted multi-map strategies. *IEEE Access : Practical Innovations, Open Solutions*, 7, 153086–153101. DOI: 10.1109/ACCESS.2019.2947699

Pei, F., Tang, W., & Cheng, K. (2006). Review on research and application of customer satisfaction. *World Standardization and Quality Management*, 10, 4–7.

Peng, Z., Shen, S., Lu, Q., & Perch, S. (2010). *''Climate Change and Transportation''*, M. Kutz (eds.), Handbook of Transportaion Engineering Volume II, New York: McGraw Hill, pp. (35.1-35.24).

Peng, Z., Wang, R., Liu, L., & Wu, H. (2020, June 19). Exploring Urban Spatial Features of COVID-19 Transmission in Wuhan Based on Social Media Data. *ISPRS International Journal of Geo-Information*, 9(402), 1–14. DOI: 10.3390/ijgi9060402

Perone, G. (2021, February 10). The determinants of COVID-19 case fatality rate (CFR) in the Italian regions and provinces: An analysis of environmental, demographic, and healthcare factors. *The Science of the Total Environment*, 755, 142523. DOI: 10.1016/j.scitotenv.2020.142523 PMID: 33022464

Phillips, A., Da Schio, N., Canters, F., & Khan, A. Z. (2023). 'A living street and not just green": Exploring public preferences and concerns regarding nature-based solution implementation in urban streetscapes'. *Urban Forestry & Urban Greening*, 86, 128034. DOI: 10.1016/j.ufug.2023.128034

Phiri, D., Salekin, S., Nyirenda, V. R., Simwanda, M., Ranagalage, M., & Murayama, Y. (2021). Spread of COVID-19 in Zambia: An assessment of environmental and socioeconomic factors using a classification tree approach. *Scientific African*, 12(e00827), 1–12. DOI: 10.1016/j.sciaf.2021.e00827 PMID: 34250321

Plummer, L. (2016) France unveils world's first solar panel road: 0.6-mile stretch to provide energy to power entire village's street lights. Daily Mail. https://www.dailymail.co.uk/sciencetech/article-4058752/France-unveils-world-s-solar-panel-road-0-6-mile-stretch-provide-energy-power-entire-village-s-street-lights.html

Poojari, S. S., & Uday, K. V. (2014). PRT systems. *International Journal of Earth Sciences & Engineering*, 7(4), 720–723.

POSCO. (2024). Korea's first personal rapid transit (PRT) Skycube. POSCO Newsroom. https://newsroom.posco.com/en/koreas-first-personal-rapid-transit-prt-skycube/

Pred, A. (1977). The Choreography of existence: comments on Hägerstrand' s time-geography and its usefulness. Economic Geography, 53(2). 7 Mart 2013, https://www.jstor.org/stable/142726?seq=1#page_scan_tab_contents

Priemus, H. (2004). Changing urban housing markets in advanced economies. *Housing, Theory and Society/Housing. Theory and Society*, 21(1), 2–16. DOI: 10.1080/14036090410032872

Qiao, J. (2022). Smart city and intelligent upgrading of urban transportation system: based on sustainable innvestment strategy. In *Proceedings of the Second International Conference on Advanced Technologies in Intelligent Control, Environment, Computing & Communication Engineering (ICATIECE)* (pp. 1-6), Bangalore, India. DOI: 10.1109/ICATIECE56365.2022.10047675

Qiu, J., Li, R., Han, D., Shao, Q., Han, Y., Luo, X., & Wu, Y. (2021, December). A multiplicity of environmental, economic and social factor analyses to understand COVID-19 diffusion. *One Health*, 13, 100335. DOI: 10.1016/j.onehlt.2021.100335 PMID: 34632042

Qiu, W., Li, W., Liu, X., Zhang, Z., Li, X., & Huang, X. (2023). 'Subjective and objective measures of streetscape perceptions: Relationships with property value in Shanghai'. *Cities (London, England)*, 132, 104037. DOI: 10.1016/j.cities.2022.104037

Rantakokko, M., Mänty, M., ve Rantanen, T. (2013). Mobility decline in old age. Exercise and Sport Sciences Reviews, 41(1), 19–25. 1 Ocak 2018, DOI: 10.1097/JES.0b013e3182556f1e

Rath, A. K., & Mohapatra, H. (2021). An IoT based efficient multi-objective real-time smart parking system. *International Journal of Sensor Networks*, 37(4), 219. DOI: 10.1504/IJSNET.2021.119483

Rathore, R., Kaushik, P., Sikarwar, S. S., Joshi, H., Mishra, A. K., & Hudda, Y. (2024). Intelligent transportation systems make use of fog and edge computing for navigation. *2024 IEEE International Conference on Interdisciplinary Approaches in Technology and Management for Social Innovation (IATMSI)*, 1-6. https://doi.org/DOI: 10.1109/IATMSI60426.2024.10502525

Razavi-Termeh, S. V., Sadeghi-Niaraki, A., Farhangi, F., & Choi, S.-M. (2021, September 14). COVID-19 risk mapping with considering socio-economic criteria using machine learning algorithms. *International Journal of Environmental Research and Public Health*, 18(9657), 9657. DOI: 10.3390/ijerph18189657 PMID: 34574582

Redman, L., Friman, M., Gärling, T., & Hartig, T. (2013). Quality attributes of public transport that attract car users: A research review. *Transport Policy*, 25, 119–127. DOI: 10.1016/j.tranpol.2012.11.005

Rehan, R. M. (2014). Urban branding as an effective sustainability tool in urban development. *HBRC Journal/HBRC Journal, 10*(2), 222–230. https://doi.org/DOI: 10.1016/j.hbrcj.2013.11.007

Rehman, F. U., Islam, M. M., & Miao, Q. (2023). Environmental Sustainability via Green Transportation: A Case of the Top 10 Energy Transition Nations. *Transport Policy*, 137, 32–44. DOI: 10.1016/j.tranpol.2023.04.013

REN21. (2022). Renewables Global Status Report 2022. https://www.ren21.net/reports/global-status-report/

Renne, J. L., Hamidi, S., & Ewing, R. (2016). Transit commuting, the network accessibility effect, and the built environment in station areas across the United States. *Research in Transportation Economics*, 60, 35–43. DOI: 10.1016/j.retrec.2017.02.003

Revathi, G., & Dhulipala, V. R. S. (2012). Smart parking systems and sensors: A survey. *2012 International Conference on Computing, Communication and Applications*, 1–5. DOI: 10.1109/ICCCA.2012.6179195

Rice, L. (2019). A health map for architecture: The determinants of health and wellbeing in buildings. Designing for Health & Wellbeing: Home, City [within Vernon]. *Society*, •••, 155–184.

Rios, V., & Gianmoena, L. (2021). On the link between temperature and regional COVID-19 severity: Evidence from Italy. *Regional Science Policy & Practice*, 13(S1), 109–137. DOI: 10.1111/rsp3.12472 PMID: 38607900

Road Traffic Technology. (n.d.). SolaRoad project: Krommenie. Retrieved August 28, 2024, from https://www.roadtraffic-technology.com/projects/solaroad-project-krommenie/

Robaina, M., & Neves, A. (2021). Complete decomposition analysis of CO2 emissions intensity in the transport sector in Europe''. *Research in Transportation Economics*, 90(101074), 101074. DOI: 10.1016/j.retrec.2021.101074

Robles, A. (2006). Masdar City: The soon-to-be eco city. Environment & Society Portal. Rachel Carson Center for Environment and Society. https://www.environmentandsociety.org/tools/keywords/masdar-city-soon-be-eco-city

Ronael, M., & Baycan, T. (2022, September 10). Place-based factors affecting COVID-19 incidences in Turkey. *Asia-Pacific Journal of Regional Science*, 6(3), 1053–1086. DOI: 10.1007/s41685-022-00257-4

Rosenbloom, S. (2001). Sustainability and automobility among the elderly: An international assessment. Transportation, 28(4), 375–408. 30 Haziran 2013, DOI: 10.1023/A:1011802707259

Rout, S., Patra, S. S., Mohanty, J. R., Barik, R. K., & Lenka, R. K. (2020). Energy aware task consolidation in fog computing environment. *Advances in Intelligent Systems and Computing*, 195-205. https://doi.org/DOI: 10.1007/978-981-15-5679-1_19

Rouwendal, J., ve Rietveld, P. (1994). Changes in commuting distances of Dutch households. Urban Studies, 31(9), 1545–1557. 26 Ocak 2018, DOI: 10.1080/00420989420081421

Royko, Y., Fornalchyk, Y., Koda, E., Kernytskyy, I., Hrytsun, O., Bura, R., Osinski, P., Markiewicz, A., Wierzbicki, T., Barabash, R., Humenuyk, R., & Polyansky, P. (2023). Public Transport Prioritization and Descriptive Criteria-Based Urban Sections Classification on Arterial Streets. *Sustainability (Basel)*, 15(3), 2363. DOI: 10.3390/su15032363

Ruck, D. J., Bentley, R. A., & Borycz, J. (2021, May). Early warning of vulnerable counties in a pandemic using socio-economic variables. *Economics and Human Biology*, 41, 100988. DOI: 10.1016/j.ehb.2021.100988 PMID: 33636583

Ruiz, E. C., De La Cruz, E. R. R., & Vázquez, F. J. C. (2019). Sustainable Tourism and Residents' Perception towards the Brand: The Case of Malaga (Spain). *Sustainability (Basel)*, 11(1), 292. DOI: 10.3390/su11010292

Ruiz-Vanoye, J. A., Diaz-Parra, O., Marroquín-Gutiérrez, F., Rodríguez-Lara, B. M., & Aguilar-Ortiz, J. (2023). Smart and Sustainable Cities. In Ruiz-Vanoye, J. A. (Ed.), *Management, Technology, and Economic Growth in Smart and Sustainable Cities*. IGI Global., DOI: 10.4018/979-8-3693-0373-3.ch001

Saarika, P. S., Sandhya, K., & Sudha, T. (2017). Smart transportation system using IoT. *2017 International Conference On Smart Technologies For Smart Nation (SmartTechCon)*. DOI: 10.1109/SmartTechCon.2017.8358540

Sabouri, S., Hamidi, S., & Ewing, R. (2020). Does Density Aggravate the COVID-19 Pandemic? *Journal of the American Planning Association*, 86(4), 495–509. DOI: 10.1080/01944363.2020.1777891

Saharan, S., Kumar, N., & Bawa, S. (2020). An efficient smart parking pricing system for smart city environment: A machine-learning based approach. *Future Generation Computer Systems*, 106, 622–640. DOI: 10.1016/j.future.2020.01.031

Samantaray, M., Biswal, A. K., Singh, D., Samanta, D., Karuppiah, M., & Joseph, N. P. (2021). Optical Character Recognition (OCR) based Vehicle's License Plate Recognition System Using Python and OpenCV. 2021 5th International Conference on Electronics, Communication and Aerospace Technology (ICECA), 849–853. DOI: 10.1109/ICECA52323.2021.9676015

Sami, Z., & Dincer, M. E. E. (2020). Study of physical, social and functional structures in Bagdat Street (Istanbul) as public space. *Civil Engineering and Architecture*, 8(3), 224–236. DOI: 10.13189/cea.2020.080305

Sardar, M. S., Asghar, N., & Rehman, H. U. (2022). *''Moderation of competitiveness in determining environmental sustainability: Economic growth and transport sector carbon emissions in global perspective''.Environment, Development and Sustainability*, 26(1), 1–23. DOI: 10.1007/s10668-022-02771-z

Sari, A. R., Sarie, F., Suparni, S., Nadi, M. A. B., & Karunia, M. N. (2024). The Effect of Environmentally Friendly Transportation Implementation and Supporting Policies on Air Pollution Reduction in Central Kalimantan. *West Science Interdisciplinary Studies*, 2(3), 605–614. DOI: 10.58812/wsis.v2i03.739

Sarmadi, M., Moghanddam, V. K., Dickerson, A. S., & Martelletti, L. (2021, April). Association of COVID-19 distribution with air quality, sociodemographic factors, and comorbidities: An ecological study of US states. *Air Quality, Atmosphere & Health*, 14(4), 455–465. DOI: 10.1007/s11869-020-00949-w PMID: 33078068

Schaie, K., & Boron, J. ve Willis, S. (2005). Everyday competence in older adults. Cambridge: Cambridge University Press. DOI: 10.1017/CBO9780511610714

Schinkel, U., Becker, N., Trapp, M., & Speck, M. (2022). Assessing the contribution of innovative technologies to sustainable development for planning and decision-making processes: A set of indicators to describe the performance of sustainable urban infrastructures (isi). *Sustainability (Basel)*, 14(4), 1966. DOI: 10.3390/su14041966

Schmitt, R. (1988). The Materialist Dialectic. Science & Society, Vol. 52 (4) (Winter, 1988/1989), 441-456.

Schwanen, T., Dieleman, F. M., ve Dijst, M. (2001). Travel behaviour in Dutch monocentric and policentric urban systems. Journal of Transport Geography, 9(3), 173–186. 20 Mayıs 2013, DOI: 10.1016/S0966-6923(01)00009-6

Schwanen, T., ve Páez, A. (2010). The mobility of older people – an introduction. Journal of Transport Geography, 18(5), 591–595. 20 Mayıs 2013, DOI: 10.1016/j.jtrangeo.2010.06.001

Sedlak, O., & Ćirić, Z. (2018). Sustainability development of smart public transportation in globalization. *The Academic Research Community Publication*, 2(1), 10. DOI: 10.21625/archive.v2i1.235

Şekercioğlu, S., & İncekara, B. (2017). *''Transforming Transport Policies: Planning for Greenhouse Gas Reduction''*. 12th Transportation Congress (Transportation Policies), 181-190.

Senvar, O., & Ünver, S. (2022). An Overview to Unmanned Aerial Vehicles from Perspectives of Quality and Safety Management in Aviation. *International Journal of Engineering Research and Development*, 14(2), 917–940. DOI: 10.29137/umagd.1064242

Shaaban, K. (2019). Assessing Sidewalks and Corridor Walkability in Developing Countries. *Sustainability (Basel)*, 11(14), 3865. DOI: 10.3390/su11143865

Shaheen, S., Cohen, A., Chan, N., & Bansal, A. (2020). *''Sharing Strategies: Carsharing, Shared Micromobility (bikesharing and scooter sharing), Transportation Network Companies, Microtransit, and Other İnnovative Mobility Modes''*. Transportation, Land Use, and Environmental Planning, 237–262.

Shaheen, S. A., & Finson, R. (2013). *Intelligent transportation systems*. Reference Module İn Earth Systems And Environmental Sciences.

Shahgholian, M., & Gharavian, D. (2018). Advanced traffic management systems: an overview and a development strategy. *arXiv preprint arXiv*:1810.02530.

Shah, I. H., Manzoor, M. A., Jinhui, W., Li, X., Hameed, M. K., Rehaman, A., Li, P., Zhang, Y., Niu, Q., & Chang, L. (2024). Comprehensive Review: Effects of Climate Change and Greenhouse Gases Emission Relevance to Environmental Stress on Horticultural Crops and Management. *Journal of Environmental Management*, 351, 119978. DOI: 10.1016/j.jenvman.2023.119978 PMID: 38169258

Shah, K. J., Pan, S. Y., Lee, I., Kim, H., You, Z., Zheng, J. M., & Chiang, P. C. (2021). Green Transportation for Sustainability: Review of Current Barriers, Strategies, and İnnovative Technologies''. *Journal of Cleaner Production*, 326, 129392. DOI: 10.1016/j.jclepro.2021.129392

Sharifi, A., & Khavarian-Garmsir, A. R. (2020). The COVID-19 pandemic: Impacts on cities and major lessons for urban planning, design, and management. *The Science of the Total Environment*, 749(142391), 2. DOI: 10.1016/j.scitotenv.2020.142391 PMID: 33370924

Sharifi, A., Lak, A., Badr, S., Zali, A., Maher, A., & Mostafavi, E. (2021). Eylül). Spatio-temporal patterns of the COVID-19 pandemic, and place-based influential factors at the neighborhood scale in Tehran. *Sustainable Cities and Society*, 72(103034), 1–13.

Sharma, S., & Bali, R. S. (2024). An edge computing framework for data analysis in intelligent transportation systems. *2024 3rd International Conference for Innovation in Technology (INOCON)*, 1-6. https://doi.org/DOI: 10.1109/INOCON60754.2024.10511983

Sharma, P. K., & Park, J. H. (2021). Blockchain-based secure mist computing network architecture for intelligent transportation systems. *IEEE Transactions on Intelligent Transportation Systems*, 22(8), 5168–5177. DOI: 10.1109/TITS.2020.3040989

Shi, X., Zhao, L., Lin, N., Xu, L., Hawbani, A., & Bi, Y. (2023). Constructing future vehicle digital twins with insights from cloud-edge collaboration. *2023 IEEE Intl Conf on Parallel & Distributed Processing with Applications, Big Data & Cloud Computing, Sustainable Computing & Communications, Social Computing & Networking (ISPA/BDCloud/SocialCom/SustainCom)*, 110-117. https://doi.org/DOI: 10.1109/ISPA-BDCloud-SocialCom-SustainCom59178.2023.00047

Shieh, A. (2022). Smart cities for technological and social innovation: Case studies, current trends, and future steps. *Urban Policy and Research*, 40(3), 281–284. DOI: 10.1080/08111146.2022.2104441

Silva, J., & Alves, M. R. (2021). Social inequalities and the pandemic of COVID-19: The case of Rio de Janeiro. *Epidemiology and Community Health*, 75(10), 975–979. DOI: 10.1136/jech-2020-214724 PMID: 33811129

Silver, K., Lopes, A., Vale, D., & da Costa, N. M. (2023). The inequality effects of public transport fare: The case of Lisbon's fare reform. *Journal of Transport Geography*, 112, 103685. DOI: 10.1016/j.jtrangeo.2023.103685

Simonds, J.O. (1998) Landscape Architecture: A Manual of Land Planning andDesign, McGraw-Hill Professional; p155,158

Singh, M., Kumar, R., & Chana, I. (2019). CDFM-based secure & Efficient architecture for data management in cloud computing. *2019 2nd International Conference on Intelligent Communication and Computational Techniques (ICCT), 13*, 207-212. https://doi.org/DOI: 10.1109/ICCT46177.2019.8969017

Singh, B., & Gupta, A. (2015). Recent trends in intelligent transportation systems: A review. *Journal of Transport Literature*, 9(2), 30–34. DOI: 10.1590/2238-1031.jtl.v9n2a6

Smirnov, A., Kashevnik, A., Shilov, N., Mikhailov, S., Gusikhin, O., & Martinez, H. (2019). Intelligent content management system for tourism smart mobility: Approach and cloud-based Android application. *Proceedings of the 5th International Conference on Vehicle Technology and Intelligent Transport Systems*, 426-433. https://doi.org/DOI: 10.5220/0007715300002179

Smith, G. C., ve Sylvestre, G. M. (2001). Determinants of the travel behavior of the suburban elderly. Growth and Change, 32(3), 395–412. 30 Mayıs 2015, DOI: 10.1111/0017-4815.00165

Solar Energy Industries Association (SEIA). (n.d.). Solar Panels. Retrieved from https://www.seia.org/solar-101/solar-panels

Solar Panels Network. (n.d.). Walkable Solar Panels: A New Era in Solar Technology. Retrieved from https://www.solarpanelsnetwork.com/walkable-solar-panels

SolaRoad. (2015). The world's first solar cycle path. [Online] Available at: https://sola-road.com/

Solomon, M., & Liang, C. (2022, October 2). Human coronaviruses: The emergence of SARS-CoV-2 and management of COVID-19. *Virus Research*, 319(198882), 198882. DOI: 10.1016/j.virusres.2022.198882 PMID: 35934258

Şolt, B. H. (2021, January 5). Changes Experienced in Cities and Urban Services During the COVID-19 Pandemic. *International Journal of Management and Social Research*, 8(15), 39–47.

Southern Aegean Development Agency. (2023). TR32 düzey 2 bölgesi 2024-2028 bölge planı taslağı [TR32 NUTS 2 region 2024-2028 draft regional plan]. GEKA. https://geka.gov.tr/tr/sayfa/planlar

Spinney, J. E. L., Scott, D. M., ve Newbold, K. B. (2009). Transport mobility benefits and quality of life: A time-use perspective of elderly Canadians. Transport Policy, 16(1), 1–11. 1 Ocak 2018, DOI: 10.1016/j.tranpol.2009.01.002

Spinoza, B. (2012). *Ethika*. Kabalcı Yayınları.

Spotswood, E. N., Benjamin, M., Stoneburner, L., Wheeler, M. M., Beller, E. E., Balk, D., & McDonald, R. I. (2021, January 11). Nature inequity and higher COVID-19 case rates in less-green neighborhoods in the United States. *Nature Sustainability*, 4(12), 1092–1098. DOI: 10.1038/s41893-021-00781-9

Srichuae, S., Nitivattananon, V., ve Perera, R. (2016). Aging society in Bangkok and the factors affecting mobility of elderly in urban public spaces and transportation facilities. IATSS Research, 40(1), 26–34. 22 Mayıs 2017, DOI: 10.1016/j.iatssr.2015.12.004

Sridhar, K. S. (2021). Urbanization and COVID-19 Prevalence in India. *Regional Science Policy & Practice*, •••, 1–13.

Srinivasan, A. (2018). IoT cloud based real time automobile monitoring system. *2018 3rd IEEE International Conference on Intelligent Transportation Engineering (ICITE)*. https://doi.org/DOI: 10.1109/ICITE.2018.8492706

Statista, (2023). ''*Global transportation sector CO_2 emissions 2000-2022 [online]*'',(19.08.2024), https://www.statista.com/statistics/1387814/carbon-dioxide-emissions-transport-subsector-worldwide/

Stefanello, F., Buriol, L. S., Hirsch, M. J., Pardalos, P. M., Querido, T., Resende, M. G., & Ritt, M. (2017). On the minimization of traffic congestion in road networks with tolls. *Annals of Operations Research*, 249(1-2), 119–139. DOI: 10.1007/s10479-015-1800-1

Steger-Vonmetz, D. C. (2005). Improving modal choice and transport efficiency with the virtual ridesharing agency. [). IEEE.]. *IEEE Intelligent Transportation Systems*, 2005, 994–999. DOI: 10.1109/ITSC.2005.1520186

Steven, R. (1997). *Constraints on elderly daily travel behavior, Master Theses*. The University of Westerm Ontario.

Storchmann, K. H. (2001). The impact of fuel taxes on public transport—An empirical assessment for Germany. *Transport Policy*, 8(1), 19–28. DOI: 10.1016/S0967-070X(00)00033-0

Su, F., ve Bell, M. G. H. (2009). Transport for older people: Characteristics and solutions. Research in Transportation Economics, 25(1), 46–55. 26 Ocak 2018, DOI: 10.1016/j.retrec.2009.08.006

Sullivan, W. C., & Lovell, S. T. (2006). Improving the visual quality of commercial development at the rural-urban fringe. *Landscape and Urban Planning*, 77(1-2), 152–166. DOI: 10.1016/j.landurbplan.2005.01.008

Tacken, M. (1998). Mobility of the elderly in time and space in the Netherlands: An analysis of the Dutch National Travel Survey. Transportation, 25(4), 379–393. 30 Haziran 2013, DOI: 10.1023/A:1005042614848

Taç, Ş. G. (2018). Karayolu ulaşımında meydana gelen trafik kazalarının önlenmesinde akıllı ulaşım sistemlerinin etkisi. *Akıllı Ulaşım Sistemleri ve Uygulamaları Dergisi*, 1(2), 12–21.

Tamam, I., Wang, S., & Djahel, S. (2020). An IoT-based Eco-Parking System for Smart Cities. 2020 IEEE International Smart Cities Conference (ISC2), 1–6. DOI: 10.1109/ISC251055.2020.9239041

Tayal, A., Bhatia, G., Singh, J., Bansal, M., & Anitha, G. (2023). Security challenges and future privacy perspectives in cloud computing for smart city environments. *2023 International Conference on Power Energy, Environment & Intelligent Control (PEEIC)*, 473-477. https://doi.org/DOI: 10.1109/PEEIC59336.2023.10451628

Taylor, N. (2003). The Aesthetic Experience of Traffic in the modern city [Carfax Publishing.]. *Urban Studies (Edinburgh, Scotland)*, 40(8), 1609–1625. DOI: 10.1080/0042098032000094450

Tchicaya, A., Lorentz, N., Leduc, K., & Lanchy, G. (2021, September 17). COVID-19 mortality with regard to healthcare services availability, health risks, and socio-spatial factors at department level in France: A spatial cross-sectional analysis. *PLoS One*, 16(9), e0257060. DOI: 10.1371/journal.pone.0256857 PMID: 34534226

Tektaş M., Korkmaz K., Erdal H. (2016). Akıllı ulaşım sistemlerinin geleceği ekonomik ve çevresel faydaları. *Balkan Journal of Social Sciences*, 561-577.

The Government of the Hong Kong Special Administrative Region. (2019, April 15). *Three restaurants in Western District and Wanchai fined for illegally emitting excessive cooking fumes.* The Government of the Hong Kong Special Administrative Region - Press Releases. https://www.info.gov.hk/gia/general/201904/15/P2019041500362.htm?fontSize=1

The Presidency of the Republic of Turkey, the Presidency of Strategy and Budget. (2019). On birinci kalkınma planı (2019-2023) [Eleventh development plan (2019-2023)]. The Presidency of Strategy and Budget, the Directorate General for Administrative Services, the Department of Information and Document Management. https://www.sbb.gov.tr/kalkinma-planlari

The Presidency of the Republic of Turkey, the Presidency of Strategy and Budget. (2023). On ikinci kalkınma planı (2024-2028) [Twelfth development plan (2024-2028)]. The Presidency of Strategy and Budget, the Directorate General for Administrative Services, the Department of Information and Document Management. https://www.sbb.gov.tr/kalkinma-planlari

The Republic of Turkey, the Ministry of Environment, Urbanization and Climate Change. (2019). 2020-2023 Ulusal akıllı şehirler stratejisi ve eylem planı [National smart cities strategy and action plan]. The Republic of Turkey, the Ministry of Environment, Urbanization and Climate Change. https://akillisehirler.gov.tr/wp-content/uploads/EylemPlani.pdf

The Republic of Turkey, the Ministry of Environment, Urbanization and Climate Change. (2024a). Akıllı şehirler portalı, proje envanteri [Smart cities portal, project inventory]. The Directorate General of Geographic Information Systems, the Department of Smart Cities. https://akillisehirler.gov.tr/wp-content/uploads/ProjeEnvanteri/map.html

The Republic of Turkey, the Ministry of Environment, Urbanization and Climate Change. (2024b). Akıllı şehir ekosistemi [Smart city ecosystem]. The Directorate General of Geographic Information Systems, the Department of Smart Cities. https://akillisehirekosistem.csb.gov.tr/

The Republic of Turkey, the Ministry of Industry and Technology. (2023). Kalkınma planlamasında istatistiki bölge birimleri sınıflandırması [Nomenclature of territorial units for statistics in development planning]. The General Directorate of Development Agencies. https://ka.gov.tr/sayfalar/kalkinma-planlamasinda-istatistiki-bolge-birimleri-siniflandirmasi--24

The Republic of Turkey, the Ministry of Industry and Technology. (2024). Bölgesel kalkınmanın yönetişiminde kalkınma ajansları [Development agencies in the governance of regional development]. The General Directorate of Development Agencies. https://ka.gov.tr/sayfalar/bolgesel-kalkinmanin-yonetisiminde-kalkinma-ajanslari--23

The Republic of Turkey, the Prime Ministry General Directorate of Legislation Development and Publication. (2004). 5216 sayılı büyükşehir belediyesi kanunu [Law no. 5216 on metropolitan municipalities]. Resmi Gazete [Official Gazette], 25531. https://www.resmigazete.gov.tr/eskiler/2004/07/20040723.htm

The Resolution of the Cabinet of Ministers of Turkey. (2002). İstatistiki bölge birimleri sınıflandırması [Nomenclature of territorial units for statistics]. Decree of the Council of Ministers dated 28/08/2002 and numbered 2002/4720. http://www.die.gov.tr/nuts/bolgekarar.htm

Tian, J., Tang, T., Lin, S., & Liang, H. (2022). A potential game based load balancing approach for mobile edge computing enabled intelligent rail construction system. *2022 IEEE 25th International Conference on Intelligent Transportation Systems (ITSC), 39*, 3461-3465. https://doi.org/DOI: 10.1109/ITSC55140.2022.9922554

Tokey, A. I. (2021). Eylül). Spatial association of mobility and COVID-19 infection rate in the USA: A county-level study using mobile phone location data. *Journal of Transport & Health*, 22(101135). PMID: 34277349

Transport & Regional Affairs Department of the Environment. (2000). *By Design, UrbanDesign in the Planning System: Towards Better Practice*. Thomas Telford Ltd.

Transport Advisory Committee. (2014). Report on Study of Road Traffic Congestion in Hong Kong. In *Transport Advisory Committee*. https://www.tlb.gov.hk/eng/boards/transport/land/Full_Eng_C_cover.pdf

Tuğaç, Ç. (2022). *''Green Transportation in the Context of Climate Change and Urban Resilience''*. İdealkent, 13(36), 545-575.

TUIK. (2024). Adrese Dayalı Nüfus Kayıt Sistemi Sonuçları, 2023. https://data.tuik.gov.tr/Bulten/Index?p=Adrese-Dayali-Nufus-Kayit-Sistemi-Sonuclari-2023-49684

Turkish Statistical Institute. (2024a). Geographic statistics portal. Turkish Statistical Institute. https://cip.tuik.gov.tr

Turkish Statistical Institute. (2024b). Regional statistics. Turkish Statistical Institute. https://biruni.tuik.gov.tr/bolgeselistatistik/anaSayfa.do?dil=en

Turnbull, K. F. (1992). *High Occupancy Vehicle Project Case Studies: Historical Trends and Project Experiences (No. DOT-T-94-18)*. Department of Transportation.

U.S. Department of Energy. (2020). *''Environmental Sustainability''*, (15.06.2024), https://www.energy.gov/lm/listings/environmental-sustainability

Ultra. (2009, June 1). Corporate brochure. Archived from the original (PDF) on July 11, 2012. Retrieved August 24, 2024, from https://web.archive.org/web/20120711192906/http://www.ultraprt.net/cms/PerimeterPRT_11x17_2_side_brochure.pdf

United Nations Economic Commission for Europe (UNECE). (2024) Smart sustainable cities. UNECE. https://unece.org/housing/smart-sustainable-cities#:~:text=A%20smart%20sustainable%20city%20is,as%20well%20as%20cultural%20aspects

United Nations. (2014). *"Department of Economic and Social Affairs, Population Division [online]''*,(17.07.2024), https://www.urbanet.info/world-urban-population/

University of Bristol. (n.d.). Heathrow personal rapid transport. Retrieved August 24, 2024, from https://www.bristol.ac.uk/research/impact/heathrow-personal-rapid-transport/

Urban, R. C., & Nakada, L. Y. (2021, April 1). GIS-based spatial modelling of COVID-19 death incidence in São Paulo, Brazil. *Environment and Urbanization*, 33(1), 229–238. DOI: 10.1177/0956247820963962 PMID: 38603029

Van Den Berg, P., Arentze, T., ve Timmermans, H. (2011). Estimating social travel demand of senior citizens in the Netherlands. Journal of Transport Geography, 19(2), 323–331. 30 Haziran 2013, DOI: 10.1016/j.jtrangeo.2010.03.018

Van Lierop, D., & El-Geneidy, A. (2016). Enjoying loyalty: The relationship between service quality, customer satisfaction, and behavioral intentions in public transit. *Research in Transportation Economics*, 59, 50–59. DOI: 10.1016/j.retrec.2016.04.001

Van Renterghem, T., & Botteldooren, D. (2016). 'View on outdoor vegetation reduces noise annoyance for dwellers near busy roads'. *Landscape and Urban Planning*, 148, 203–215. DOI: 10.1016/j.landurbplan.2015.12.018

Verma, R., Yabe, T., & Ukkusuri, S. V. (2021, May 26). Spatiotemporal contact density explains the disparity of COVID-19 spread in urban neighborhoods. *Scientific Reports*, 11(10952), 1–11. DOI: 10.1038/s41598-021-90483-1 PMID: 34040093

Vicente, P., Sampaio, A., & Reis, E. (2020). Factors influencing passenger loyalty towards public transport services: Does public transport providers' commitment to environmental sustainability matter? *Case Studies on Transport Policy*, 8(2), 627–638. DOI: 10.1016/j.cstp.2020.02.004

Vij, A., Ryan, S., Sampson, S., & Harris, S. (2020). Consumer preferences for on-demand transport in Australia. *Transportation Research Part A, Policy and Practice*, 132, 823–839. DOI: 10.1016/j.tra.2019.12.026

Vijayan, D. S., Koda, E., Sivasuriyan, A., Winkler, J., Devarajan, P., Kumar, R. S., Jakimiuk, A., Osinski, P., Podlasek, A., & Vaverková, M. D. (2023). Advancements in solar panel technology in civil engineering for revolutionizing renewable energy solutions—A review. *Renewable & Sustainable Energy Reviews*, 183, 113206.

Wallander, J., & Mäkitalo, M. (2012). Data mining in rail transport delay chain analysis. *International Journal of Shipping and Transport Logistics*, 4(3), 269–285. DOI: 10.1504/IJSTL.2012.047492

Wang, J., Zeng, F., Tang, H., Wang, J., & Xing, L. (2022, January). Correlations between the urban built environmental factors and the spatial distribution at the community level in the reported COVID-19 samples: A case study of Wuhan. *Cities (London, England)*, 129, 103932. DOI: 10.1016/j.cities.2022.103932 PMID: 35975194

Wang, Q., & Sun, H. (2019). Traffic Structure Optimization İnhistoric Districts Based on Green Transportation and Sustainable Development Concept. *Advances in Civil Engineering*, 2019(1), 9196263. Advance online publication. DOI: 10.1155/2019/9196263

Wang, R., Lu, T., Wan, C., Sun, X., & Jiang, W. (2023). 'Measuring the Effects of Streetscape Characteristics on Perceived Safety and Aesthetic Appreciation of Pedestrians'. *Journal of Urban Planning and Development*, 149(3), 05023020. DOI: 10.1061/JUPDDM.UPENG-4314

Watkins, K. (2012). Safe and sustainable roads. *Make Roads Safe The Campaign For Global Road Safety*, 1, 40.

Watts, M. J. (2022, September). Macro-level drivers of SARS-CoV-2 transmission: A data-driven analysis of factors contributing to epidemic growth during the first wave of outbreaks in the United States. *Spatial and Spatio-temporal Epidemiology*, 43, 100539. DOI: 10.1016/j.sste.2022.100539 PMID: 36460448

Waygood, E. O. D., Sun, Y., & Susilo, Y. (2014). Transportation carbon dioxide emissions by built environment and family lifecycle: Case study of the Osaka metropolitan area. *Transportation Research Part D, Transport and Environment*, 31, 176–188. DOI: 10.1016/j.trd.2014.06.001

Whelan, M., & Langford, J. Oxley, J.; Koppel, S., ve Charlton, J.; (2006). The elderly and mobility: a review of the literatre. 30 Haziran 2013, https://www.monash.edu/muarc/research/our-publications/muarc255

White Paper. (2011). Roadmap to a Single European Transport Area – Towards a competitive and resource efficient transport system. European Commission, Brussels, 28.3.2011.

WHO. (2020). Coronavirus Disease 2019 (COVID-19) Situation Report. World Health Organization. Retrieved April 03, 2023, from https://www.who.int/docs/default-source/coronaviruse/situation-reports/20200229-sitrep-40-covid-19.pdf?sfvrsn=849d0665_2

WHO. (March 22, 2023). WHO Coronavirus (COVID-19) Dashboard. Retrieved from WHO Health Emergency Dashboard: https://covid19.who.int/

Willer, C. J. (2021). Rebranding place "to build community": Neighborhood branding in Buffalo, NY. *Urban Geography*, 43(9), 1350–1371. DOI: 10.1080/02723638.2021.1927323

Wolch, J. R., Byrne, J., & Newell, J. P. (2014). Urban green space, public health, and environmental justice: The challenge of making cities 'just green enough.'. *Landscape and Urban Planning*, 125, 234–244. DOI: 10.1016/j.landurbplan.2014.01.017

World Bank. (2022). ''*The World Bank Open Data*'', (15.07.2024), https://data.world bank.org/

World Bank. (2024). Global smart city partnership program: Supporting cities to become 'smart' – efficient, innovative, inclusive, and resilient – using data and technology. World Bank. https://www.worldbank.org/en/programs/global-smart-city-partnership-program#:~:text=Smart%20Cities%20%26%20Urban%20Development,improve%20urban%20planning%20and%20management

World Health Organization (2002). Active ageing: A policy framework. The Aging Male, 5(1), 1–37. 24 Haziran 2014, DOI: 10.1080/713604647

World Health Organization. (2011). Global health and aging. NIH Publication No 117737, 1(4), 273–277. 10 Haziran 2015, https://doi.org/DOI: 11-7737

Worth, M. Creating significance through public places art an inclusive+interdisciplinary practice, Public Art and Urban Design. Interdisciplinary and SocialPerspectives Waterfronts of Art III, web version on the w@terfront nr, 4.http://www .ub.edu/escult/epolis/WaterIII.pdf. p51.

Wu, X., & Zhang, J. (2021, March 4). Exploration of Temporal-Spatially Varying Impacts on COVID-19 Cumulative Case in Texas Using Geographically Weighted Regression (GWR). Environmental Science and Pollution Reserch, 1-21.

Wu, Y., Xiao, W., & Wei, Y. D. (2022). Neighborhood, built environment and resilience in transportation during the COVID-19 pandemic. Transportation Research Part, 110(103428), 1-15.

Wu, I., & Pojani, D. (2016). Obstacles to the creation of successful bus rapid transit systems: The case of Bangkok. *Research in Transportation Economics*, 60, 44–53. DOI: 10.1016/j.retrec.2016.05.001

Wu, S., Dong-qiang, G., Wu, Y., & Wu, Y. (2018). Future development of taiwan's smart cities from an information security perspective. *Sustainability (Basel)*, 10(12), 4520. DOI: 10.3390/su10124520

Wu, Y., Hu, Z., Su, M., Xie, L., Zhang, A., Lin, X., & Nie, Y. (2021). Population migration, spread of COVID-19, and epidemic prevention and control: Empirical evidence from China. *BMC Public Health*, 21(529), 1–12. PMID: 33731053

Xiao, J., Zhao, J., Luo, Z., Liu, F., & Greenwood, D. (2022, September). The impact of built environment on mental health: A COVID-19 lockdown perspective. *Health & Place*, 77, 102889. DOI: 10.1016/j.healthplace.2022.102889 PMID: 36027740

Xiong, Z., Sheng, H., Rong, W., & Cooper, D. E. (2012). Intelligent transportation systems for smart cities: A progress review. *Science China. Information Sciences*, 55(12), 2908–2914. DOI: 10.1007/s11432-012-4725-1

Xu, B., & Thakur, G. (2021). Introduction to the special issue on smart transportation. *GeoInformatica*, 25(3), 417–418. DOI: 10.1007/s10707-021-00432-3

Xu, G., Jiang, Y., Wang, S., Qin, K., Ding, J., Liu, Y., & Lu, B. (2022, January). Spatial disparities of self-reported COVID-19 cases and influencing factors in Wuhan, China. *Sustainable Cities and Society*, 76, 103785. DOI: 10.1016/j.scs.2021.103485 PMID: 34722132

Xu, W., Yan, C., Jia, W., Ji, X., & Liu, J. (2018). Analyzing and Enhancing the Security of Ultrasonic Sensors for Autonomous Vehicles. *IEEE Internet of Things Journal*, 6(6), 5015–5029. DOI: 10.1109/JIOT.2018.2867917

Yang, G., Wang, D., & Mao, X. (2020). Modelling the modal shift effects of converting a general traffic lane into a dedicated bus lane. *Promet (Zagreb)*, 32(5), 625–637. DOI: 10.7307/ptt.v32i5.3375

Yang, Z., Li, J., Huang, X., Zhang, A., Lu, Y., Zhao, X., & Yang, X. (2023). The impact of urban spatial environment on COVID-19: A case study in Beijing. *Frontiers in Public Health*, •••, 1287999. PMID: 38259769

Yan, H., Xu, X., Bilal, M., Xia, X., Dou, W., & Wang, H. (2024). Customer centric service caching for intelligent cyber–physical transportation systems with cloud–edge computing leveraging digital twins. *IEEE Transactions on Consumer Electronics*, 70(1), 1787–1797. DOI: 10.1109/TCE.2023.3326969

Yavuz, H. R. (2024). *"Istanbul Example in Smart Cities Regarding Transportation"*. Premium e-. *Journal of Social Sciences*, 8(38), 91–96.

Yazıcı, A. B., & Ünverdi, N. Ö. (2024). LoRa Technology Overview and Smart Parking System Design. 2024 32nd Signal Processing and Communications Applications Conference (SIU), 1–4. DOI: 10.1109/SIU61531.2024.10600948

Yeh, S., Paltsev, S., Reilly, J. M., Daniels, D., & Linares, P. (2024). *''Designing Resilience for Multi-System Dynamics of Future Transportation''*. MIT Joint Program on the Science and Policy of Global Change, 371.

Yigitcanlar, T., & Kamruzzaman, M. (2014). Investigating the İnterplay Between Transport, Land Use and The Environment: A Review of the Literature''. *International Journal of Environmental Science and Technology*, 11(8), 2121–2132. DOI: 10.1007/s13762-014-0691-z

Yook, D., & Heaslip, K. (2015). Effective modeling for a distance-based fare structure with a time-expanded network. *Journal of Public Transportation*, 18(1), 1–13. DOI: 10.5038/2375-0901.18.1.2

Yoshino, D., Fujiwara, A., & Zhang, J. (2010). Environmental Efficiency Model based on data Envelopment Analysis and its Application to Environmentally Sustainable Transport Policies. *Transportation Research Record: Journal of the Transportation Research Board*, 2163(1), 112–123. DOI: 10.3141/2163-13

You, L., Zhao, F., Cheah, L., Jeong, K., Zegras, C., & Ben-Akiva, M. (2018). Future mobility sensing: An intelligent mobility data collection and visualization platform. *21st International Conference on Intelligent Transportation Systems (ITSC)* (pp. 2653-2658). IEEE. DOI: 10.1109/ITSC.2018.8569697

Yücel, G. F. (2013). Street Furniture and Amenities: Designing the User-Oriented Urban Landscape. In Özyavuz, M. (Ed.), *Advances in Landscape Architecture*. IntechOpen., DOI: 10.5772/55770

Zahedi, S. (2012). Green Transportation a More Eco-efficient Option''. *Journal of Social and Development Sciences*, 3(7), 223–228. DOI: 10.22610/jsds.v3i7.706

Zawieska, J., & Pieriegud, J. (2018). Smart city as a tool for sustainable mobility and transport decarbonisation. *Transport Policy*, 63, 39–50. DOI: 10.1016/j.tranpol.2017.11.004

Zefreh, M. M., Hussain, B., & Sipos, T. (2020). In-depth analysis and model development of passenger satisfaction with public transportation. *KSCE Journal of Civil Engineering*, 24(10), 3064–3073. DOI: 10.1007/s12205-020-1871-7

Zelyut, S. (2021). *Spinoza*. Say Yayınları.

Zeng, Y., Zhang, Z., Ye, Z., & Li, L. (2023). Regional innovation effect of smart city construction in China. *PLoS One*, 18(2), e0281862. DOI: 10.1371/journal.pone.0281862 PMID: 36795669

Zenker, S., Braun, E., & Petersen, S. (2017). Branding the destination versus the place: The effects of brand complexity and identification for residents and visitors. *Tourism Management*, 58, 15–27. DOI: 10.1016/j.tourman.2016.10.008

Zhang, Z., Mao, B., Liu, M., Chen, J., ve Guo, J. (2007). Analysis of Travel Characteristics of Elders in Beijing. Journal of Transportation Systems Engineering and Information Technology, 7(6), 11–20. 24 Eylül 2016, .DOI: 10.1016/S1570-6672(08)60001-X

Zhang, C., Zhu, L., Xu, C., Zhang, C., Sharif, K., Wu, H., & Westermann, H. (2020). BSFP: Blockchain-Enabled Smart Parking With Fairness, Reliability and Privacy Protection. *IEEE Transactions on Vehicular Technology*, 6(6), 6578–6591. DOI: 10.1109/TVT.2020.2984621

Zhang, M., Li, Z., Si, H., Cheng, L., Zhou, X., & Wang, B. (2023). Urban travel time and residential location choice: The impacts of traffic congestion. *Sustainable Cities and Society*, 99, 104975. DOI: 10.1016/j.scs.2023.104975

Zhang, P., Chen, N., Xu, G., Kumar, N., Barnawi, A., Guizani, M., Duan, Y., & Yu, K. (2024). Multi-target-Aware dynamic resource scheduling for cloud-fog-Edge multi-tier computing network. *IEEE Transactions on Intelligent Transportation Systems*, 25(5), 3885–3897. DOI: 10.1109/TITS.2023.3330419

Zhang, S., Lu, C., Jiang, S., Shan, L., & Xiong, N. N. (2020). An unmanned intelligent transportation scheduling system for open-pit mine vehicles based on 5G and big data. *IEEE Access : Practical Innovations, Open Solutions*, 8, 135524–135539. DOI: 10.1109/ACCESS.2020.3011109

Zhang, X., Liu, H., Xu, M., Mao, C., Shi, J., Meng, G., & Wu, J. (2020). Evaluation of passenger satisfaction of urban multi-mode public transport. *PLoS One*, 15(10), e0241004. DOI: 10.1371/journal.pone.0241004 PMID: 33079972

Zhu, L., Liu, X., Huang, H., Avellán-Llaguno, R. D., Lazo, M. M., Gaggero, A., & Zhu, Y. G. (2020, November). Meteorological impact on the COVID-19 pandemic: A study across eight severely affected regions in South America. *The Science of the Total Environment*, 744, 140881. DOI: 10.1016/j.scitotenv.2020.140881 PMID: 32674022

Zurob, C., Allard, J. M., Macário, R., Garcia, B., & Garcia, C. (2016). Passenger information systems. *Restructuring Public Transport through Bus Rapid Transit*, 247-260.

About the Contributors

Görkem Gülhan graduated from Dokuz Eylül University, Faculty of Architecture, Department of Urban and Regional Planning. He completed his master's degree at Dokuz Eylül University, Department of Urban and Regional Planning. He completed his PhD at Pamukkale University, Department of Civil Engineering, specialising in transport planning. His research interests focus on transport master plans, accessibility and transport planning. Apart from transport, he is interested in urban design, urban sociology, numerical methods in planning. He has publications in many national and international indexes. He is currently working as an associate professor at Pamukkale University, Faculty of Architecture and Design, Department of Urban and Regional Planning.

Olcay Polat received his Industrial Engineering PhD degree from the Technical University of Berlin (Germany) with the scholarship of German Academic Exchange Service (DAAD) in 2013. He has been working as an Professor in the Department of Industrial Engineering at Pamukkale University (Turkey). He has a special interest in supply chain management, logistics and transportation optimization, production management and risk management. He has been published more that 100 papers in respected journals, conference proceedings and books.

***

Metin Mutlu AYDIN works as full-time professor at Ondokuz Mayıs University. He completed his undergraduate studies at the Civil Engineering Department at Dokuz Eylül University in 2010, followed by a master's degree in Transportation Engineering from the same university in 2013. He later earned his Ph.D. from Akdeniz University in 2017, with a research focus on road geometry optimization. Dr. AYDIN is currently a faculty member at Ondokuz Mayıs University, where he serves as a member of the Department of Civil Engineering. He has significantly contributed to the Intelligent Transportation Systems (ITS) field and marketing through his research, writing, and teaching. His scholarly work has been widely published in many academic journals and presented at national and international conferences. He worked as a coordinator and researcher in many national and international projects related to sustainability and intelligent transportation.

Divya Bhardwaj is an Undergraduate student at KIIT University.

Dickson K.W. Chiu received the B.Sc. (Hons.) degree in Computer Studies from the University of Hong Kong in 1987. He received the M.Sc. (1994) and Ph.D. (2000) degrees in Computer Science from the Hong Kong University of Science and Technology (HKUST). He started his own computer consultant company while studying part-time. He has also taught at several universities in Hong Kong. His teaching and research interest is in Library & Information Management, Service Computing, and E-learning with a cross-disciplinary approach involving library and information management, e-learning, e-business, service sciences, and databases. The results have been widely published in nearly 400 international publications (most of them have been indexed by SCI/-E, SSCI, and EI, such as top journals MIS Quarterly, Computer & Education, Government Information Quarterly, Decision Support Systems, Information Sciences, Knowledge-Based Systems, Expert Systems with Application, Information Systems Frontiers, IEEE Transactions, including many taught master and undergraduate project results and around 20 edited books. He received a best paper award at the 37th Hawaii International Conference on System Sciences in 2004. He is an Editor (-in-chief) of Library Hi Tech, a prestigious journal (SSCI IF=3.4), and a Senior Editor of the Journal of Organizational Computing and Electronic Commerce (SCIE IF=2.9). He is the Editor-in-chief Emeritus of the International Journal on Systems and Service-Oriented Engineering (founding) and the International Journal of Organizational and Collective Intelligence, and serves on the editorial boards of several international journals. He co-founded several international workshops and co-edited many journal special issues. He also served as a program committee member for over 300 international conferences and workshops. Dr. Chiu is a Senior Member of both the ACM and the IEEE and a life member of the Hong Kong Computer Society. According to Google Scholar, he has over 9,600 citations, h-index=51, i-10 index=205, and ranked worldwide 1st in "LIS," "m-learning," and "e-services." He received nearly 1,000 citations in 2022 and 2,600 in 2023.

K. Mert Cubukcu is a Professor of City and Regional Planning at Dokuz Eylül University, Izmir, Turkey. He holds a Bachelor of City Planning degree from Middle East Technical University, and a Ph.D. in City and Regional Planning from The

Ohio State University. He was the runner-up for the Regional Science Association International (RSAI) Dissertation Contest. Cubukcu's research interests include quantitative planning techniques, GIS, transport planning, and spatial statistics. Cubukcu's research has appeared in international journals including Journal of Transport Geography, Applied Geography, Land-Use Policy, Environment and Planning A, The Annals of Regional Science, Spatial Information Research, Journal of Architectural and Planning Research, Studies in Regional and Urban Planning, Management Research and Practice, and International Journal of Education Policy and Leadership. He is the author of two books and the recipient of Stollman Planning Award.

Eren Dagli works as a lecturer at Selçuk University. After completing his undergraduate education at Akdeniz University, Department of Civil Engineering in 2018, he completed his master's degree at Gümüşhane University. Now he is continuing his doctorate at Ondokuz Mayıs University. He is conducting studies on transportation and traffic engineering and smart transportation.

Miguel Ángel Ruiz Jaimes, estudió la Licenciatura en Informática, Maestría y el Doctorado en Sistemas Computacionales. Actualmente, labora como Profesor de Tiempo Completo en la Universidad Politécnica del Estado de Morelos (UPEMOR) actualmente es el Director de la ingeniería en Informática y Electrónica en la misma institución y contacto principal de la academia CISCO Networking. El Doctor Miguel Ángel, ha sido responsable de proyectos de investigación y cuenta con diversas publicaciones de los resultados. Tiene publicados capítulos de libros nacionales e internacionales. Ha participado en diferentes congresos como ponente. Ha dirigido tesis de nivel Licenciatura y Maestría. Ha participado también, como Asesor en el diseño de proyectos tecnológicos. Forma parte del Comité Internacional del Congreso México-Colombia. Fue Dictaminador de libro en la Benemérita Universidad Autónoma de Puebla. Ha sido Miembro del Sistema Estatal de Investigadores del Estado de Morelos. Y es revisor de revistas internacionales: en estudios en Inteligencia Artificial (IA),Ciudades Inteligentes(Smart Cites), Internet de las Cosas (IoT).

Anuj Jaiswal is an Assistant professor in the Department of Architecture and Planning MANIT Bhopal. He has wide experience in the field of transportation planning and modeling. He has delivered several expert lectures/talks,especially in the fields of public transportation, traffic management, ITS, Active transport and road safety, etc.Dr Anuj Jaiswal has about 20 years of professional experience in Transport Planning and Traffic Engineering. He has been teaching transport planning, modeling, trafficengineering, geometric design of highways and Transport

Economics at the graduate and postgraduate levels in MANIT Bhopal since 2005. He has also been involved in interdisciplinary research including traffic safety, and public transport systems since 2003 as member of various Transportation organization. Dr Jaiswal has done Doctor of Philosophy in Public Transportation (Transport Planning), from MANIT Bhopal, in 2013. He had done Master's in Transport Planning from, the School of Planning and Architecture, New Delhi, India and graduated from the School of Architecture, I.E.P.T Indore, India.

Ezgi Karaçoban, M.Sc., is a human resources specialist in the textile industry with a diverse educational background. She holds a master's degree in City and Regional Planning from Pamukkale University (2023), where she focused on smart cities and regional development strategies. Ms. Karaçoban earned her Bachelor of Arts in Sociology from Pamukkale University (2019), concentrating on violence against healthcare workers. She also completed an Associate Degree in Social Services from Anadolu University (2018). Her academic interests include urban sociology, smart cities, and regional development, highlighting her interdisciplinary approach to understanding social and urban dynamics.

Shivam Kumar is an Undergraduate student at KIIT University.

Vicky Lalramsangi is an Assistant Professor in Department of Architecture and Planning, Maulana Azad National Institute of Technology, Bhopal. Having completed a B.Arch and Masters in Urban Design, she is currently pursuing PhD investigating the utility of public open spaces in hill cities, focussing on the less documented north-eastern hill region of India. Her primary areas of interest include urban design, public spaces, gender dynamics, socio-cultural and behavioral aspects, pedestrian mobility, and inclusive and safe urban environments.

Amlan Lenka Undergraduate student at KIIT University.

Josue Martinez-Mireles Master of Science in Mechatronics Engineering, specialized in the design of multidisciplinary engineering solutions, has more than 17 years as a full time researcher at UPP, teaches undergraduate and graduate level, has had the opportunity to hold management positions at UPP.

Ayush Mohapatra is an Undergraduate student at KIIT University.

Hitesh Mohapatra received the B.E. degree in Information Technology from Biju Patnaik University of Technology (BPUT), Odisha in 2006, and the MTech. Degree in CSE from Odisha University of Technology and Research (OUTR) University, Odisha in 2009. He received his Ph.D. in Computer Science &

Engineering in 2021 from Veer Surendra Sai University of Technology (VSSUT), Burla, India. He has contributed 30 SCI and Scopus indexed research papers, 21 international/national conferences and books on Software Engineering and C Programming respectively. He has 15 years of teaching experience both in industry and academia. He has served the research community with various capacities like session chair, technical chair, keynote speaker etc. His research interests include wireless sensor network, smart city, smart grid and smart water. Currently he is working as Associate professor at School of Computer Engineering, KIIT DU, Bhubaneswar, and Odisha.

Oğuz Özbek is a Turkish academic specializing in City and Regional Planning at Pamukkale University. He holds a Ph.D. from Middle East Technical University, focusing on Financial Geography and the Geography of Banking. His research addresses regional development strategies, spatial planning, and the application of new technologies in urban and regional contexts. Prof. Özbek's career progression spans from Research Assistant to Professor, with positions held at Selcuk University and Pamukkale University. He has served in administrative roles, including Department Head and Vice Dean. His academic work encompasses teaching, research, and international collaboration. Prof. Özbek has participated in exchange programs, delivering lectures at various European universities on topics such as spatial planning systems and regional development policies. Prof. Özbek has published in international and national journals, contributed chapters to books, and presented at conferences. Linguistically, he is proficient in Turkish and English, with elementary skills in Polish, Russian, and Norwegian.

Aditya Panda is an Undergraduate student at KIIT University.

Arshi Parashar is an Assistant Professor in the Department of Architecture and Planning at Maulana Azad National Institute of Technology Bhopal, India. Before joining India, she taught in the Department of Architecture at Prince Sultan University in Riyadh, Saudi Arabia, from the year 2014 to 2019. She has a Bachelors degree in Architecture, a Masters degree in Urban Planning, and a Ph.D. in Urban Design from IIT Roorkee, India. She has ten years of experience in teaching and academia. Her research interests are situated at the intersection of urban design and human behavior, gender-inclusive cities, public spaces, Walkability, and behavioral urbanism. Her published works including research articles, viewpoints, book chapters and conference papers are majorly about urban design and its effect on users.

Ayush Parida is an Undergraduate student at KIIT University.

Supriya Priyadarshi is an Undergraduate student at KIIT University.

Bikram Rath is an Undergraduate student at KIIT University.

Jazmín Rodríguez Flores Master of Science in Computer Science, specializing in systems analysis and design, has more than 17 years as a full time researcher at UPP, teaches undergraduate and graduate courses.

Jorge A. Ruiz-Vanoye was born in D.F. Mexico in 1975. He obtained his Ph.D. degree in Computer Science in 2008 from CENIDET. He has worked at Electric Research Institute of Mexico government (IIE) and in diverse companies more. He has given classes in diverse Mexican Universities since 1996. He is reviewer of Computers & Security, Neural Computing & Applications journal, Journal of Computer Information Systems, Computers & Electrical Engineering, Journal of Computational and Applied Mathematics and others journals and conferences; for publications and more information see www.ruizvanoye.com.

Sanjana Subudhi is an Undergraduate student at KIIT University.

Aylin Tepecik completed his primary, secondary and high school education in Aydın-Nazilli. He completed his associate degree from Celal Bayar University, Department of Map and Cadastre in 2016. Then, in 2021, she completed her undergraduate education at Erciyes University, Department of City and Regional Planning. She completed her master's degree at Erciyes University Institute of Science and Technology in 2023. Conducts research on a wide range of topics; Rural Development, Regional Planning, Sustainable Planning and Urban Sustainability.

Index

A

Aging 174, 175, 179, 193, 195, 196, 197, 199, 200, 351
alternative transportation modes 14
Autonomous Vehicles 52, 88, 107, 108, 112, 114, 117, 124, 125, 129, 134, 171, 291, 292, 293, 294, 298, 299, 300, 303, 304, 325, 326

B

Bagdat Street 201, 203, 204, 205, 206, 207, 208, 209, 211, 212, 218, 219, 220, 222, 223, 227
branding 244, 268, 269, 272, 287, 288, 289, 290

C

Circularity in Transportation 291
City rebranding 268, 272, 289
climate change 1, 2, 3, 4, 7, 8, 9, 11, 12, 13, 14, 16, 17, 18, 19, 20, 24, 26, 39, 42, 52, 57, 58, 59, 60, 66, 78, 80, 255, 298, 302, 304, 324
Covid-19 329, 330, 331, 332, 333, 334, 335, 336, 337, 338, 339, 340, 341, 342, 343, 344, 345, 346, 347, 348, 349, 350, 351, 352, 353, 361, 362, 363, 364, 365, 366, 367, 368, 369, 370, 371, 372, 373, 374, 375, 376, 377, 378, 379, 380

E

Energy Efficiency 11, 12, 13, 24, 26, 27, 29, 30, 31, 32, 33, 35, 36, 41, 42, 52, 86, 147, 151, 152, 292, 300, 303, 305, 306, 320, 321, 322, 323
environmental sustainability 1, 3, 4, 12, 14, 18, 19, 20, 21, 25, 29, 30, 33, 52, 54, 77, 227, 294, 298, 305, 321, 323

G

Good road environment 261
greenhouse gas emissions 1, 2, 3, 4, 5, 6, 7, 8, 9, 11, 13, 14, 15, 28, 37, 84, 108, 167
green transportation systems 1, 11, 12, 15

I

Intelligent Transportation System 126, 132, 138, 139, 226
Internet of Things (IoT) 25, 26, 27, 44, 53, 58, 68, 90, 91, 105, 106, 122, 128, 129, 137, 143, 144, 155, 169, 170, 171

M

Marxist Critiques 291, 296
microsimulation 201

O

one-way analysis of variance 187, 190

P

public transportation 10, 11, 12, 13, 14, 15, 18, 24, 27, 52, 53, 55, 56, 57, 59, 65, 66, 70, 71, 72, 73, 76, 78, 79, 92, 97, 98, 99, 100, 102, 104, 105, 108, 109, 110, 111, 112, 119, 126, 128, 173, 174, 175, 178, 179, 180, 181, 183, 184, 192, 193, 194, 195, 196, 201, 202, 228, 285, 293, 298, 310, 324, 331, 344, 351, 353, 368

Q

Quantitative Studies 329, 332

R

Raspberry Pi 156, 158, 163
regional development 47, 48, 49, 50, 51, 52, 53, 54, 55, 56, 57, 58, 60, 61, 62, 63, 64, 66, 67, 68, 69, 70, 74, 75, 76, 78, 80

Road Aesthetics 233, 234, 235, 238, 242, 243, 244, 249, 261, 262
Road Safety 37, 72, 88, 91, 95, 119, 122, 135, 136

S

Smart Cities 21, 23, 24, 25, 26, 27, 28, 30, 35, 38, 39, 42, 44, 45, 47, 48, 49, 50, 51, 52, 53, 54, 56, 57, 58, 59, 60, 61, 62, 63, 66, 70, 77, 78, 79, 80, 85, 114, 117, 119, 121, 137, 165, 169, 171
Smart Parking 13, 65, 66, 70, 71, 72, 73, 126, 139, 144, 146, 147, 148, 150, 151, 155, 156, 165, 166, 167, 168, 169, 170, 171
Solar Energy 27, 40, 291, 293, 294, 297, 298, 300, 301, 306, 311, 320, 321, 326, 327
Spinoza's Views 291
Sustainable Transport 21, 78, 85, 179, 296, 298, 304, 309, 323
sustainable transportation 1, 7, 9, 10, 11, 14, 19, 49, 52, 55, 60, 70, 71, 72, 74, 92, 94, 101, 103, 110, 116, 134, 202, 300, 301, 302, 324

T

Technological Transformation 293
telematics services 122
TR32 Region 47, 48, 49, 50, 54, 55, 61, 62, 63, 64, 65, 66, 67, 68, 69, 70, 71, 72, 73, 74, 75, 76
traffic management 9, 11, 27, 28, 35, 36, 53, 55, 56, 57, 59, 65, 66, 70, 72, 76, 83, 88, 89, 90, 91, 92, 93, 95, 101, 102, 111, 113, 114, 118, 122, 124, 125, 126, 129, 135, 136, 166
transportation behavior 174, 176
transportation planning 52, 70, 72, 85, 89, 90, 91, 92, 94, 95, 96, 97, 98, 100, 101, 102, 103, 105, 107, 108, 109, 110, 111, 115, 118, 174, 176, 194, 195, 198, 201, 234
transportation policies 1, 3, 9, 10, 11, 14, 20, 47, 48, 49, 51, 53, 54, 55, 56, 61, 67, 193, 202, 222
Transport Circularity 291, 294, 297

U

Urban development 10, 25, 26, 27, 28, 29, 30, 33, 36, 52, 55, 72, 144, 146, 168, 178, 235, 288, 289, 335
Urban Environments 23, 27, 28, 32, 34, 42, 48, 52, 122, 125, 134, 144, 148, 201, 223, 245, 263, 298, 306, 311, 320, 321, 326, 329, 332
Urban Factors 329, 332, 333, 334, 336, 340, 343, 349, 353, 375
Urban Features 329, 332
urban traffic congestion 152

V

vehicle capacity 202, 223

www.ingramcontent.com/pod-product-compliance
Lightning Source LLC
LaVergne TN
LVHW080346011026
849161LV00009B/2207